COMPUTATIONAL ACCELERATOR PHYSICS

The proceedings of this meeting were made possible by the support of the Office of Energy Research of the U.S. Department of Energy and the Office of Naval Research.

COMPUTATIONAL ACCELERATOR PHYSICS

Williamsburg, Virginia   September 1996

EDITORS
Joseph J. Bisognano
*Thomas Jefferson
National Accelerator Facility*

Alfred A. Mondelli
*Science Applications
International Corporation*

AIP CONFERENCE
PROCEEDINGS 391

American Institute of Physics   Woodbury, New York

Authorization to photocopy items for internal or personal use, beyond the free copying permitted under the 1978 U.S. Copyright Law (see statement below), is granted by the American Institute of Physics for users registered with the Copyright Clearance Center (CCC) Transactional Reporting Service, provided that the base fee of $6.00 per copy is paid directly to CCC, 222 Rosewood Drive, Danvers, MA 01923. For those organizations that have been granted a photocopy license by CCC, a separate system of payment has been arranged. The fee code for users of the Transactional Reporting Service is: 1-56396-671-9/ 97 /$6.00.

© 1997 American Institute of Physics

Individual readers of this volume and nonprofit libraries, acting for them, are permitted to make fair use of the material in it, such as copying an article for use in teaching or research. Permission is granted to quote from this volume in scientific work with the customary acknowledgment of the source. To reprint a figure, table, or other excerpt requires the consent of one of the original authors and notification to AIP. Republication or systematic or multiple reproduction of any material in this volume is permitted only under license from AIP. Address inquiries to Office of Rights and Permissions, 500 Sunnyside Boulevard, Woodbury, NY 11797-2999; phone: 516-576-2268; fax: 516-576-2499; e-mail: rights@aip.org.

L.C. Catalog Card No. 97–70181
ISBN 1-56396-671-9
ISSN 0094-243X
DOE CONF- 9609256

Printed in the United States of America

CONTENTS

Preface ix
List of Participants xi
Photographs xv

STATUS OF COMPUTATIONAL ACCELERATOR PHYSICS

Computational Tools and Lattice Design for the PEP-II B-Factory 3
    Y. Cai, J. Irwin, Y. Nosochkov, and Y. Yan
Numerical Modeling of Beam-Environment Interactions
in the PEP-II B-Factory 9
    C.-K. Ng, K. Ko, Z. Li, and X. E. Lin
Understanding the Nonlinear Beam Dynamics of the Advanced
Light Source 15
    D. Robin and J. Laskar
TESLA FEL Gun Simulations with PARMELA and MAFIA 21
    M. Zhang and P. Schuett
Manipulation of High-Current Pulses for Heavy-Ion Fusion 27
    W. M. Sharp, D. A. Callahan, A. Friedman, and D. P. Grote

ELECTROMAGNETIC AND PIC SIMULATION

Modeling Large Heterogeneous RF Structures 39
    Z. Li, K. Ko, V. Srinivas, and T. Higo
Simulation of High-Brightness Electron Photoinjectors 45
    E. R. Colby
Methods Used in WARP3d, a Three-Dimensional PIC/Accelerator Code 51
    D. P. Grote, A. Friedman, and I. Haber
MAGY: Time Dependent, Multifrequency, Self-Consistent Code
for Modeling Electron Beam Devices 59
    M. Botton, T. M. Antonsen, and B. Levush
MAFIA Version 4 65
    T. Weiland, M. Bartsch, U. Becker, M. Bihn, U. Blell, M. Clemens,
    M. Dehler, M. Dohlus, M. Drevlak, X. Du, R. Ehmann, A. Eufinger,
    S. Gutschling, P. Hahne, R. Klatt, B. Krietenstein, A. Langstrof,
    P. Pinder, O. Podebrad, T. Pröpper, U. van Rienen, D. Schmidt,
    R. Schuhmann, A. Schulz, S. Schupp, P. Schütt, P. Thoma,
    M. Timm, B. Wagner, R. Weber, S. Wipf, H. Wolter, and Z. Min
QUICKSILVER—A General Tool for Electromagnetic PIC Simulation 71
    D. B. Seidel, R. S. Coats, W. A. Johnson, M. L. Kiefer, L. P. Mix,
    M. F. Pasik, T. D. Pointon, J. P. Quintenz, D. J. Riley, and C. D. Turner
High Accuracy 3D Electromagnetic Finite Element Analysis 77
    E. M. Nelson
Electromagnetic PIC Modeling with a Background Gas 83
    J. P. Verboncoeur and D. Cooperberg

Higher Order Modes in Tapered Disc-Loaded Structures 89
    U. van Rienen

**Effects of Space Charge on the Current-Voltage Characteristics
of Field Emitter Arrays** 95
    K. L. Jensen, E. G. Zaidman, and M. A. Kodis

Enhancements to the Opera-3d Suite 101
    C. P. Riley

Code Update: MicroWaveLab 107
    J. F. DeFord

Status and Perspectives of the PRIAM/ANTIGONE Codes. 113
    G. Le Meur and F. Touze

PISCES II: 2.5-D RF Cavity Code. 119
    Y. Iwashita

**"Thick-Slice" Simulation of Short Longitudinal-Scale Phenomena
on a Space-Charge-Dominated Beam** 125
    I. Haber, D. A. Callahan, A. Friedman, and D. P. Grote

BPERM Version 3.0—A 2D Wakepotential/Impedance Code 131
    T. Barts and W. Chou

**Renoir, a Numerical Simulation Code for the Study of Halo
in Intense Charged Particle Beams** 137
    A. Piquemal

**GATOR: A Hybrid Spectral/PIC Formulation of the Interaction
of Electron Beams with Slow-Wave Structures** 143
    H. P. Freund and E. G. Zaidman

**Large-Timestep Techniques for Particle-In-Cell Simulation of Systems
with Applied Fields That Vary Rapidly in Space.** 149
    A. Friedman and D. P. Grote

3-D Electromagnetic Modeling of Wakefields in Accelerator Components 155
    B. R. Poole, G. J. Caporaso, W. C. Ng, C. C. Shang, and D. Steich

Beamtracking in Cylindrical and Cartesian Coordinates. 161
    B. Schillinger and T. Weiland

Space Charge Effects with Periodic Focusing 167
    N. Brown

**SCD-Beam Main Regularities in Beginning Part of High-Current
Proton Linac** 173
    B. I. Bondarev and A. P. Durkin

High-Order Space Charge Effects Using Automatic Differentiation. ... 179
    M. F. Reusch and D. L. Bruhwiler

A Tracking Code for Injection and Acceleration Studies in Synchrotrons 185
    E. Lessner and K. Symon

The Dynamics of Space Charge Compensation 191
    R. Becker

**Generalized Time-Domain Method for Solution of Maxwell's
Integral Equations** 197
    M. Hano

**Space-Charge-Dominated Beam Dynamics Simulations Using
the Massively Parallel Processors (MPPs) of the Cray T3D** 203
    H. Liu

PARTICLE TRACKING AND BEAM TRANSPORT

Optimization of Dynamic Aperture for the KEKB B-Factory 215
    K. Oide, H. Koiso, and K. Ohmi
Differential Algebras with Remainder and Rigorous Proofs
of Long-Term Stability 221
    M. Berz
Longitudinal Wake Field Corrections in Circular Machines 229
    K. R. Symon
A C++ Implementation of the Differential-Algebraic Model
for the TASCC Superconducting Cyclotron 235
    S. R. Douglas, W. G. Davies, and G. E. Lee-Whiting
MXYZPTLK and BEAMLINE: Status and Future 241
    L. Michelotti
TRACE 3-D Code Improvements 247
    W. P. Lysenko, D. P. Rusthoi, K. C. D. Chan, G. H. Gillespie,
    and B. W. Hill
COSY INFINITY Version 7 253
    K. Makino and M. Berz
Zlib: A Numerical Library for Optimal Design of Truncated Power
Series Algebra and Map Parameterization Routines 259
    Y. T. Yan
The Particle Beam Optics Interactive Computer Laboratory 264
    G. H. Gillespie, B. W. Hill, N. A. Brown, R. C. Babcock, H. Martono,
    and D. C. Carey
Automatic Differentiation and Lattice Function Matching 270
    J.-F. Ostiguy, L. Michelotti, and J. A. Holt
Computation and Analysis of Spin Dynamics 276
    V. Balandin, M. Berz, and N. Golubeva
Hamiltonian Methods for the Study of Polarized Proton Beam
Dynamics in Accelerators and Storage Rings 282
    V. Balandin and N. Golubeva

SIMULATIONS FOR CONTROL SYSTEMS

Evaluation of a Server-Client Architecture for Accelerator Modeling
and Simulation 291
    B. A. Bowling, W. Akers, H. Shoaee, W. Watson, J. van Zeijts,
    and S. Witherspoon
Accelerator Physics Computing in a Control System Environment 297
    J. A. Holt, A. Braun, L. Michelotti, and M. Martens
Abductive Model Refinement for Accelerator Control 303
    C. Stern, W. Klein, G. Luger, and M. Kroupa
A Simple Tool for Beamline Commissioning and Transport
Optimization 309
    L. Catani

Comparison between Various Beam Steering Algorithms
for the CEBAF Lattice.. 317
    M. Chowdhary, Y-C Chao, and S. Witherspoon

NEW COMPUTER TECHNIQUES AND ENVIRONMENTS

The Classic Project.. 325
    F. C. Iselin
A Consistent Interface between PIC-Simulations 331
    U. Becker, M. Dohlus, and T. Weiland
Unified Accelerator Libraries 337
    N. Malitsky and R. Talman
An Object Oriented C++ Class Library for Solving Electromagnetic
Time Domain Problems... 343
    H. Abe
Multilevel Codes RFQ.3L for RFQ Designing 349
    B. I. Bondarev, A. P. Durkin, and S. V. Vinogradov
A Matrix Representation of Lie Algebraic Methods for Design
of Nonlinear Beam Lines... 355
    S. N. Andrianov
An Object Model for Beamline Descriptions 361
    B. W. Hill, H. Martono, and J. S. Gillespie
Mapa—an Object Oriented Code with a Graphical User Interface
for Accelerator Design and Analysis 366
    S. G. Shasharina and J. R. Cary
Recent Developments in the Accelerator System Model Code................ 369
    S. Mendelsohn, D. H. Berwald, M. H. Hughes, T. J. Myers, C. C. Paulson,
    M. A. Peacock, C. M. Piaszczyk, E. M. Piechowiak, J. W. Rathke,
    G. H. Gillespie, and B. W. Hill

HIGH-PERFORMANCE COMPUTING

Parallel Beam Dynamics Calculations on High-Performance Computers 377
    R. Ryne and S. Habib
Parallel Computation of Transverse Wakes in Linear Colliders.............. 389
    X. Zhan and K. Ko
Recent Fortran 90 Developments in 3D Electric Fields Calculations
and Applications Related to the Spiral Project at GANIL................... 395
    P. Bertrand
Author Index... 401

PREFACE

The sixty-two papers appearing in this volume were presented at CAP96, the Computational Accelerator Physics Conference held September 24–27, 1996, in Williamsburg, Virginia. Science Applications International Corporation (SAIC) and the Thomas Jefferson National Accelerator Facility (Jefferson Lab) jointly hosted CAP96, with financial support from the U.S. Department of Energy's Office of Energy Research and the Office of Naval Research. The conference followed 1988, 1990, and 1993 meetings in La Jolla, California, in Los Alamos, New Mexico, and in Pleasanton, California. Stanford Linear Accelerator Center will host CAP98.

Some 97 people attended CAP96: 70 from the United States, 7 from Germany, 5 each from France and Switzerland, 4 from Japan, 2 each from Russia and Canada, and 1 each from Italy and the United Kingdom. The program included 42 talks and 31 poster presentations.

Topics ranged from descriptions of specific codes to advanced computing techniques and numerical methods. Update talks were presented on nearly all of the accelerator community's major electromagnetic and particle tracking codes. Like CAP93, the conference included sessions on forward-looking computer methods and technologies such as high-performance computing. A session was added on simulation codes in accelerator control systems, in recognition of their increasingly important role. CAP96 also saw increased attention to the development of object-oriented techniques and specialized class libraries. Scripting languages and their role in building interactive, programmable application software were addressed as well. At the final session, William McCurdy, Thomas Kitchens, Robert Ryne, and Paul Dubois served as panelists for a discussion titled "The Future of Computational Accelerator Physics."

Code demonstrations held throughout the conference included MAPA (John Cary), MAFIA (Martin Timm), OPERA-3D (Chris Riley), UAL/SMF (Nicolay Malitsky), QUICKSILVER (David Seidel), OOPIC (John Verboncoeur), COSY INFINITY 7 (Martin Berz), LIDOS (Alexander Durkin), Space-Charge-Dominated Beam Codes (Boris Bondarev), and MICROWAVELAB (John DeFord).

The agenda also afforded opportunities for break-out sessions—smaller meetings on topics of special interest. Robert Ryne organized a break-out session on high-performance computing; the other topics and organizers were: high-brightness photo-injectors, Claudio Parazzoli; electromagnetic simulation codes, John Petillo and Alfred Mondelli; applications of the WARP code, Alex Friedman; and the CLASSIC code, John Irwin and Chris Iselin.

On the final evening, after a banquet at Evelynton Plantation in nearby Charles City County, conference participants went outside for a special program of entertainment: the last total lunar eclipse visible from North America in this millennium. The conference ended the next afternoon with a tour of the Continuous Electron Beam Accelerator Facility at Jefferson Lab, just southeast of Williamsburg in Newport News. While a nuclear physics experiment continued in one of CEBAF's three experimental halls, participants visited open areas of the facility, saw superconducting accelerating sections being assembled, and toured an experimental hall being instrumented for future experiments.

As conference organizers, we were fortunate to work with particularly effective people. The Program Committee strove tirelessly to attract the finest speakers and participants. Its members were Robert Ryne of Los Alamos National Laboratory, Richard Cooper of UCLA, John DeFord of Ansoft, Alex Dragt of the University of Maryland, Kwok Ko and John Irwin of Stanford Linear Accelerator Center, Leo Michelotti of Fermilab, Johannes van Zeijts of Jefferson Lab, Alex Friedman of Lawrence Livermore National Laboratory, and George Gillespie of G. H. Gillespie and Associates. Cela Callaghan and Marty Hightower of Jefferson Lab and Anita Mahaffey of SAIC ably handled registration, local arrangements, program support, and day-to-day details. Others contributing from Jefferson Lab included Ruth Bizot, Steve Corneliussen, Cindy Garwood, Randy Hartman, Sarah Spata, Steve Wells, and Camille White. We are grateful to both the committee and the staff for making CAP96 a success.

Alfred A. Mondelli
Science Applications International Corporation
Joseph J. Bisognano
Thomas Jefferson National Accelerator Facility

PARTICIPANTS

Abe, Hiroshi	C&C Research Laboratories
Akers, Walter	Jefferson Lab
Balandin, Vladimir	Michigan State University
Becker, Reinard	Universität Frankfurt
Becker, Ulrich	TH-Darmstadt
Bertrand, Patrick	GANIL
Berz, Martin	Michigan State University
Bisognano, Joe	Jefferson Lab
Bondarev, Boris	Moscow Radiotechnical Institute
Botton, Moti	University of Maryland
Bowling, Bruce	Jefferson Lab
Brown, Nathan A.	G. H. Gillespie Associates, Inc.
Cai, Yunhai	SLAC
Cary, John	Tech-X Corporation
Catani, Luciano	Istituto Nazionale di Fisica Nucleare
Chae, Yong	Argonne National Laboratory
Chao, Yu-Chiu	Jefferson Lab
Chou, Weiren	Fermilab
Chowdhary, Mahesh	Jefferson Lab
Colby, Eric R.	Fermilab
Davies, Walter	Chalk River Laboratories
DeFord, John F.	Ansoft Corporation
Douglas, Stephen	Chalk River Laboratories
Dragt, Alex	University of Maryland
Dubois, Paul F.	Lawrence Livermore National Laboratory
Durkin, Alexander	Moscow Radiotechnical Institute
Forest, Etienne	KEK
Freund, Henry	SAIC
Friedman, Alex	Lawrence Livermore National Laboratory
Fuchi, Kyoko	Michigan State University
Gillespie, George H.	G. H. Gillespie Associates, Inc.
Godlove, Terry	FM Technologies, Inc.
Golubeva, Nina	Institute for Nuclear Research
Grote, David P.	Lawrence Livermore National Laboratory
Grote, Hans J.	CERN
Haber, Irving	Naval Research Laboratory
Hancock, Steven	CERN
Hano, Mitsuo	Yamaguchi University
Hill, Barrey W.	G. H. Gillespie Associates, Inc.
Holt, James	Fermilab
Irwin, John	SLAC
Iselin, Christoph	CERN
Iwashita, Yoshihisa	Kyoto University

Kitchens, Tom	DOE Germantown
Ko, Kwok	SLAC
Kroupa, Mike	Vista Control Systems
Langdon, Bruce	Lawrence Livermore National Laboratory
Lebedev, Valeri	Jefferson Lab
Lee, We-li	Princeton Plasma Physics Laboratory
LeMeur, Guy	University of Paris
Lessner, Eliane	Argonne National Laboratory
Levush, Baruch	Naval Research Laboratory
Li, Rui	Jefferson Lab
Li, Zenghai	SLAC
Lidel, John	(affiliation unknown)
Liu, Hongxiu	Jefferson Lab
Lysenko, Walter	Los Alamos National Laboratory
Malitsky, Nikolay	Cornell University
Mendelsohn, Stanley	Northrop Grumman Corporation
McCurdy, William	Lawrence Berkeley National Laboratory
Millich, Antonio	CERN
Mondelli, Alfred A.	SAIC
Nelson, Eric	Los Alamos National Laboratory
Ng, Cho	SLAC
Niederer, James	Brookhaven National Laboratory
Oide, Katsunobu	KEK
Ostiguy, Jean-Francois	Fermilab
Parazzoli, Claudio Gilbert	Boeing
Peters, Jerry	U. S. Department of Energy
Petillo, John	SAIC
Piquemal, Alain	CEA
Reusch, Michael F.	Northrop Grumman Corporation
Riley, Christopher P.	Vector Fields, Ltd.
Robin, David S.	Lawrence Berkeley National Laboratory
Ryne, Robert	Los Alamos National Laboratory
Schillinger, Brigette	TH-Darmstadt
Schoessow, Paul	Argonne National Laboratory
Segré, Jacques	CEA
Seidel, David	Sandia National Laboratories
Sharp, William M.	Lawrence Livermore National Laboratory
Shasharina, Svetlana	Tech-X Corporation
Smithe, David	Mission Research Corporation
Srinivas, Vinay	SLAC
Stern, Carl	Vista Control Systems, Inc.
Symon, Keith	Argonne National Laboratory
Timm, Martin	TH-Darmstadt
Touze, Francois	University of Paris
van Rienen, Ursula	TH-Darmstadt
van Zeijts, Johannes	Jefferson Lab

Verboncoeur, John	University of California-Berkeley
Vrankovic, Vjeran	Paul Scherrer Institute
Whealton, John	Oak Ridge National Laboratory
Witherspoon, Sue E.	Jefferson Lab
Yan, Yiton T.	SLAC
Yunn, Byung C.	Jefferson Lab
Zaidman, Ernest	Naval Research Laboratory
Zhan, Xiaowei	SLAC
Zhang, Min	DESY

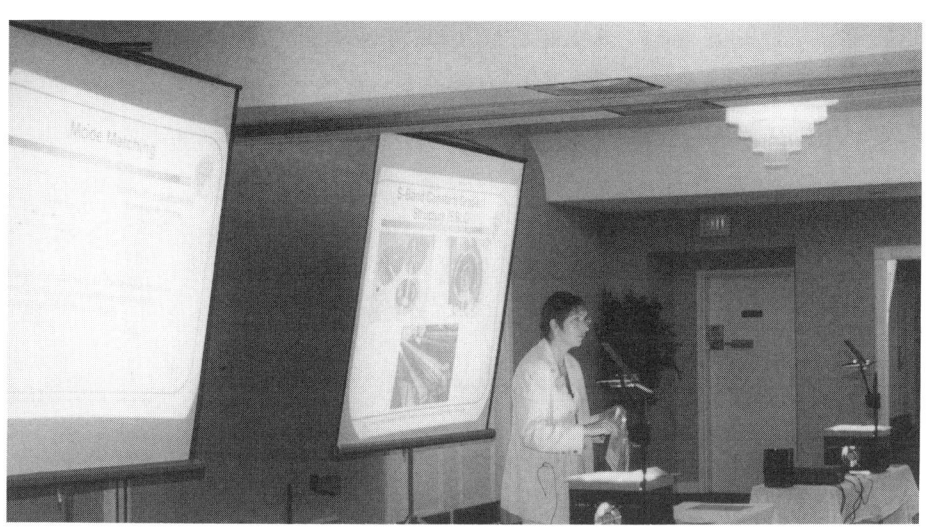

STATUS OF COMPUTATIONAL ACCELERATOR PHYSICS

Computational Tools and Lattice Design for the PEP-II B-Factory

Yunhai Cai, John Irwin, Yuri Nosochkov and Yiton Yan
Stanford Linear Accelerator Center, Stanford University, Stanford, CA 94309

Abstract

> Several accelerator codes were used to design the PEP-II lattices, ranging from matrix-based codes, such as MAD and DIMAD, to symplectic-integrator codes, such as TRACY and DESPOT. In addition to element-by-element tracking, we constructed maps to determine aberration strengths. Furthermore, we have developed a fast and reliable method(nPB tracking) to track particles with a one-turn map. This new technique allows us to evaluate performance of the lattices on the entire tune-plane. Recently, we designed and implemented an object-oriented code in C++ called LEGO which integrates and expands upon TRACY and DESPOT.

Introduction

When we designed the PEP-II B factory[1], we encountered many challenging problems in the area of chromatic optics and non-linear dynamics. The accelerator design and simulation codes played very important roles for solving those problems. First, the automated design processes speeded up turn-around time of evaluating lattices. This allowed us to try more alternatives that often resulted in a better or simpler solution of lattice[2]. Secondly, special elements, such as a dipole and quadrupole nested inside a solenoid field, implemented in the codes allowed us to make accurate modeling of the accelerator possible. Finally, the non-linear analysis capabilities with maps made diagnosis of many problems quicker and easier than the tracking method.

In this paper, we will begin with the features and issues in the design of B-Factory lattices. In the second part, we will describe the physics and approximations we used in the tracking codes. Hopefully, it will make many potential users of the codes more comfortable after understanding the physics inside. We will continue with analysis of linear optics and non-linear generalization. At the end, we will discuss a map-tracking technique called nPB. Since many concrete examples of how to use those tools had been published, we will not present them again in this paper.

Design Features and Issues

Main parameters of PEP-II B-Factory are tabulated in the Table 1. The energies and luminosity are basically determined by the mass and physics of B and $\bar{B}$ system. From the Table 1, we can see that the design luminosity is at least an order of magnitude higher than any of the existing accelerators. In order to achieve this high luminosity, we need to make the beam current as high as possible and beam size as small as possible at the collision point. This implies, in terms of optics, that

---

*Work supported by Department of Energy, contract DE-AC03-76SF00515.

the β functions should be made very small at the interaction point and the final focusing quadrupoles very strong. As a consequence, the large chromatic effects from the strong quadrupoles have to be corrected properly with nearby sextupoles. The small beta functions at the IP also requires a small bunch length as well if one wants to optimize the luminosity.

The PEP-II consists of two rings with different energies. The high energy ring(HER) is very similar to the PEP since all magnets and the tunnel of the PEP are reused. One of the main differences is the β function in the vertical plane at the IP is reduced from 5.0 cm to 2.0 cm. This change requires a better chromatic correction scheme near the interaction region. As a result, non-interleaved sextupole pairs are introduced in the two arcs that are adjacent to the interaction region. In addition, we generated beta-bumps at the locations of sextupole pairs to make the chromatic correction more orthogonal between the horizontal and vertical planes.

Table 1: Main PEP-II nominal parameters.

Parameter	LER	HER
Energy, E[Gev]	3.1	9.0
Circumference, C[km]	2.2	2.2
Emittance, ϵ_x/ϵ_y[nm-rad]	64.3/2.6	48.2/1.9
Beta function, β_x^*/β_y^*[cm]	37.5/1.5	50.0/2.0
B-beam tune shift, $\xi_{0,x}/\xi_{0,y}$	0.03/0.03	0.03/0.03
Synchrontron tune, μ_s	0.03344	0.05207
RF frequency, f_{RF}[MHz]	476	476
RF voltage, V_{RF}[MV]	5.1	18.5
Damping time, τ_E/τ_x[ms]	29.2/60.5	18.4/37.2
Bunch length, σ_l[cm]	1.0	1.0
RMS $\delta E/E$, σ_E	7.7×10^{-4}	6.1×10^{-4}
Total current, I[A]	2.14	0.99
Synch. loss, U_0[Mev/turn]	0.77	3.58
Luminosity, $\mathcal{L}[cm^{-2}s-1]$	3.0×10^{33}	

The low energy ring(LER) is resided on the top of the high energy ring. Because of its lower energy, wigglers are needed in order to match up the damping time and emittance with the HER. In addition, to obtain 1 cm bunch length without excessive high voltage on the RF cavity, the phase advance per cell is chosen to be 90^0 instead of 60^0 in the HER. The LER chromatic correction scheme is very similar to the one used in the HER. Due to the 90^0 cells, we found that it is necessary to use non-interleaved sextupoles with an optimized pattern in the remaining four arcs as well.

Tracking Codes

Several accelerator codes were commonly used to design the PEP-II lattices. MAD and DIMAD are used to design linear optics. TRACY and DESPOT are used for tolerance and tracking study. Since MAD and DIMAD are well known and documented, we will not have any further discussions about them. Instead, we will give a short introduction of TRACY and DESPOT.

TRACY[5] was first developed by Hiroshi Nishimura. The basic idea was using a PASCAL interpreter as part of user interface. As a result, the PASCAL language became its command language of the code. An user could use variables, arrays, procedures and functions in input command file to call the compiled PASCAL procedures directly without recompiling the code. We found that this feature was very powerful and user-friendly because many correction schemes would be implemented easily at the lever of an input file.

The physics part of the code was replaced completely later by J. Bengtsson and E. Forest. Now, it is called "TRACY II". One of the feature missing in the TRACY is extracting a non-linear map. Hence, its analysis is limited at the linear level. As a complementary, DESPOT contains a truncated power algebra package[6] and Lielib package by E. Forest. We used it extensively to calculate non-linear maps.

Recently, we started to design and implement an object-oriented code in C++ called LEGO which integrates and expands upon TRACY and DESPOT. In terms of physics, all three codes belong to the same family of codes called symplectic-integrator. The main expansions in LEGO are three dimensional geometry and more flexibilities in choosing different integrators for a given element.

Let's start to describe the physics in the codes with symplectic integrators. For a sector bend in cylindrical coordinate, its Hamiltonian can be expressed as

$$H = -(1 + \frac{x}{\rho})\sqrt{(1+\delta)^2 - p_x^2 - p_y^2} + b_0 x + b_0 \frac{x^2}{\rho} - \frac{q}{p_0} A_s \quad (1)$$

where $\delta = \frac{p-p_0}{p_0}$, $b_0 = \frac{qB_y}{p_0}$, p_x and p_y are momentum normalized by the design momentum p_0, and A_s is longitudinal component of vector potential. If we take out the vector potential A_s, the remaining Hamiltonian is solvable[7], which could be selected as an integrator in the LEGO. This Hamiltonian is suitable for a small ring, for example.

For a large machine, $x \ll \rho$, and high energy, $p_x \ll 1, p_y \ll 1$, we can make a Taylor expansion of square root keeping only up to quadratic part. This leads us to the Hamiltonian used in TRACY and DESPOT

$$H = \frac{1}{2}\frac{p_x^2 + p_y^2}{(1+\delta)} - \frac{x}{\rho}\delta + \frac{x^2}{2\rho^2} - \frac{q}{p_0} A_s \quad (2)$$

b_0 has been set to $\frac{1}{\rho}$ assuming that bending matches the curvature of the cylindrical coordinate system. In general, the Hamiltonian could not be solved if A_s contains terms beyond quadratic one. There are many ways to make an approximation. We choose symplectic integrators as an approximation of the Hamiltonian. One of the advantages of symplectic integrator is that symplecticity is preserved in the process of integration. This property becomes very important if the long-term stability of particles is the issue of concern. Another advantage is that one can obtain a transfer map of an arbitrary order easily by integrating a truncated power series through the element.

We can separate the Hamiltonian in eq.2 into two exact solvable parts

$$H = H_0 + H_1 \qquad (3)$$

where $H_0 = \frac{1}{2}\frac{p_x^2+p_y^2}{1+\delta}$ and $H_1 = -\frac{q}{p_0}A_s + \frac{1}{2}(\frac{x}{\rho})^2 - \frac{x}{\rho}\delta$. It can be shown easily by using the CBH theorem that

$$exp(:-LH:) = exp(:-\frac{L}{2}H_0:)exp(:-LH_1:)exp(:-\frac{L}{2}H_0:) + O(L^3) \qquad (4)$$

The result is very simple since it can be interpreted as just placing the integrated kick at the middle of the drift. The result does not depend on the specific form of H_0 or H_1. In fact, if we use $H_0 = -(1+\frac{x}{\rho})\sqrt{(1+\delta)^2 - p_y^2 - p_y^2}$, the integration process becomes the same as the one in TEAPOT code. Because the Hamiltonian is locally defined in this approach, we do not have any difficulty to integrate through a vertical bending magnet. We call this kind of integrator second-order since the residual errors are of the third order in the length.

This process could be generalized to construct higher-order integrators by make use of more drifts and kicks. A forth-order integrator has been used for strong quadrupoles inside the interaction regions of the LER.

Linear and Non-linear Analysis

There are many ways to calculate twiss functions in an accelerator. The one implemented in DESPOT is less known to the community. Because this method has a straight forward generalization of non-linear and linear coupled cases. We will give a brief discussion of this approach. For a ring, a one-turn matrix at a given point could be written as

$$M = \begin{pmatrix} cos\mu + \alpha sin\mu & \beta sin\mu \\ -\gamma sin\mu & cos\mu - \alpha sin\mu \end{pmatrix} \qquad (5)$$

where μ is tune and β, α, γ are the Courant-Snyder parameters. The matrix can be normalized into a simple rotation by a symplectic transformation

$$A^{-1}MA = R = \begin{pmatrix} cos\mu & sin\mu \\ -sin\mu & cos\mu \end{pmatrix} \qquad (6)$$

One can show easily that the lattice functions are related to the symplectic transformation in following way

$$\beta = A_{11}^2 + A_{12}^2, \alpha = -(A_{11}A_{21} + A_{12}A_{22}), \gamma = A_{21}^2 + A_{22}^2 \qquad (7)$$

In fact, the choice of A is infinite but the lattice functions do not depend on any particular choice, largely because they are closely related to the physical observables.

Furthermore, we can find a symplectic transformation that normalizes the one-turn matrix at another point of the ring by $A' = M_{trans}A$, where M_{trans} is the transfer matrix between two points. This result is obtained because the tune is invariant in the ring.

From this result, we can calculate the lattice functions simply by propagating the scripted $\mathcal{A}$ around the ring. To generalize this procedure to the nonlinear case, we simply make a non-linear symplectic transformation to normalize the Taylor map at a given point in the ring. Then propagate the non-linear scripted $\mathcal{A}$ from this point to others. In this way, we could, for instance, calculate the lattice functions as a power series expansion of the energy.

Map Analysis and Tracking

To analyze the performance of the lattices for the PEP-II, we always simulate the effects of all known alignment and multipole errors in the lattice with the TRACY. To correct those errors in the lattice, we make many routine corrections, such as orbit and coupling. At the end of the process, we tracked the particles element-by-element to evaluate the dynamic aperture at the nominal working point.

We also used maps extensively to diagnose problems in the lattices when the dynamic aperture is decreased. We calculated the strength of resonances normalized at the amplitude of dynamic aperture. We found[8] that it was very effective method to identify a large source of the errors or mistakes sometimes.

Now let's briefly discuss the techniques developed for fast tracking with maps. A Taylor map is symplectic up to the order of the truncation if the effects of the radiation damping and quantum excitation are ignored. It can be factorized into a single Lie transformation as follows

$$\mathcal{M} = \mathcal{A}^{-1}(z,\delta)\mathcal{R}(z)exp:f(z,\delta):\mathcal{A}(z,\delta) + O(N+1) \qquad (8)$$

where $\mathcal{A}$ and $\mathcal{A}^{-1}$ are the linear matrix that generate the linear normalization discussed in the last section; $\mathcal{R}$ is the linear rotation; and $f(z,\delta)$ is a polynomial from order 3 to order n+1. $f(z,\delta)$ contains all non-linear information of the map.

We can expand the factor $exp:f(z,\delta):$ in terms of Poisson brackets

$$exp:f(z,\delta):z_{in} = z_{in} + \{f, z_{in}\} + \frac{1}{2}\{f, \{f, z_{in}\}\} + ... \qquad (9)$$

We call this method nPB tracking[9]. The key is that the brackets could be evaluated directly in the action-angle basis. If we truncate the expansion at the third order, the speed of tracking is two order of magnitude faster than element-by-element tracking. The increase of speed opens doors to many interesting studies, for example, beam-beam simulation and evaluation of the dynamic aperture on entire tune plane. However, one of the drawbacks of this technique is that it is not symplectic. Therefore, it should be used with caution in studying the long-term stability of particles.

These map factorizing and tracking procedures were implemented using Zlib[10]. We use them routinely in the analysis of PEP-II lattice.

Summary

The codes are crucial tools used in the process of designing a lattice. All those codes were very useful for the problems they were designed to solve. However, every code has its own input format and command language. It is always troublesome to convert from one to another. Mistakes could be made during the translation. Sometimes, the different approximations made in the codes could make results very confusion. Based on our experience, all those codes should be used with caution. One should always check the physics and approximations made in the code before using them for the design.

Acknowledgement

We would like to thank E. Forest for explaining the physics and procedures implemented in TRACY and DESPOT.

References

[1] "PEP-II: An Asymmetric B Factory," Conceptual Design Report, SLAC-418, June 1993.

[2] Y. Cai et al., "Low Energy Ring Lattice of the PEP-II Asymmetric B-Factory," Proc. of 1995 Particle Accelerator Conference, Dallas, May 1995.

[3] M.H.R. Donald et al., "Lattice Design for the High Energy Ring of the SLAC B-Factory (PEP-II)", Proc. of 1995 Particle Accelerator Conference, Dallas, May 1995.

[4] Y. Nosochkov, Y. Cai, J. Irwin, M. Sullivan, E. Forest, "Detector Solenoid Compensation in the PEP-II B-Factory," Proc. of 1995 Particle Accelerator Conference, Dallas, May 1995.

[5] H. Nishimura, "TRACY, A Tool for Accelerator Design and Analysis," Proc. of EPAC, Rome, 1988, p-803.

[6] M. Berz *Part. Accel.* **24** 109 (1989).

[7] E. Forest, private communication.

[8] Y. Yan, J. Irwin, Y. Cai, M. Donald, and Y. Nosochkov, "Dynamic Aperture Improvement of PEP-II Lattices Using Resonance Basis Lie Generators," EPAC 1996 proceeding.

[9] Y.T. Yan, J. Irwin and T. Chen, "Resonance Basis Maps and nPB Tracking for Dynamic Aperture Studies," Particle Accelerators, 1996 Vol. 55, pp. 17-26.

[10] Y.T. Yan, "Zlib and Related Rpograms for Beam Dynamics Studies in Computational Accelerator Physics," AIP Conference Proceedings 297, p. 279 (1993),

Numerical Modeling of Beam-Environment Interactions in the PEP-II B-Factory[*]

C.-K. Ng, K. Ko, Z. Li and X. E. Lin
Stanford Linear Accelerator Center, Stanford University, Stanford, CA 94309

1 Introduction

The PEP-II B-Factory [1] is designed to operate at high currents with many bunches (1658) to achieve the luminosity required for physics studies. Interactions of a beam with its environment in a storage ring raise various issues of concern for accelerator physics, mechanical design and device performance. First, for accelerator physics, wakefields generated by interactions of a beam with beamline components, if not properly controlled, will drive single-bunch and coupled-bunch instabilities [2]. The total broad-band impedance of the ring cannot exceed a budget limited by single-bunch effects. The growth rate of a coupled-bunch mode contributed from narrow-band impedance should be smaller than the damping rate due to synchrotron radiation; otherwise, suppression by feedback control will be necessary. Second, the energy loss by a beam at a beamline component in the form of higher-order-mode (HOM) power leads to additional heating on the component, and to TE mode radiation through openings on vacuum chamber walls. Last, calculations of transfer and beam impedances of pickup and kicker devices are essential for improving their performance and for identifying trapped modes. To address these issues quantitatively requires numerical simulations of each beamline component which include the realistic geometry and the relevant physics involved in the particular beam-environment interactions.

2 Simulations of beam-environment interactions

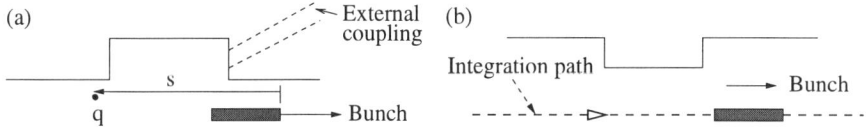

Figure 1: Schematic of beam-environment interaction.

Beam-environment interactions in complex geometries can be studied directly by solving Maxwell equations on a numerical grid in the time domain using a relativistic charged Gaussian bunch. The wakefield generated by the bunch at a beamline component affects particles within the bunch or those in subsequent bunches (see Fig. 1), and its Fourier transform is the impedance. The short-range wakefield corresponds to the broad-band impedance which is governed by excitation above the beampipe cutoff, while the long-range wakefield is associated with the narrow-band resonances below cutoff. The loss parameter, which is the overlap integral of the bunch shape with the short-range wakefield, determines the energy loss.

Many beamline components have external couplings for monitoring or damping purposes. They can be the coaxial cables of a BPM [4] or the damping waveguides of

---

[*]Work supported by Department of Energy, contract DE-AC03-76SF00515.

a damped RF cavity [5]. Numerically, these external loading can be represented by ports, and a broad-band matched condition at the ports is required if the response over the frequency range of the beam spectrum is to be evaluated. MAFIA provides this capability so that, for example, one can determine the transfer impedance of the BPM from the beam-induced signal at the coaxial port. Similarly one is able to resolve the external Q's of the HOMs of the damped RF cavity from the long-range wakefield excited by the beam.

There are two methods of wakefield integration in MAFIA. For a cavity-type structure (Fig. 1(a)) where the structure extrudes out of identical upstream and downstream beampipes, one applies the indirect method to integrate the wakefield along the beampipe wall. Then the only contribution comes from integration across the cavity opening. The beampipes can be reasonably short, thus enabling the computations of long-range wakefields. For a collimator-type structure (Fig. 1(b)) where the structure intrudes into the beampipe, one can use the direct method to integrate the wakefield along the bunch path. Now the downstream beampipe has to be sufficiently long in order that the scattered fields off the structure can catch up with the bunch. Specifically the beampipe length L_d has to satisfy $L_d \sim s/(1-v_g/c)$, where s is the distance of the wakefield to be computed and v_g the group velocity of the scattered waves at the bunch r.m.s. frequency. L_d becomes impractically large for short bunches as v_g approaches c and for long-range wakefield calculations for which s is large.

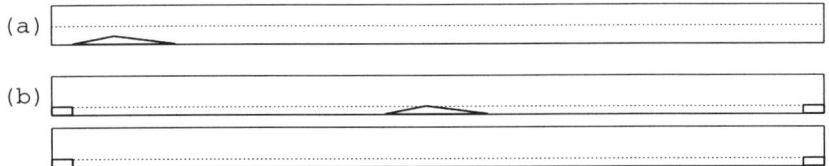

Figure 2: Wakefield calculation of a mask in the vacuum chamber (a) using the direct method; (b) by subtraction of two indirect computations. The dotted lines indicate the integration path in each case.

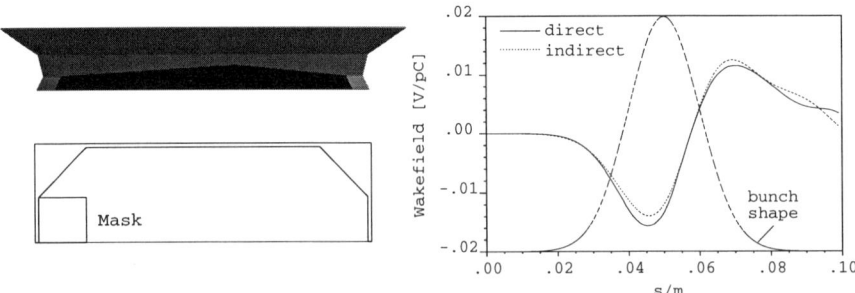

Figure 3: MAFIA model of mask.   Figure 4: Longitudinal wake of mask.

There is an alternate method to treat collimator-type structures by introducing artificial beampipes at the ends to convert them to cavity-type structures (see Fig. 2(b)). Then the indirect method is applied twice, one with the structure and

10

another without, to obtain the wakefield by subtraction. The error in this method comes from crosstalk between the beampipes and is negligible if they are far apart. We consider a mask in the PEP-II vacuum chamber (see Fig. 3) as an example, and compare the subtraction method with the direct method which requires a long downstream beampipe (Fig. 2(a)). The results from the two methods are shown in Fig. 4. Qualitatively the two wakefields are similar with a difference in the loss parameters of about 15%.

In the following sections we will cover three topics of practical interest related to PEP-II beamline component design that we have investigated with time domain numerical analyses.

3 Beam impedance and transfer impedance

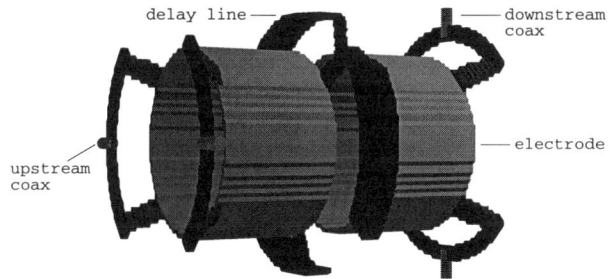

Figure 5: MAFIA model of the longitudinal feedback kicker.

We present the longitudinal feedback kicker as an example of a pickup device. The numerical modeling involves a wakefield analysis in the presence of external couplings. A MAFIA geometry of the device is shown in Fig. 5. The PEP-II kicker has two electrodes in series connected via $\lambda/2$ delay lines to generate voltages across three gaps. It is designed to operate from 952 MHz to 1071 MHz, and to provide a shunt impedance of about 400 Ω over this bandwidth with acceptable beam induced heating of the electrodes. Power in the specified frequency range is driven from a pair of coaxial feeds downstream and couples out at another pair upstream. From the simulation we would like to obtain estimates of the device performance, and to uncover potential danger with trapped modes. We address these issues by calculating the transfer and beam impedances. For numerical expediency (smaller mesh and shorter run time) a 2 cm rather than the actual 1 cm bunch length was used since their spectra differs little over the frequency range of interest. To resolve reasonably well any trapped modes the wakefield was calculated up to 5 m.

Figs. 6(a) and (b) show the monitored signal at the upstream cables and the transfer impedance obtained from its Fourier transform. The transfer impedance $Z_{transfer}$ is fairly constant over the operating bandwidth, and corresponds to a shunt impedance of 353 Ω, using the relationship $R_s = (2Z_{transfer})^2/R_o$ where $R_o = 50$ Ω, the coax impedance. This is in good agreement with the measured value of 385 Ω [6]. The longitudinal wakefield and the beam impedance spectrum are shown in Figs. 7(a) and (b). The peak beam impedance around 1 GHz is about 100 Ω which also agrees with measurement. There is a sharp peak at 2.15 GHz with an impedance of 700 Ω and a Q of 30. This trapped mode has also been observed in measurements [6]. For a total of two kickers in a ring, this HOM contributes 1.4 kΩ which is below the coupled bunch limit of 1.7 kΩ at this frequency. Since Q is low, resonance heating due to coupled bunches is not a concern.

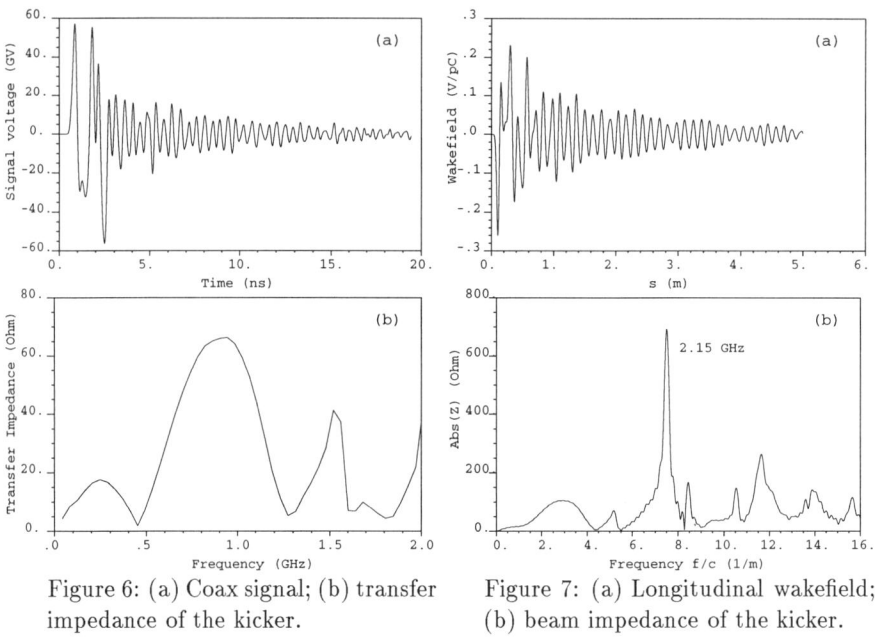

Figure 6: (a) Coax signal; (b) transfer impedance of the kicker.

Figure 7: (a) Longitudinal wakefield; (b) beam impedance of the kicker.

4 HOM power dissipation

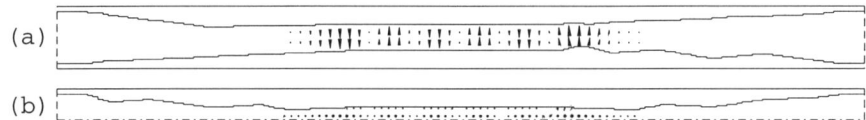

Figure 8: TE_{11} mode trapped in the IR (a) horizontal plane; (b) vertical plane (half of the structure in this plane is shown because of symmetry).

We consider the HOM heating of the beryllium chamber in the interaction region (IR). This central chamber contains the interaction point (IP) and connects at each end to a series of irregularly shaped masks for the purpose of shielding synchrotron radiation before tapering up to larger beampipes at both sides further from the IP. The masks are symmetric with respect to the beam path in the vertical plane but form a constriction at each end of the beryllium chamber (Fig. 8(b)). In the horizontal plane they are not symmetric, which means that the cross section of the vacuum chamber in the mask region varies along the beam path (Fig. 8(a)). The geometry is very large (over 4 m long) and so is the variation in dimension (radius changes from 2 cm to 6 cm over 2 m length). This, and the difficulty to properly terminate the beampipe ends, make mode analysis not a viable approach.

Alternatively we model in the time domain the heating by a single bunch. Instead of the wakefield our focus is on the beam induced currents on the beryllium chamber wall. Fig. 9(a) shows the time variation of the magnetic field at one wall location up to a time of several bunch spacings. The large initial peak is due to the beam image current, and the subsequent oscillations are a result of HOMs. The HOM spectra are shown in Fig. 9(b). The TE_{11} modes are generated because the asymmetry of the masks leads to finite longitudinal electric fields along the beam direction as seen in the mode pattern shown in Fig. 8. This is a 3D effect which previous 2D calculations cannot produce by assuming total symmetry of the masks about the beam axis. It is evident that the HOMs are trapped in the beryllium chamber due to the constrictions in the vertical plane and to mask offsets in the horizontal plane. Using the spectra sampled at many wall locations, one can integrate over the chamber surface to find the power loss. At 3 A current, the wall dissipation is estimated to be 12 W from image current and 2.5 W from HOMs. The enhancement from coupled bunches depends on the loaded Q's of the HOMs, and work is in progress to evaluate this resonant heating to ensure that the cooling specifications can handle the additional dissipation.

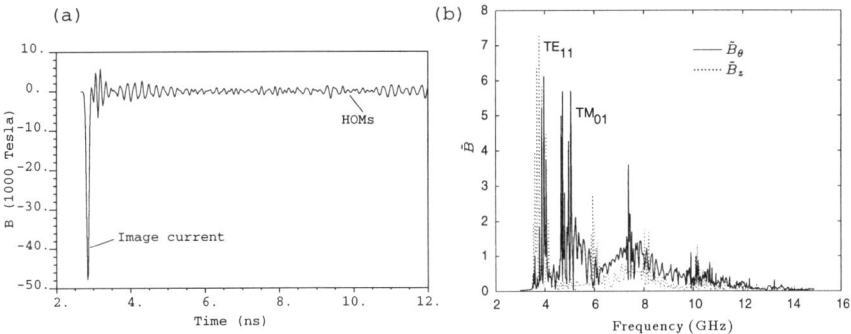

Figure 9: (a) Time variation, (b) Fourier transform of the longitudinal and azimuthal components, of the magnetic field on the beryllium chamber wall.

5 TE power radiation

HOM TE power will propagate in the PEP-II rings after being generated by the beam at asymmetric beamline components such as the RF cavities. The concern is that this power may have adverse effects on a beamline component and its surroundings if it is allowed to radiate through openings in the vaccum chamber wall. An example is the distributed-ion-pump (DIP) screen employed to separate the pump chamber from the beam chamber. The screen is primarily designed to provide adequate conductance for pumping not at the expense of generating high beam impedance. Now additional consideration to prevent TE power radiation from damaging the pumps in the pump chamber has to be included.

MAFIA models of a design with 4-cm slots and a 'microwave screen' design consisting of 3mm×3mm holes are shown in Fig. 10. We compare their effectiveness in screening out TE power by a transmission calculation of a TE_{10} mode propagating in the beam chamber. The results are shown in Fig. 11. The deviation of the transmission coefficient from unity represents radiation through openings on the chamber wall. The slotted design is found to allow non-negligible penetration over a broad range of frequencies, especially around 3.3 GHz. Apparently, the slots are long enough to interrupt the azimuthal current of the TE mode. In contrast, the screen design is basically opaque to TE penetration at all frequencies with 100% transmission. In light of this comparison, the final design consists of six continuous grooves of 5.64 m long and 3.75 mm wide, with round holes of diameter 3 mm hidden halfway inside. The broad-band impedance of this hidden hole design is extremely small and has been described in Ref. [7].

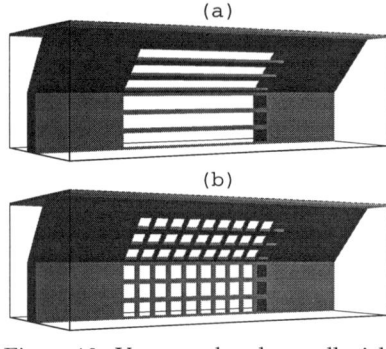

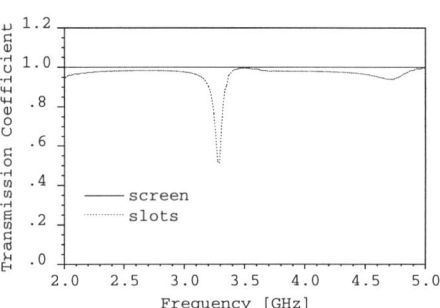

Figure 10: Vacuum chamber wall with (a) slots; (b) a screen. Only parts of the models of the beam and pump chambers are shown.

Figure 11: Transmission coefficients of TE_{10} mode for the slots and screen.

6 Summary

For the past three years, we have modeled many of the PEP-II beamline components and have helped optimize their designs to meet the requirements of beam impedance, mechanical design or device performance. We have tried to benchmark our simulation models with measured data whenever they are available. The methodology described here can be applied to other storage rings such as light sources and the damping rings in linear colliders.

Acknowledgements

We are grateful to many members of the PEP-II collaboration at SLAC, LBNL and LLNL for their collaboration on modeling the PEP-II beamline components. We thank Sam Heifets for theoretical discussions, and Thomas Weiland and Martin Dohlus for their expertise with MAFIA.

References

[1] An Asymmetric B Factory, Conceptual Design Report, SLAC-418, June 1993.
[2] S. Heifets et. al., Impedance Study for the PEP-II B-Factory, SLAC-PUB-6989, 1996.
[3] MAFIA User's guide Version 3.20, CST GmbH, Darmstadt, Germany.
[4] N. Kurita et. al., Numerical Simulation of the PEP-II Beam Position Monitor, SLAC-PUB-7006, 1995.
[5] E. Lin, K. Ko and C. Ng, Impedance Spectrum for the PEP-II RF Cavity, SLAC-PUB-6902, 1995.
[6] J. Corlett and J. Byrd, PEP-II Longitudinal Feedback Kickers- Prototype Measurement Results, 1996.
[7] C.-K. Ng and T. Weiland, Impedance of the PEP-II DIP Screen, SLAC-PUB-7005 1995.

Understanding the Nonlinear Beam Dynamics of the Advanced Light Source[1]

David Robin

Lawrence Berkeley National Laboratory
Berkeley, CA 94720 ; e-mail: robin@lbl.gov

Jacques Laskar

CNRS, Astronomie et Systèmes Dynamiques, Bureau des Longitudes
3, rue Mazarine, 75006 Paris ; e-mail: laskar@bdl.fr

Abstract. Frequency map analysis is used to study the single particle transverse beam dynamics in the Advanced Light Source. The maps, which provide details about the diffusion of orbits and limits on long term stability, are generated by a postprocessor attached to a tracking code. In this paper we describe the method and show how the map is changed when the 12-fold symmetry of the linear lattice is perturbed by including measured magnetic field imperfections. Also the long term stability of orbits that reside in regions of large diffusion is studied.

INTRODUCTION

In a particle storage ring, the stability of the particle motion affects both the beam lifetime and the injection efficiency. If the motion of particles at large amplitudes is unstable, then particles which are scattered to large amplitudes via gas or intrabeam scattering may become lost. If the motion of particles is chaotic, then particles may also become lost due to orbit diffusion. Similarly, particles injected at large amplitude may not survive. So for long beam lifetimes and good injection efficiencies it is necessary to design and operate rings which have large stable phase space areas.

Typically, ring designers rely heavily on short term tracking studies to determine the size of a ring's stable area. In these studies, a model of the ring exists in which particles can be tracked. Particles are launched in the model with a variety of different initial conditions and then tracked around the ring for a few hundred to a few thousand terms or until the particles become lost. The short term dynamic aperture is then drawn as a boundary that separates the region of space where particles survived from the region where they were lost. The short term dynamic aperture is then judged to be sufficiently large based on certain criteria which are necessary for long beam lifetimes and high injection efficiencies. If the short term dynamic aperture is not large enough, then the lattice design needs to be modified.

---

[1] This work was supported by the Director, Office of Energy Research, Office of Basic Energy Sciences, Material Sciences Division of the U.S. Dept. of Energy, under Contract No. DE-AC03-76SF00098.

Short term dynamic aperture studies by themselves provide minimal information about the beam dynamics. The results of these studies only specify whether a particle with a given initial condition will survive over a short time period. These studies provide no information about how unstable particles are lost, and thus about the long term stability of particles. As a result, short term dynamic aperture studies are of limited help for improving the design of the lattice.

Frequency map analysis is a numerical method that is useful in studying the stability of dynamical systems. Frequency map analysis was developed by Jacques Laskar (1–6) who initially applied the method towards problems in celestial mechanics. According to the KAM theorem (7–9), in the phase space that is sufficiently close to integrable conservative system, many invariant tori will persist. Trajectories starting on one of these tori remain on it thereafter, executing quasiperiodic motion with a fixed frequency vector depending only on the torus. The frequency analysis method will numerically computes the frequency vectors associated with each of these invariant tori.

Although the frequencies are strictly speaking only defined and fixed on these tori, the frequency analysis algorithm (NAFF) will numerically compute over a finite time span a frequency vector for any initial condition. One can thus construct a map, called the frequency map, which will associate the frequency vector to each set of action-like initial conditions. On the set of inital conditions corresponding to KAM tori, the frequency vector will be a very accurate approximation of the actual frequencies, and the frequency map will be smooth as the restriction to this set of a true diffeomorphism (11), which will not be the case in the chaotic regions Therefore one can then distinguish between regular and nonregular orbits by just looking to the frequency map. Moreover, the frequency vector of regular orbits will remain fixed within numerical accuracy. Those that do not remain fixed are also revealed as nonregular and their orbits are chaotically diffusing.

The NAFF algorithm is described in detail in other papers (2,4,11). The focus of this paper is to study the dynamics of the Advanced Light Source (ALS) storage ring lattice using frequency map analysis. In the paper, the procedure to generate the frequency map is outlined. First the map of the ideal lattice (no errors) is presented. Second, measured linear focusing errors are included in the lattice which shows enhanced resonance excitation and its effects on the injection rate. Finally the results of long term tracking studies are presented.

GENERATING FREQUENCY MAPS

Two ingredients are necessary in order to generate the frequency map. The first is a model of the lattice in which a particle can be tracked, turn-by-turn, around the ring. The second is a frequency analysis postprocessor which takes the turn-by-turn tracking data and computes the fundamental frequencies (NAFF package).

For the study of the ALS, a new computer code is written that combines a particle tracking code, DESPOT (14), with the NAFF postprocessor. Particles are launched on a regular mesh in action space. (In all cases presented in this paper, the particles were launched with an initial position offset but no initial angle.) Each particle is then tracked for N turns ($N = 1024$ for the results in this paper) or until the particle is lost. Each surviving particle is mapped into the frequency plane by computing the fundamental frequencies or tunes of that particle by means of frequency analysis. For each run, 40,000 particles (400×100 mesh) are tracked.

In the case of the ALS, more than 99% of the computational time time is spent in the tracking loop. There is little overhead in the computation of the frequencies.

The particles are tracked in series. On a dedicated SUN SPARC 10 workstation the map takes several days to generate. However, since all the particles are completely independent of each other, the particles could in practice be tracked in parallel. Therefore the generation of the frequency map is ideally suited to run on a massively parallel processing (MPP) machine where the map would take minutes instead of days to generate.

ADVANCED LIGHT SOURCE

Dumas and Laskar first applied frequency map analysis techniques in accelerator dynamics to the study of the global dynamics and diffusion of a modelized cell of the ALS (1). Later Laskar and Robin studied the full 12-cell ALS lattice and looked at the effect of artificial symmetry breaking (15). In that study one quadrupole was detuned by 5%. The detuned lattice showed strong excitation of low-order resonances that were not excited in the ideal ring. This increase in resonance activity caused a reduction in the dynamic aperture. An inspection of the map clearly showed that the nominal working point was not the optimal one in order to maximize the size dynamic aperture. By just looking at the map it was possible to predict the position of a better working point where the map is cleaner and the dynamic aperture is larger.

In this paper frequency maps are generated for lattices with measured gradient errors. Also the issue of long term stability is explored through long term tracking of certain particles that lie in chaotic regions of the map. In all cases presented in this paper, particles are tracked on-energy and with no synchrotron oscillations.

In Figure 1, the results of short term tracking of the ideal lattice is presented. Each of the points plotted in Figure 1 (left plot) represent the initial coordinates of all particles that survived 1024 turns. As can be seen from Figure 1, the dynamic aperture extends to more than 22 mm horizontally and 11 mm vertically at the tracking point.

Ideal Ring

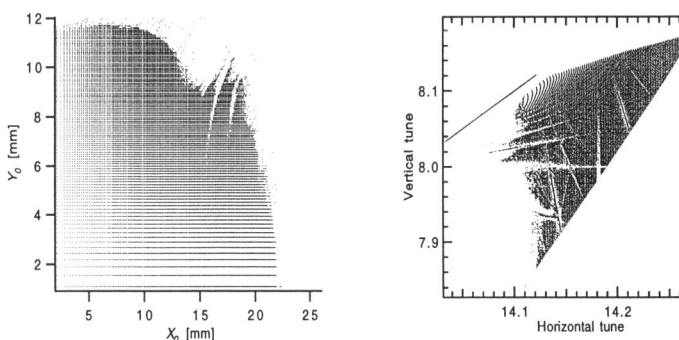

FIG. 1. Ideal lattice: Initial coordinates of all particles that survived 1024 turns (left) and the corresponding frequency map in frequency space (right).

The frequency map is generated by computing the tunes of each surviving particle (using the first 1024 turns) and mapping them into the frequency plane. The frequency map for the ideal lattice is presented in Figure 1 (right plot). The working point of the machine is $\nu_x = 14.28$ and $\nu_y = 8.18$ (upper right corner of Figure 2).

Small amplitude particles tend to oscillate with tunes close to the working point. Particles with large amplitudes oscillate with tunes that are shifted left and downward from the working point. This is due to the negative tune-shift with amplitude that is generated by the sextupoles. In particular, particles that are oscillating at large horizontal amplitudes are shifted more downwards, and particles that are oscillating at large vertical amplitudes are shifted more to the left.

We also see the influence of resonances in the map. In regions of tunespace where there are no strongly excited resonances, the tunes are evenly spaced. In regions of the map where there are strongly excited resonances the tunes are unevenly spaced. In these regions the motion may be chaotic and the particles may have large orbit diffusion rates.

Lattice with Measured Errors

The ALS lattice is designed with 12 sectors which in the ideal ring are all identical (12). The beam dynamics in the ALS lattice is very sensitive to linear focusing errors which perturb the natural 12-fold symmetry of the ring. When the symmetry of the lattice is broken, many low-order resonances, that were previously not exited in the ideal ring, may become excited. This increased resonance excitation has been observed experimentally in the beam tail distributions (13). As a result gradient errors coming from mispowered quadrupoles or orbit misteering in sextupoles can greatly effect the dynamic aperture and beam profile.

Because of this fact, in the spring of 1996, the gradient distribution of the ring was measured (16). The measurements revealed that four quadrupoles were more that 1% lower than the rest. The errors created an rms perturbation of the betatron function that is as large as 6% horizontally and 19% vertically. These measured gradient errors were included in the lattice and the map was generated. The resultant map can be seen in Figure 2 (left plot).

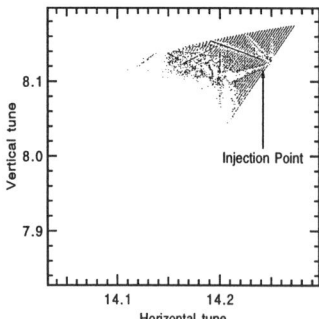 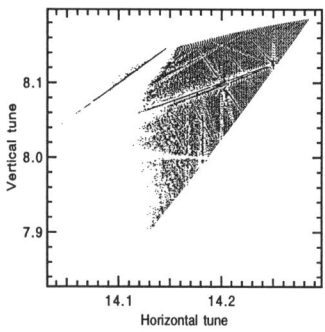

FIG. 2. Frequency map with measured errors (left) and with symmetry partially restored (right)

Comparing the frequency map for the ideal ring (Figure 1) with this map it is clear that the dynamic aperture is smaller in the ring with broken symmetry. Also there is an enhancement of low-order resonance excitation. In particular, one sees that many excited resonances intersect in the region near the tune of 14.25 horizontally and 8.125 vertically. This is a critical area because this is close to where particles are injected into the ring (see Figure 2). Because of the enhanced resonance excitation, one might imagine that as particles are injected into the ring, they may chaotically diffuse out of the beam. In other words, this increase in resonance activity may have an impact on our injection efficiency.

The ring's symmetry was partially restored by adjusting individual quadrupole gradients in the lattice. The model was then refit to match the new ring. The resulting betatron beating was measured and found to be less than 1% both horizontally and vertically (16). A map for this new lattice was generated and the results can be seen in Figure 2 (right plot). Comparing the left and right plots in Figure 3, it is clear that the more symmetrical lattice has a much cleaner map than the old lattice. As a result one would expect that the injection rates would be better in that lattice. A measurement of the injection efficiency was made for the two lattices. The injection efficiency was found to be a factor of two larger in the more symmetrical lattice (16).

Orbit Diffusion

In a previous paper (3), Laskar used frequency map analysis to study the diffusion of orbits in the case of a four dimensional symplectic map. In that study, orbits were followed over an extended time with the frequencies being calculated at regular intervals. This allowed their orbits to be tracked in the frequency plane. As expected, in regular regions of the map there was practically no diffusion. In nonregular regions of the map the diffusion rate was very rapid and some particles escaped. A similar study is made here for the ALS system. It is shown that the map provides an intuitive understanding of the long term stability of the particles.

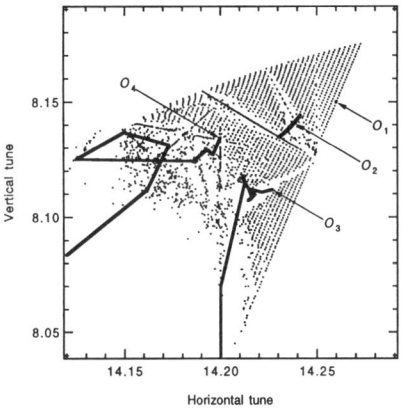

FIG. 3. Orbit diffusion of four orbits over 500,000 turns

In particular, let's look at diffusion in the lattice in the case of large symmetry breaking (left plot in Figure 2). In Figure 3 the background points are just the same as the points plotted Figure 2. Several distinct orbits (O_1, O_2, O_3 and O_4) are followed over an extended time (500,000 turns). On each interval of 1024 turns the frequencies are computed and plotted as points connected by lines. For the orbit, O_1, in a regular region, there is practically no diffusion. For orbits in nonregular regions (O_2, O_3 and O_4) the diffusion is large. Take the orbit O_2 for example. The diffusion is large but perpendicular to the $3\nu_x + 2\nu_y = 59$ resonance. However the diffusion of the orbit is bounded by regular regions of the map that border that resonance. Because of the direction of the diffusion and the location of the regular regions, it is unlikely that particles in the region will escape in a reasonable time.

On the contrary, orbits O_3 and O_4 escape in a relatively short number of turns ($< 100,000$). Unlike the diffusion of the orbit O_2, the related nonregular region of the frequency map is not bounded, but widely open on the outside of the beam. Thus many resonances overlap, and the direction of the diffusion for both O_3 and O_4 randomly changes. Therefore one suspects that many particles in this nonregular area will escape.

SUMMARY

This study demonstrates how frequency map analysis uncovers the global dynamics of the ALS. The method clearly shows how resonance excitation increases and the injection efficiency suffers when measured errors are included in the model. The method also provides strong intuition about orbit diffusion and particles long term stability.

ACKNOWLEDGMENTS

One of the authors (DR) would like to thank the staff of the ALS and in particular Alan Jackson for supporting this work.

REFERENCES

1. Dumas, H. S., Laskar, J.: Global Dynamics and Long-Time Stability in Hamiltonian Systems via Numerical Frequency Analysis, *Phys. Rev. Lett.* 70, (1993), 2975–2979
2. Laskar, J.: The chaotic motion of the solar system. A numerical estimate of the size of the chaotic zones, *Icarus*, **88**, (1990), 266–291
3. Laskar, J., Frequency analysis for multi-dimensional systems. Global dynamics and diffusion, *Physica D* **67** (1993) 257–281
4. Laskar, J., Froeschlé, C., Celletti, A., The measure of chaos by the numerical analysis of the fundamental frequencies. Application to the standard mapping, *Physica D*, **56**, (1992) 253-269
5. Laskar, J. Robutel, P.: The chaotic obliquity of the planets, *Nature*, **361**, (1993), 608–612
6. Laskar, J., Joutel, F., Robutel, P., Stabilization of the Earth's obliquity by the Moon, *Nature* **361** (1993) 615-617
7. Kolmogorov, A.N.: 1954, On the conservation of conditionally periodic motions under small perturbation of the Hamiltonian *Dokl. Akad. Nauk. SSSR*, **98**, (1954), 469
8. Arnold V., Proof of Kolmogorov's theorem on the preservation of quasi-periodic motions under small perturbations of the Hamiltonian, *Rus. Math. Surv.* **18**, **N6** (1963) 9–36
9. Moser, J.K., *Nachr. Akad. Wiss. Göttingen, Math. Phys. Kl.* II (1962), 1–20.
10. Laskar, J., Frequency map analysis of an Hamiltonian system, *Workshop on Non-Linear dynamics in Particle accelerators, Arcidosso Sept. 1994 AIP Conf. Proc.* (1995) **344** 130–159
11. Laskar, J.: Introduction to Frequency Map analysis.*in the proceedings of 3DHAM95 NATO Advanced institute, S'Agaro, June 1995* (1996) in press
12. 1-2 GeV Synchrotron Radiation Source Conceptual Design Report, LBNL publication *PUB-5172 Rev.* (1986)
13. Robin, D. et. al., Observation on Nonlinear Resonances in the Advanced Light Source, *Workshop on Non-Linear dynamics in Particle accelerators, Arcidosso Sept. 1994 AIP Conf. Proc.* (1995) **344** 170–175
14. DESPOT is a symplectic integrating tracking code which was written by Etienne Forest.
15. Laskar, J. and Robin, D., Application of Frequency Map Analysis to the ALS, *to be published in a special addition of Particle Accelerators dedicated to the Workshop on Single-Particle Effects in Large Hadron Colliders, Montreux, Switzerland (1996)*
16. D. Robin, J. Safranek, G. Portmann and H. Nishimura; Model Calibration and Symmetry Restoration of the Advanced Light Source, *to be published in the proceedings of the 1996 European Accelerator Conference. (1996)*

TESLA FEL Gun Simulations with PARMELA and MAFIA

Min Zhang and Petra Schuett*

*Deutches Elektronen Synchrotron (DESY) -MPY-,
Notkestr. 85, D-22603 Hamburg, Germany
* FB 18, Fachgebiet Theorie Elektromagnetischer Felder, TH Darmstadt,
Schlossgartenstr. 8, D-64289 Darmstadt, Germany*

Abstract The most recent simulation results of the DESY TESLA FEL gun are presented. Two codes are used: PARMELA and MAFIA. Since the two use different schemes in particle simulations, we will address their differences and try to give an explanation for them.

INTRODUCTION

In recent years, a great deal attention has been aroused to using the state-of-the-art linear accelerator technologies to build a high performance Angstrom regime FEL light sources. DESY is now pushing forward to building an FEL using its on-going TESLA Test Facility (TTF) with its wave lengths tunable from 20 nm to 6 nm (VUV) (1). The lasing is based on the Self-Amplified Spontaneous Emission (SASE) scheme, where no resonant optical mirrors are needed. For an efficient SASE lasing, the two most important beam parameters, transverse and longitudinal emittances, must be reasonably small. The former describes the parallelism of a beam ray and the latter a measure of beam monochromaticity.

For the TTF FEL, λ_{ph}=6 nm, γ=2000 (1 GeV), it requires ε_t^n=1 mm-mrad according to $\varepsilon_t^n=\lambda_{ph}\gamma/(4\pi)$, if we keep the undulator length to 30 meters. This number becomes even more strict, if a factor of two emittance dilution in the bunch compressor is taken into account. In the undulator section, longitudinal and transverse emittances are coupled. Corresponding to 1 mm-mrad ε_t^n, ε_s should be no greater than 32 deg-keV for 1nC charge. The goal of the simulations is to find a set of gun and drive laser parameters with which the two emittances are fulfilled.

PARMELA AND MAFIA SIMULATIONS

The overall structure to be simulated is shown in Fig. 1. The gun has a

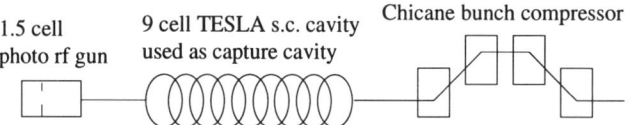

Figure 1. Overall TESLA FEL injector setup

symmetric rf coupler which accommodates an easy adjustment of the focusing solenoid. The capture cavity (CC) is the TESLA L-band 9 cell super-conducting cavity with E_{acc} = 15~25 MV/m. The overall simulation length is about 5~6 m.

Parameter Space for Simulations

Given numbers: klystron power = 5 MW, frequency = 1.3 GHz, charge = 1nC. Automatic variables: E_h - electric field at cathode, E_h/E_f - field balance, B_{zmax} - maximum B field on axis, Z_B - location of B_{zmax}, φ_{rf} - launch phase, r_b - bunch radius, τ_1 - rising and falling time of laser pulse, τ_2 - FWHM of laser pulse.

Two goals are to achieve: find the minimum transverse emittance and minimize the transverse emittance with practical constraints.

The Minimum Transverse Emittance

We used the mesh scheme in PARMELA to simulate space charge effects, since the point-to-point scheme was very noise. For all simulations below, 10,000 particles were used.

In order to reduce transverse emittance, the laser temporal profile should be as close to a square as possible, for the radial outward space charge forces are uniform over the bunch length, making the solenoidal focusing forces uniformly compensated (2). Generally speaking, a large aspect ratio of bunch length over radius is more favorable for getting a low transverse emittance. But a low longitudinal emittance requires a short bunch. Considering common laser capability, we prefer to use a flat-top laser pulse, which is technically realized by adding several gaussian-shaped laser pulse subsequently together.

Trajectories of selected particles are shown in Fig. 2. Most particles (in core) are "reflected" away from the axis while a few (in head and tail) are traveling across the axis due to lack of space charges near the axis.

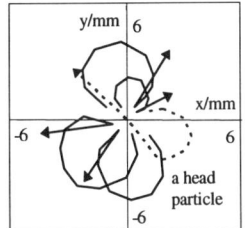

Figure 2. Traces of selected particles in transverse plane

Low emittance means small bunch radius and small divergence. To achieve small radius, we need an inward impact to the particles by a strong solenoidal field. But this will increase transverse potential energy. The particles will be bounced more quickly from the axis, resulting in a large divergence.

The minimum ε_x^n obtained for the gun itself was 1.3 mm-mrad, the longitudinal emittance ε_s = 39.5 deg-keV at z = 200 cm from the cathode with φ_{rf} = 32^0, B_{zmax} = 2200 G, Z_B = 12 cm, r_b = 1.5 mm, τ_1 = 2.2 ps, τ_2 = 8.76 ps, E_h = 50 MV/m, E_h/E_f = 1.0. We refer to this set of optimum

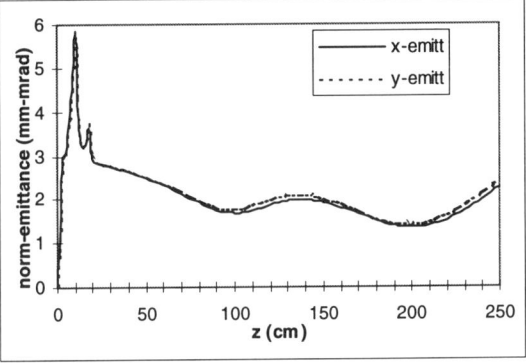

Figure 3. Transverse emittances (gun only)

emittances as case-A. $\varepsilon_x^n(z)$ is plotted in Fig. 3. When the capture cavity was included, we got ε_{xmin}^n = 1.02 mm-mrad, ε_s = 40 deg-keV at z = 407 cm with E_{acc} = 15 MV/m, Z_{cc} = 115 cm, and φ_{cc} = -10^0. This is called case-B optimum.

The above results are fine if all of the parameters can be realized. It is, unfortunately, not the case. The issue is Z_{cc}, which is too close to the cathode. The minimum distance from the cathode plane to the first iris of the capture cavity is 150 cm, which covers 40 cm gun plus rf coupler, laser port, ICT, valve, BPM, Farady cup, and thermal transition of the cryomodule. Another limit is the laser bandwidth (σ_t). The minimum σ_t of a currently available laser system is 3.5 ps. Furthermore, the ideal capture cavity was replaced by the DESY C19 cavity. It has an uneven field profile. We redid the case-B with Z_{cc} = 160 cm, σ_t = 3.5 ps, and C19 cavity's field profile.

It was found that for this practical case (referred to as case-C), it was almost impossible to use multi-pulse scheme to achieve a flat-top laser pulse in order to get a low ε_x^n, because this would increase ε_s to an untolerable value.

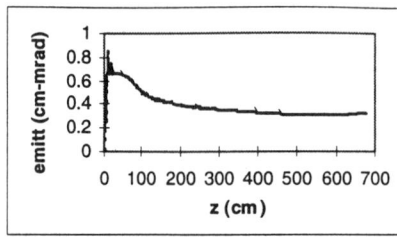

Figure 4. Normalized emittance (case-C)

Parameters used for this case were $\varphi_{rf} = 32^0$, $B_{zmax} = 2110$ G, $Z_B = 13$ cm, $r_b = 1.5$ mm, $\sigma_t = 3.5$ ps, $E_h = 50$ MV/m, $E_h/E_f = 1.0$, $E_{acc} = 14.6$ MV/m, $Z_{cc} = 160$ cm, $\varphi_{cc} = -14^0$. We got $\varepsilon_{xmin}^n = 3.02$ mm-mrad, $\varepsilon_s = 42$ deg-keV at z = 544 cm. With $\varphi_{cc} = -14^0$ we got α_s (α parameter of longitudinal phase space) = -2.68. According to (3), compression ratio is equal to 2.9. Figure 4 shows ε_x^n versus z for case-C.

Bunch Collimations

In the case-C, we find that both ε_x^n and ε_s are too high with respect to their design values. It has been found that once Z_{cc} is set to 160 cm, it is impossible to achieve the desired emittance values according to extensive PARMELA simulations. A straightforward passive way to improve the emittance is to apply collimators to the beam. We simulated both transverse and longitudinal collimations with MAFIA. It was found that in certain circumstances longitudinal collimation is more effective than transverse one. This can be demonstrated by collimating the case-A bunch at the minimum ε_x^n (Fig. 5).

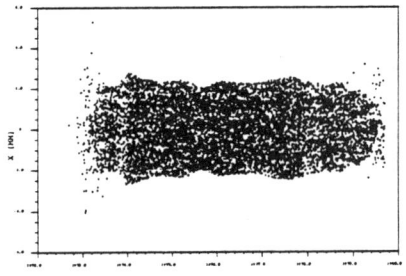

Figure 5. Longitudinal bunch shape at minimum transverse emittance (case-A)

By clipping off head and tail, we loose ~ 10% particles but improve ε_x^n and ε_s by 27% and 31%, respectively. If we apply transverse collimator to the same bunch at the same loss, we get 9.3% gain in ε_x^n but 1% loss in ε_s. The deterioration of ε_s can be understood by studying Fig. 6, which shows energy spread in the transverse plane. Clipping off outside a certain x range will make statistic energy deviation bigger due to statistical deconcentration of energy values. From this it can be deduced that transverse collimation is more effective only when the bunch head and tail part extend more than its core does in the radial direction, as the bunch shape Fig. 7 shows. It is the bunch at $\varepsilon_x^n = \varepsilon_{xmin}^n$ of case-C (i.e. z = 544 cm).

A longitudinal collimation of 23% particles has the same effect as a transverse one of 16% particles, both resulting in a 50% reduction of ε_x^n, i.e. $\varepsilon_x^n = 1.5$ mm-

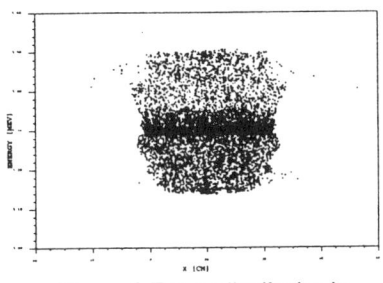
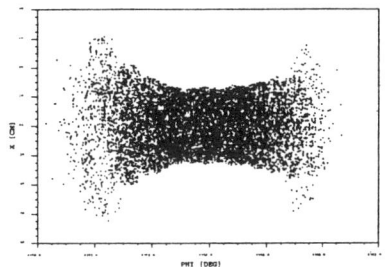

Figure 6. Energy distribution in transverse plane (case-A)

Figure 7. Longitudinal bunch shape at minimum transverse emittance (case-C)

mrad. The former gives ε_s = 15.2 deg-keV and the latter 28.4 deg-keV. This example demonstrates that the transverse collimation can also be superior to the longitudinal one. There seems no reason to reject either of them.

Another more active way to reduce ε_x^n is to increase E_{acc} of the capture cavity. We simulated E_{acc} = 25 MV/m with the case-A gun settings. We got ε_{xmin}^n = 1.1 mm-mrad when Z_{cc} = 140 cm. Without any collimations, we had nearly reached the ε_x^n = 1.0 mm-mrad design value. So for the TTF FEL phase II (1GeV), the design goal is within reach, though largely depending on the laser performance.

MAFIA Simulations

MAFIA TS2 module was used for the PIC simulation of the gun. It is a 2.5D r-z self-consistent leap-frog code. The dynamics is of second order.

Since the TTF FEL gun has a very short bunch (σ_s = 1~2 mm), TS2 needs a very fine mesh near the cathode, typically ~ 10~20 µm. Thus we only simulated the gun itself with no radial boundaries, i.e. no wakefields included. It was a mere point-to-point space charge simulation. We present the MAFIA results here only for a crosscheck with PARMELA's.

The simulation was carried out using 230,000 mesh points and 4,950 macro-particles. The smallest mesh steps in r and z were 100 µm and 20 µm, respectively. Total computing time was 4 days CPU (SUN Sparc 20). $\varepsilon_x^n(z)$ is given in Fig. 8.

We find that the MAFIA ε_x^n is quite different from the PARMELA's (Fig. 3). Up to 25 cm, ε_x^n has three peaks. Their locations are the same for the two plots. The major difference lies in the second peak, which is just within the solenoid.

In this range particles undergo a strong rotation. A possible explanation would be that MAFIA uses radial momentum p_r in ε_x^n evaluation while PARMELA may just use cartesian momentum p_x, in which p_φ is implicitly included. We crosschecked the 6D coordinates for the two codes and found no significant differences. So the only thing that may differ is at the final emittance evaluation step. In general, the two codes deliver comparable emittance results in rotation-free regions. We plan to use even finer mesh ($\Delta z_{min} = 10$ μm) in MAFIA simulations in the near future.

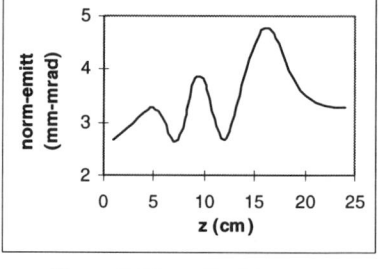

Figure 8. Normalized transverse emittance obtained with MAFIA

SUMMARY

Numerical simulations of the DESY TESLA FEL gun and capture cavity using PARMELA was presented. Two sets of boundary conditions were used, one for optimum design, the other for practical design. The simulation showed that the TESLA FEL gun design goals of ε_x^n and ε_s could be achieved, depending largely on the laser performance. High E_{cc} is favorable for lowering ε_x^n. Both transverse and longitudinal bunch collimations are effective ways to passively reduce ε_x^n and/or ε_s. Under different circumstances, one of them is more effective.

MAFIA simulation of the gun was possible but very time-consuming. The PARMELA and MAFIA results generally agreed with each other. The discrepancy in ε_x^n was mainly in the solenoid range, which was probably caused by the presence of particles angular momenta. To speed up MAFIA simulations, further improvements are planned, e.g. a 2D-2D interface, new Δt (particle) = n × Δt (EM field) scheme, etc.

REFERENCES

1. "A VUV free electron laser at the TESLA test facility at DESY, Conceptual design report", DESY Print, June 1995, TESLA-FEL 95-03
2. Carlsten, B.E., "New photoelectric injector design for the Los Alamos National Laboratory XUV FEL accelerator", Nucl. Instr. and Meth. A285 (1985) 313
3. Rosenzweig, J. B., "Pulse compression in TTF Injector II", DESY Print, April 1995, TESLA 95-05

Manipulation of High-Current Pulses for Heavy-Ion Fusion*

W. M. Sharp, D. A. Callahan, A. Friedman, and D. P. Grote

Lawrence Livermore National Laboratory, University of California
Livermore, California 94550, U. S. A.

Abstract. For efficient induction-driven heavy-ion fusion, the current profile along a pulse must be modified in a non-selfsimilar manner between the accelerator and the target. In the accelerator, the pulse should have a duration of at least 50 ns in order to make efficient use of the induction cores, and the current should by nearly uniform along the pulse to minimize the aperture. In contrast, the optimal current profile on target consists of a main pulse of about 10 ns preceded by a longer low-current "foot." This pulse-shape manipulation must be carried out at the final pulse energy (5-10 GeV for 200 amu ions) in the presence of a large nonlinear longitudinal space-charge field. A straightforward method is presented here for doing the required pulse shaping. Induction-cell voltages are generated using idealized beam profiles both in the accelerator and on target, and they are verified and checked for error sensitivity using the fluid/envelope code CIRCE.

1. Introduction

For a heavy-ion beam to be an effective inertial-fusion driver, the energy deposited on the target must be carefully controlled. Although the details of energy deposition depend on the target design, a typical indirect-drive target requires a main pulse of about 10 ns duration, with several beams delivering a total of 5-10 MJ. This pulse is normally preceded by a low-current "foot" containing about a tenth of the main-pulse energy, and through the entire pulse, ion velocities should be nearly uniform to minimize chromatic aberration in the final focus. For induction accelerators, which work best with a small number of high-current beams, the peak current on target can be as high as 10 kA per beam, and in the absence of charge neutralization, the longitudinal space-charge fields can be as high as 10 MV/m.

Producing a beam like this at the target is problematic because the optimal beam in an accelerator has a very different current profile. For induction drivers, the aperture is minimized for beams with a uniform line-charge density and at most a small head-to-tail velocity increase to give slow compression. Furthermore, the pulsed-power is used most efficiently when the beam duration is much longer than the pulse rise and fall times, leading to pulses that are ten to twenty times longer in the accelerator than on target. These longer beams also reduce the longitudinal space-charge fields at the beam ends to a manageable magnitude. Considerable manipulation is needed to shape this optimal accelerator pulse into the form needed on target, and the large space-charge field of the final beam prevents this shaping from being done during the final stages of compression, since induction accelerators can presently produce an average gradient of about 1 MV/m.

The conventional scenario for generating the current profile needed on target is to give a pulse carefully controlled current and velocity profiles before it enters a "drift-compression" lattice, where alternating-gradient (AG) focusing is applied, but the beam is not accelerated. The imposed head-to-tail velocity variation, or "tilt," is as large as 5%, and it is "tailored" so that the beam compresses to the desired final profile just as the space-charge field causes longitudinal-velocity "stagnation" at all points along the pulse. This plan requires intricate manipulation of the beam after the main acceleration sequence, probably in a straight "shaping" lattice between the accelerator and the drift-compression section.

---

* Work performed under the auspices of the U. S. Department of Energy by Lawrence Livermore National Laboratory under contract W-7405-ENG-48.

To date, the principal published work addressing the question of shaping a constant-current beam into a form suitable for final compression is a paper by Ho, Brandon, and Lee[1]. That paper used analysis and one-dimensional numerical simulations to model a beam-compression sequence that produced uniform current and velocity on target. The effects of longitudinal space charge were ignored during shaping, on the grounds that shaping could be done in a sufficiently short time. The present paper presents an alternate procedure for shaping intense heavy-ion pulses that extends the initial work in three major ways. First, we look at the problem of producing a final beam with a prescribed profile,rather than one with uniform current. Second, we include the longitudinal space charge during compression and compensate for it by modifying the shaping fields. Finally, the effect of the changing radial structure of the beam is included approximately in the model. The new procedure is detailed in Section 2 of this paper, and the numerical simulation of a simple case is presented in Section 3. The concluding section gives a brief summary of conclusions.

2. PROCEDURE

The procedure here for calculating the pulse-shaping fields has four main steps, each of which is detailed below. The current and velocity profiles needed at the beginning of the drift-compression lattice are first determined by simulating the target beam drifting backward through the section. A corresponding beam with the same total charge and with user-specified current and velocity profiles is then constructed at the start of the shaping section, and a set of shaping fields to be applied in that lattice is calculated, ignoring the effects of the longitudinal space-charge field. Finally, longitudinal-control fields, referred to here as "ears," are added to the shaping fields to approximately balance the space-charge field. The combined fields are validated and checked for error sensitivities by numerical simulations using the code CIRCE[2], which models the transverse beam dynamics using envelope equations and treats longitudinal dynamics using a Lagrangian fluid approximation. The beam is divided lengthwise into $I-1$ "slices," each characterized by an enclosed charge δq_i, and the positions and velocities of the slice boundaries are calculated from fluid-like equations. The code has been benchmarked against the three-dimensional particle-in-cell code WARP3d[3] and is found to give reliable results, provided that the normalized emittance does not increase significantly and there is little longitudinal mixing.

2.1 Calculate beam profile entering drift section

The current and velocity profiles on target must be selected first. This choice obviously depends on the details of the target, and the target in turn must be designed with limitations of the accelerator and final focus in mind. Here, we use a simple "generic" target beam, and we sidestep problems of the final focus and transport to the target by assuming that there is little change in the current and velocity after final focus. Instead, we set up a beam that is matched to the quadrupole lattice in the drift-compression section.

We would like to obtain the appropriate beam at the beginning of the drift-compression section by numerically running the chosen target beam backward in time through the compression lattice. However, due to limitations of the modeling code CIRCE, we instead run the equivalent problem of the target beam expanding forward in time through the spatially reversed compression lattice. If the scaled longitudinal velocity $\beta \equiv v_z/c$ along the target beam deviates from the charge-density weighted average velocity $\bar{\beta}$, the the sign of $\beta - \bar{\beta}$ must also initially be reversed. After this beam is allowed to expand, the current is equated to the desired current I_{bf} at the endpoint z_f of the shaping section, and the desired velocity there β_f is set to $2\bar{\beta} - \beta$. Here, the f subscripts denote final values after beam shaping, and c is the speed of light in vacuum.

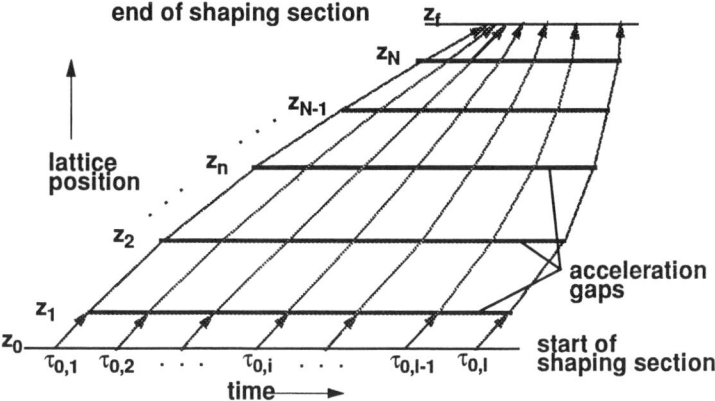

Figure 1. Cartoon of the beam-shaping sequence.

2.2 Calculate beam profile entering shaping section

To calculate an appropriate initial distribution, the shaping lattice must first be specified. In most cases, this lattice consists of AG focusing magnets with single or clustered acceleration gaps at one or more locations in a half-lattice period. In principle, the lattice could also include bends and higher-order focusing elements, but these are not considered here. The minimum number of acceleration cells is set by breakdown electric field across a cell and the velocity tilt required at the start of drift compression. If we assume instantaneous acceleration in gaps, the velocity change in the nth gap due to a voltage $V_n(\tau)$ is approximately by

$$\Delta \beta_{i,n}^2 \equiv \beta_{i,n}^2 - \beta_{i,n-1}^2 \approx \frac{2qe}{\gamma_{i,n}^3 M c^2} V_{i,n}, \qquad (1)$$

where q is the ion charge state, e is the elementary charge, M is the ion mass, and $V_{i,n} \equiv V_n(\tau_{i,n})$ is the voltage at the time $\tau_{i,n}$ that the ith beam slice traverses the gap. The Lorentz factor here $\gamma_{i,n} \approx \left[1 - \frac{1}{2}(\beta_{i,n}^2 + \beta_{i,n-1}^2)\right]^{-1/2}$ complicates the relation between $\Delta\beta_{i,n}$ and $V_{i,n}$, but since it varies by less than 0.4% during shaping and compression, we replace the factor here by the nominal value $\bar{\gamma} \approx (1 - \bar{\beta}^2)^{-1/2}$. According to Eq. (1), the average voltage that must be applied to the ith beam slice in each of the N acceleration modules is

$$\bar{V}_i \approx \frac{1}{\alpha N} \left(\beta_{i,N}^2 - \beta_{i,0}^2\right), \qquad (2)$$

where $\alpha \equiv 2qe/\bar{\gamma}^3 M c^2$. The subscript $n = 0$ here denotes values at the start of the shaping section z_0. We choose to apply the same voltage to the beam head in all cells, and likewise, to the tail, so from Eq. (2), the required head and tail voltages are $V_{head} = \bar{V}_1$ and $V_{tail} = \bar{V}_I$ respectively. We take the maximum electric field that can be held off by the gap insulators to be $E_{max} \approx 10^7$ V/m. Although advanced insulators can hold off twice this voltage for up to 50 ns, the longer beams required for heavy-ion fusion (HIF) and the need to minimize cost lead us to use the more conservative figure. If all acceleration gaps are taken to have the same length L_{gap}, then the minimum number of cells needed in the shaping section is $N_{min} \approx \max(V_{head}, V_{tail})/(E_{max} L_{gap})$.

Due to the low average gradient in induction accelerators, the shaping section must typically be longer that 100 m, so the beam duration is expected to decrease measurably as the beam traverses the section. To construct a suitable beam profile at z_0 to be manipulated into the calculated profile at z_f, we first estimate this change in beam duration, again assuming that the beam head and tail respectively see V_{head} and V_{tail} in each cell. In the cartoon of Fig. 1, this calculation corresponds to finding the trajectories $\tau_{1,n}$ and $\tau_{I,n}$ of the head and tail slices back to z_0. The transit time between cells in the absence of longitudinal space charge is given by

$$\Delta \tau_{i,n} \equiv \tau_{i,n+1} - \tau_{i,n} \approx \frac{\Delta z_n}{c\beta_{i,n}}, \qquad (3)$$

where $\Delta z_n \equiv z_{n+1} - z_n$ is the distance between the centers of the nth and $(n+1)$th gap, and from Eq. (1),

$$\beta_{i,n} = \left(\beta_{i,0}^2 + \alpha \sum_{m=1}^{n} V_{i,n}\right)^{1/2}. \qquad (4)$$

Defining z_{N+1} to be z_f and $\tau_{i,N+1}$ to be τ_i there, we iterate Eqs. (3) and (4) for the head and tail slices backward to z_0, obtaining $\tau_{i,n}$ and $\beta_{i,n}$ along those trajectories. The desired current profile $I_{b0}(\tau)$ at z_0 is then scaled to have the appropriate duration $\tau_{I,0} - \tau_{1,0}$ and the correct total charge $Q = \sum_{i=1}^{I} \delta q_i$.

To complete the specification of the initial beam, we must choose times $\tau_{i,0}$ for $1 < i < I$ such that each slice contains the same current δq_i as the beam at z_f. These intermediate τ values are found numerically by requiring that the τ integral of I_{b0} from $\tau_{1,0}$ to $\tau_{i,0}$ equals the charge in the same section of beam $Q_i = \sum_{j=1}^{i} \delta q_j$.

2.3 Calculate pulse-shaping fields

The values of $V_{i,n}$ and $\tau_{i,n}$ now must be specified for $1 < i < I$ and $1 < n < N$. We would like $V_{i,n}$ to be close to $\bar{V}_i$ so that all cells have to generate approximately the same voltage, but taking $V_{i,n} = \bar{V}_i$ does not, in general, give the correct slice arrival times at z_f. Instead, we write $V_{i,n} = \bar{V}_i + \delta V_{i,n}$ and choose the voltage difference $\delta V_{i,n}$ to satisfy two constraints:

$$\sum_{n=1}^{N} \delta V_{i,n} = 0, \qquad (5a)$$

$$\frac{1}{c} \sum_{n=1}^{N-1} \frac{\Delta z_n}{(\beta_{i,0}^2 + n\alpha \bar{V}_i + \alpha \sum_{m=1}^{n} \delta V_{i,m})^{1/2}} = \tau_{i,N} - \tau_{i,1}. \qquad (5b)$$

The first constraint guarantees that the final slice velocity is unchanged, and the second, from Eq. (3), ensures that the total transit time from z_1 to z_N is correct. We simplify Eq. (5b) by assuming that the contribution of the voltage perturbation to $\beta_{i,n}$ is small, so that the expression can be linearized in $\delta V_{i,n}$:

$$-\frac{\alpha}{2c} \sum_{n=1}^{N-1} \frac{\Delta z_n}{(\beta_{i,0}^2 + n\alpha \bar{V}_i)^{3/2}} \sum_{m=1}^{n} \delta V_{i,m} \approx \tau_{i,N} - \tau_{i,N}^0. \qquad (6)$$

Here, $\tau_{i,n}^0$ denotes the slice arrival times along constant-voltage trajectories, given explicitly by

$$\tau_{i,n}^0 \equiv \tau_{i,1} + \frac{1}{c} \sum_{m=1}^{n-1} \frac{\Delta z_m}{(\beta_{i,0}^2 + m\alpha \bar{V}_i)^{1/2}}. \qquad (7)$$

As a sample solution of these equations, we take $\delta V_{i,n}$ to vary linearly with n, so that

$$\delta V_{i,n} = A_i \left(\frac{n-1}{N-1} - \frac{1}{2} \right). \tag{8}$$

This expression satisfies Eq. (5a) for any value of A_i, and an appropriate A_i approximately satisfying Eq. (5b) is found by substituting Eq. (8) into Eq. (6). We then calculate the voltage adjustments $\delta V_{i,n}$ from Eq. (8) and find the corresponding $\tau_{i,n}$ values by iterating Eq. (3), using the new voltages $\bar{V}_i + \delta V_{i,n}$ in the $\beta_{i,n}$ expression of Eq. (4). The resulting shaping voltages and time data are then written to an external file for use in CIRCE. This procedure for calculating the shaping voltages is used in the sample problem of Section 3, but it is in no sense optimum. Performance measures for global optimization can readily be formulated, but the large number of shaping cells and beam slices typically used makes the problem unwieldy. Various optimization schemes are under study.

The pulse-shaping method here is predicated on the assumption of instantaneous acceleration in the induction cells. We have compared the results of this approximation with an analytic expression obtained for a finite-length gap, and we find that the relative error in $\tau_{i,n}$ due to the approximation is, to lowest order,

$$\frac{L_{gap}}{\Delta z_n} \left[\frac{\Delta \beta_{i,n}^2}{4 \bar{\beta}_{i,n}^2} \right]^2, \tag{9}$$

where $\bar{\beta}_{i,n} c$ is the average velocity of the ith slice going through the nth cell. Even when the gap length is a significant fraction of Δz_n, this error is still small due to the low fractional energy change in a single induction cell of a heavy-ion driver. For example, the factor $\Delta \beta_{i,n}^2 / \bar{\beta}_{i,n}^2$ is approximately 10^{-5} for a 10 GeV driver with conventional acceleration modules. We conclude that instantaneous acceleration is a very good model for the cases under consideration.

2.4 Calculate ear fields

The straight-line trajectories assumed in calculating the shaping fields are, of course, only valid in the absence of longitudinal space-charge fields. To calculate appropriate ear fields, we use CIRCE to simulate the beam transit through the shaping section, applying the shaping fields, but with the longitudinal component of the space-charge field artificially set to zero. When the beam enters each shaping cell, the beam profile is used to calculate the space-charge field at that point, and this field is then used to construct the appropriate ear field there.[4] This ear field is added to the shaping field, and the combined field is written along with timing data to an external file for use on later runs.

3. SAMPLE CASE

The shaping procedure is illustrated here using an idealized analytic beam-current profile on target and a simple lattice for shaping and compression. The purpose is primarily didactic, since no effort has been made either to optimize the lattice or to choose a target profile appropriate for a specific target design.

3.1 Target-beam characteristics

The beam on target is deliberately chosen to stress the pulse-shaping algorithm. An ion mass of 200 amu with a final energy of 10 GeV is assumed to give the beam high rigidity, and the 10 kA peak current on target, which is appropriate when there are four beams there,

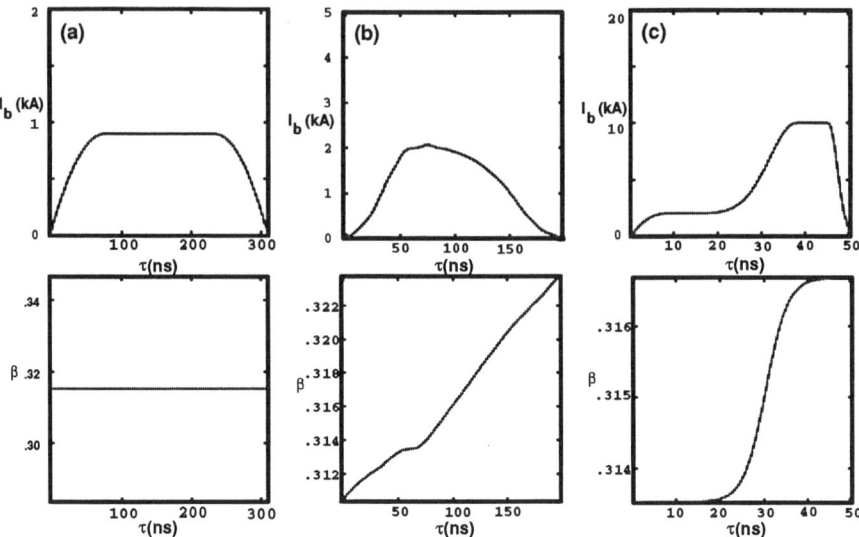

Figure 2. The current and velocity profiles of an idealized heavy-ion beam (a) after acceleration, (b) after shaping, and (c) at target. Profiles are plotted as functions of arrival time τ relative to the beam head.

exaggerates the effect of space charge. The final normalized x and y emittances are taken to be 10^{-5} m-rad, which should allow the beams to be focused to a 3 mm radius spot on the target. The functional form of the current has the general features needed for a target with two end plugs, similar to that specified by Ho[5], but the actual analytic form has been chosen for its simplicity and convenience.

It is found that the current profile on target cannot be specified arbitrarily. For beams in which all ions have the same mass and momentum, several classes of profiles are unusable due to pathological dynamics when they expand. For example, the expansion wave from the leading edge of the main pulse tends to form a steep density gradient as it expands into the lower-density foot, provided that the space-charge force is sufficiently high. If "wave-breaking" occurs, ions in the expanding wave overlap the slower ions in the foot, producing a multi-valued velocity distribution that is impossible to produce using external fields. Similarly, if the target distribution has two peaks separated by a lower current region, a beam distribution with two distinct velocity components in the same spatial region is formed if the two expansion waves meet. In the fluid-like longitudinal dynamics model of CIRCE, where the charge in every beam slice is constant, the density develops nearly singular spikes at the points where a multi-valued distribution would form in reality. Consequently, we disqualify a target profile when such spikes form during expansion.

The tendency to form a multi-valued distribution is reduced by increasing the current in the foot, which increases the velocity of electrostatic waves there, or by lengthening the rise time of the main pulse, which reduces the space-charge force driving the main-pulse expansion. A third approach is to give the main pulse a slightly higher velocity, so that it "catches up" with the foot as the beam compresses. All three strategies have been used to generate the target profile shown in Fig. 2c. Here, the 2 kA foot current is about a factor of two higher than would be optimal, but choosing the foot velocity to be about 1% lower

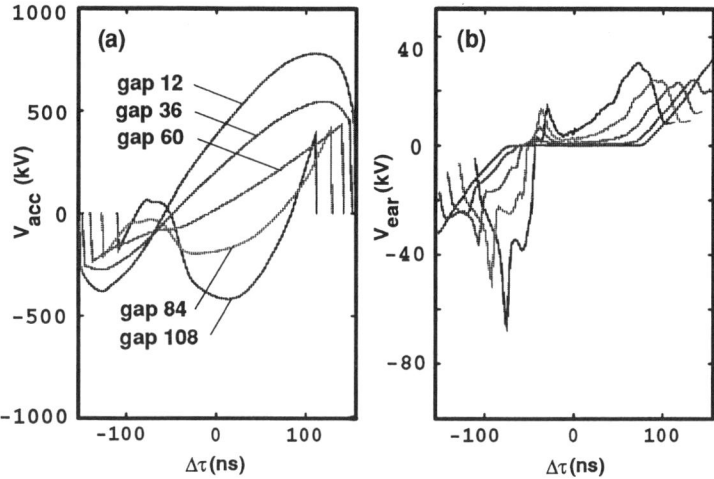

Figure 3. The (a) shaping fields and (b) ear fields at selected acceleration modules in the shaping section. No optimization has been done to smooth ear fields.

than in the main pulse partially compensates for the higher current by increasing the focal spot size, so that some current misses the capsule end plugs. Although this profile expands in a reversible fashion, the deliberate defocusing of the foot wastes energy and lowers the capsule gain. Careful optimization is clearly needed here.

3.2 Beam characteristics before and after shaping

The 45-period drift-compression section used here is a simple AG lattice with a half period of 4.18 m, with the quadrupole strength and occupancy chosen to give an maximum undepressed phase advance of 75°. When the target-beam in Fig. 2c is allowed to propagate backwards without longitudinal confinement through this lattice, it develops the current and velocity profiles shown in Fig. 2b. Here, as in Fig. 2c, we have reversed the velocity so the plot is appropriate for a compressing beam. For this case, the beam duration has approximately quadrupled to 200 ns, and the rarefaction waves from the ends of the main pulse have met, leaving the velocity roughly linear. We also note that the beam tail has spread out more than the head during expansion because the short fall time of the main pulse leads to a large space-charge field there.

The 60-period shaping section for this case is also a simple AG lattice with the same nominal phase advance and 4.18 m half period as the drift-compression section. However, the beam-pipe radius is smaller, reflecting the lower line-charge density, and most of the space between quadrupoles is filled with acceleration gaps. A cluster of ten 10 cm gaps is inserted between each quadrupole doublet, allowing an average gradient in excess of 1 MV/m. We assume that the beam from the accelerator has a uniform longitudinal velocity, although allowing a small tilt would reduce the velocity change that must be imposed in the shaping section. When the beam ends are followed backward through this lattice, we find that the beam duration increases to about 300 ns at the starting point. Choosing the rise and fall times to be 25% of the beam duration, we can then set up the initial beam profile shown in Fig. 2a. We see by comparing Figs. 2a and 2b that a head-to-tail velocity tilt in excess of 4% must be imposed by the shaping section.

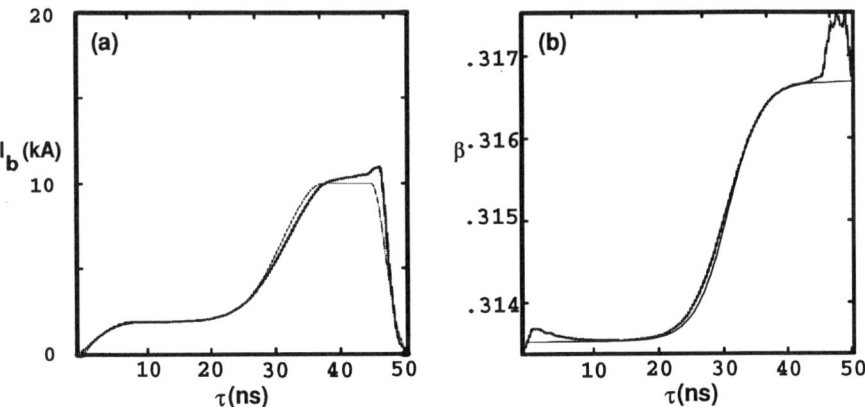

Figure 4. Comparison of the (a) current and (b) velocity of the simulated beam (wide line) and assumed beam (narrow line) at the target.

3.3 Shaping and ear fields

A selection of waveforms for this illustrative case, taken at five equally spaced modules, is shown in Fig. 3a. Here, the duration of the waveform is seen to decrease at successive z positions as the beam compresses. The procedure in Section 2c generates acceptably smooth shaping voltages, but the corresponding ear voltages, shown in Fig. 3b, are seen to have considerable high-frequency content. This feature is a consequence of the shaping voltages being adjusted along each slice trajectory, without regard for voltages along neighboring trajectories. An erratic current profile results when $\tau_{i,n}$ at any z_n are irregularly spaced, producing longitudinal fluctuations in the space-charge field there. This irregularity in the ear fields would be reduced by a suitable optimization procedure.

3.4 Dynamics

When the idealized accelerator beam in Fig. 2a is transported through the shaping and drift-compression sections with ears and shaping fields applied, the final beam has the basic features of the intended target beam, as shown in Fig. 4. The wide lines in the figure give the current and velocity profiles of this test pulse, and the narrow lines are the intended profiles from Fig. 2c. The greatest departures from the intended profiles occur at the tail of the main pulse, where the longitudinal space-charge field is largest. Presumably, these departures occur because the high space-charge fields make the beam particularly sensitive to errors in the current or velocity at the start of the drift-compression section. The ear fields are found to be important to replicating details of the the intended current profile. When the ears in Fig. 3b are not added to the shaping fields, the final current profile of the main pulse assumes a nearly triangular shape, and the rising and falling portions stretch out by a total of 13 ns.

A preliminary study of the sensitivity of the final profiles to errors in the shaping fields has been made. It is found that random 2.5% rms errors in the magnitudes of the shaping fields produce final profiles that are not significantly worse than the error-free case in Fig. 4, but progressively larger discrepancies from the intended profiles are seen for random 5% and 10% errors. Systematic errors of the same magnitude cause much worst discrepancies in each case and are probably unacceptable even at the 2.5% level. Random shaping-field timing errors with an rms magnitude of 10 ns are found to cause a head-to-tail current increase exceeding 10% across the tops of both the main pulse and the foot. This deviation from the optimum current profile is probably acceptable in practice, but for a random 25 ns timing jitter, the beam no longer compresses successfully, due to the formation of large density spikes near the beam tail. These sensitivities illustrate the need for high-precision pulsed power in the shaping section.

A much more thorough examination of errors sensitivities is obviously needed. In particular, the effects of limiting the bandwidth of the shaping voltage should be checked. Also, these cases should be simulated using WARP3d to verify that the fluid-like longitudinal dynamics of CIRCE has not introduced any unphysical effects here.

4. SUMMARY

The work presented here suggests that an ion pulse from a heavy-ion driver can be systematically shaped by a series of small changes. The procedure here has not been optimized or coordinated with target simulations, but it nonetheless illustrates that predetermined beam profiles can in principal be produced. Generating the required pulses with adequate precision may prove challenging because existing pulse-generation technology imposes limits on both the magnitude and frequency content of the shaping signal. A more fundamental problem is the limited choice of target profiles that can be formed by drift-compression in vacuum. Various ways to circumvent this problem as being considered, such as the partially neutralized transport, the application of "mid-course corrections" during compression, and the use of separate beams, possibly with different ion masses, for the foot and main pulse.

5. REFERENCES

[1] D. D.-M. Ho, S. T. Brandon, and E. P. Lee, *Part. Accel.* **35**, 15-42 (1991).

[2] W. M. Sharp, J. J. Barnard, D. P. Grote, S. M. Lund, and S. S. Yu, "Envelope Model of Beam Transport in ILSE" in *Proceedings of the 1993 Conf. on Computation Accelerator Physics*, Pleasanton, CA, February 1993, pp. 540-548.

[3] A. Friedman, D. P. Grote, and I. Haber, "Three Dimensional Particle Simulation of Heavy Ion Beams," *Phys. Fluids B* **4**, 2203-2210 (1992).

[4] W. M. Sharp, D. A. Callahan, and D. P. Grote, "Longitudinal Dynamics and Stability in Beams for Heavy-Ion Fusion," in *Proceedings of the 1995 Heavy-Ion Fusion Symposium* Princeton, NJ, 6-9 September 1995, in press.

[5] D. D.-M. Ho, private communication.

ELECTROMAGNETIC AND PIC SIMULATION

Modeling Large Heterogeneous RF Structures[1]

Zenghai Li, Kwok Ko, Vinay Srinivas and Toshiyasu Higo[†]

SLAC, Stanford University, Stanford, CA 94309
[†]KEK, Tsukuka, Ibaraki, 305 Japan

Abstract

Large heterogeneous structures are difficult to model on a numerical grid because of the limitations on computing resources, so that alternate approaches such as equivalent circuits and mode-matching have been developed to treat this problem. This paper will describe the three methods and will analyze a structure representative of the SLAC and JLC detuned structures to compare the efficacy of each approach.

1. Introduction

Large heterogeneous structures have been in use for many years in accelerator systems such as the Stanford Linear Collider (SLC) which employs 3-m long 86-cell tapered disk-loaded waveguides (DLWGs) to provide constant acceleration. The SLC experienced beam breakup due to cumulative deflection of the bunch by transverse wakefields in the linac. In the TeV scale linear colliders proposed by SLAC and KEK, wakefield effects will be important since multi-bunch operation (versus single bunch in the SLC) is required to reach desired luminosity. A primary goal of the accelerator structure design has been the suppression of wakefields to preserve the low-emittance of the long bunch trains (90 bunches) during their delivery to the interaction point.

2. Detuned Structure

A promising candidate being considered for SLAC's Next Linear Collider (NLC) [1] and KEK's Japan Linear Collider (JLC) [2] is the detuned structure whereby the cell dimensions are varied by design to detune the most dangerous dipole modes in a Gaussian manner. This detuning decoheres the dipole modes so that the aggregate wakefield is reduced to a safe level (a factor of 100 from peak) over the length of the bunch train. SLAC has fabricated and tested a detuned structure to confirm the wakefield reduction [3]. Meanwhile, considerable effort has been devoted to the analysis of wakefields to assess the effectiveness of this suppression mechanism.

---

[1]This work was supported by the U.S. Department of Energy, under contract No. DE-AC03-76SF00515.

3. Modeling Approaches

The NLC detuned structure is a 1.8-m 206-cell DLWG while the JLC prefers a 1.3-m 150-cell design, both operating at X-band. Due to the size of the geometry and the small variation in cell to cell dimensions (of the order of 10th's of microns), direct numerical simulation is beyond most computing capabilities, except on multi-processor supercomputers [4]. Alternative approaches such as equivalent circuits and mode-matching have been developed, and calculations based on these methods have been documented in the literature.

This paper discusses three methods for modeling large heterogeneous structures: (1) equivalent circuits with two circuit chains [5], (2) mode-matching with open modes [6], and (3) grid-based simulation using code MAFIA [7]. Due to limited space, the reader is referred to the original work for theoretical details. The emphasis here is to apply these methods to a test structure small enough that's numerically practical, yet has properties representative of the NLC and JLC detuned structures. More importantly this enables for the first time a direct comparison between the three approaches.

The wakefield analysis will assume the structure is closed at both ends since the dipole modes couple weakly to the input/output waveguides. Then one is looking at essentially a large standing-wave cavity in which the eigenmodes in the dipole band of frequencies are of interest. The transverse wakefield is obtained from the eigenmodes by summing their contributions as follows

$$W(s) = 2 \sum_{m=1}^{M} K_m sin(\frac{\omega_m s}{c}) \tag{1}$$

where M is the number of modes, ω_m and K_m are the eigenfrequency and the transverse kick of the mth mode respectively. In all the models described below the cell geometry is taken to be from disk to disk, and cell data such as dispersion curves and mode patterns are obtained with MAFIA.

4. Two-chain Model

A two-chain circuit model comprises two rows of LC circuits with coupling between neighbors as well as cross-couplings between rows. To parametrize each LC circuit, the corresponding cell is considered as part of a periodic chain and one uses selected passband frequencies from the Brilloun diagram to determine the circuit frequency and the coupling coefficient. Accordingly two chains of circuits can be parametrized by considering two passbands.

The motivation for the two-chain model is because the dipole modes consist of a mixture of TE111 and TM110 modes. The two circuit chains incorporate the cell response in the lower and upper dipole passbands, and therefore include both TE and TM components. The two-chain result in the periodic limit is found to approximate far better the actual dispersion curves than the single-chain case. Once the parameters of the $2 \times N$ circuits are determined and the boundary conditions at the two end cells are fixed, the eigenvalue problem is completely specified. One then solves for the eigenfrequencies and computes the transverse kick of each eigenmode from the kick factors of the individual cells by following the results derived in [5].

5. Open-mode Model

While conventional mode-matching separates the cell geometry into regions of uniform waveguides, the open-mode model considers the whole cell as one region and expands the electric field in the cell as a superposition of cell modes generated with open (or magnetic) boundary conditions at the disks. In principle, an expansion in closed modes with electrically shorted planes can also be used. The open modes are appropriate here because the dipole mode of interest is synchronous with the beam near π phase which is specified by magnetically shorted planes. The advantage this method has over the usual waveguide mode expansion is superior convergence since only the cell mode close in frequency to the eigenmode of the structure will predominate.

The coupled set of equations relating the mode expansions in one cell to the next have been derived in [6], and each of them has the conservative form which balances the stored energy in the cell against the power flux through the disk openings. It turns out that the closed modes are needed as well (otherwise power flow is zero). The parameters in the coupled system are determined by overlapping volume and surface integrals of the open and closed modes. Setting the boundary conditions at the end cells then completes the eigensystem for the whole structure. There are $j \times N$ eigenvlaues to solve for if the expansion includes j number of modes, and the kickfactors in Eq. (1) can be obtained directly by integrating the fields of the open modes in each cell.

6. Test Structure with Linear Taper

The test structure in this comparison is a linearly tapered DLWG at X-band with flat-ended disks so a structured-grid code like MAFIA can model exactly. Available computing resources limit the length of the structure that can be simulated accurately to around a maximum of forty cells. The taper is chosen to cover a group velocity span v_g/c from 6.46% to 3.91% for the lower dipole mode, which is within the NLC and JLC range. This span of v_g also simplifies the analysis by excluding any overlap with the upper dipole passband. The slope of the taper as measured by $dv_g/dN/c$ with $N = 41$ is close to the average NLC and JLC values of .04% and .05% respectively. A stronger taper is also considered by reducing N to 21 cells. A schematic of the 21-cell linearly tapered DLWG is shown in Fig. 1.

Figure 1: 21-cell linearly tapered structure.

7. Wakefield Results

The 21-cell ($dv_g/dN/c = .12\%$) results from MAFIA and two-chain model on dispersion curves, eigenfrequencies and kickfactors are shown in Fig. 2. Fig. 3 compares the eigenfrequencies and kickfactors from MAFIA and open-mode model. The number of modes included in the open-mode expansion is 8. The wakefields from all three methods are plotted in Fig. 4. Finally the same comparison for a 41-cell structure ($dv_g/dN/c = .061\%$) is shown in Fig. 5.

Taking the MAFIA results close to being exact one can summarize the comparison with the following observations. In the limit of a gradual taper (41 cells), both the two-chain and open-mode model agree well with MAFIA. For a less gradual taper (21 cells), the two-chain model is less accurate than the open-mode model. The difference from MAFIA in the two-chain model can be attributed to how well the dispersion curve and the coupling coefficient are approximated and to terminations at the ends. In the open-mode model the mismatch at the disk openings and the end boundaries are sources of errors.

8. Conclusion

The NLC and JLC detuned structures have tapers more gradual than the test structure, so one can expect less errors from approximations applied at the disk openings. Furthermore, because the modes in these structures extend over many cells, and because those touching the ends contribute less to the wakefields (they are at the tails of the Gaussian distribution), the effects of the end cells are reduced. Therefore both the two-chain and open-mode methods can be considered as viable alternatives to modeling these large heterogeneous structures when numerical simulations become impractical.

References

[1] The NLC Design Group, *Zeroth-order Design Report for the Next Linear Collider*, LBNL-PUB-5424, SLAC Report 474, UCRL-ID-124161, 1996.
[2] JLC Group, "JLC-I," KEK Report 92-16, 1992.
[3] C. Adolphsen *et. al.*, "Measurement of wakefield suppression in a detuned X-band accelerator structure," SLAC-PUB-6629, 1994.
[4] X. Zhan *et. al.*, "Parallel computation of transverse wakes in linear colliders," this proceedings.
[5] K. L. F. Bane and R. L. Gluckstern, "The transverse wakefields of a detuned X-band accelerator structure," SLAC-PUB-5783, 1992.
[6] M. Yamamoto, *Study of Long-Range Wake Field in Accelerating Structure of Linac*, KEK Report 94-9, 1995.
[7] MAFIA User's Guide Version 3.20, CST GmbH, Darmstadt, Germany.

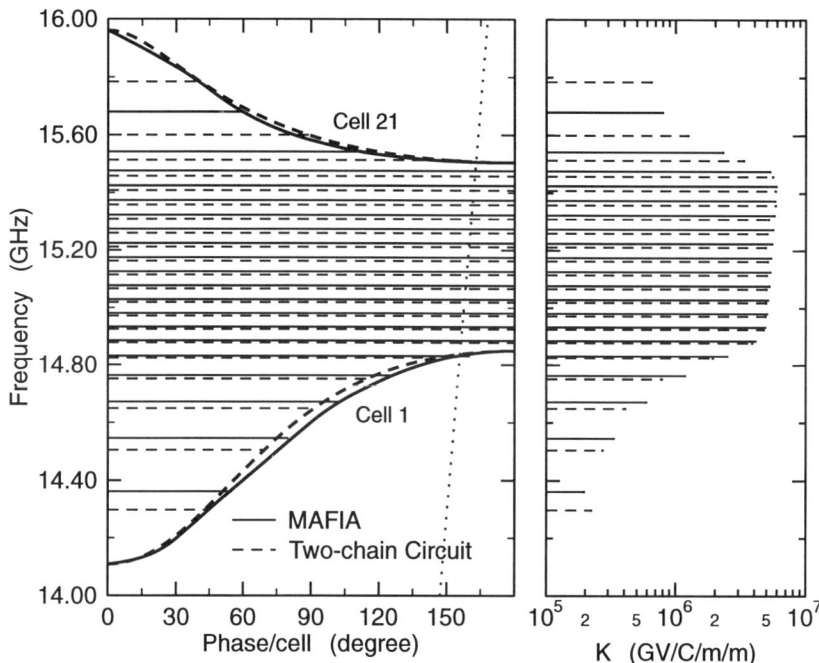

Figure 2: Coupled frequencies and kick factors of a 21-cell structure, MAFIA and two-chain circuit results.

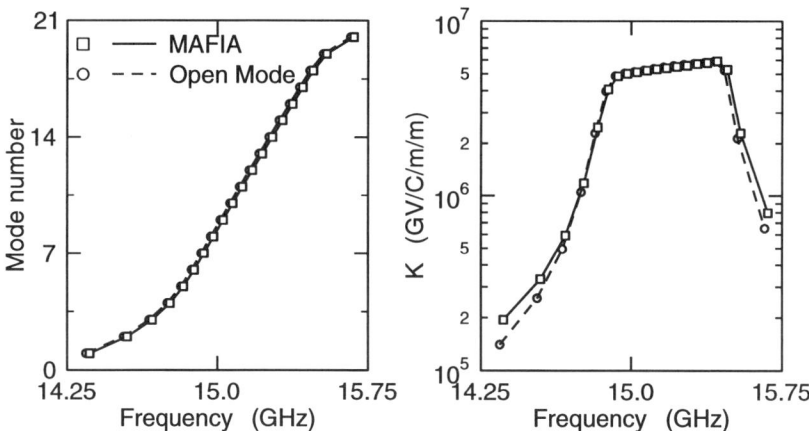

Figure 3: Coupled frequencies and kick factors of a 21-cell structure, MAFIA and open-mode results.

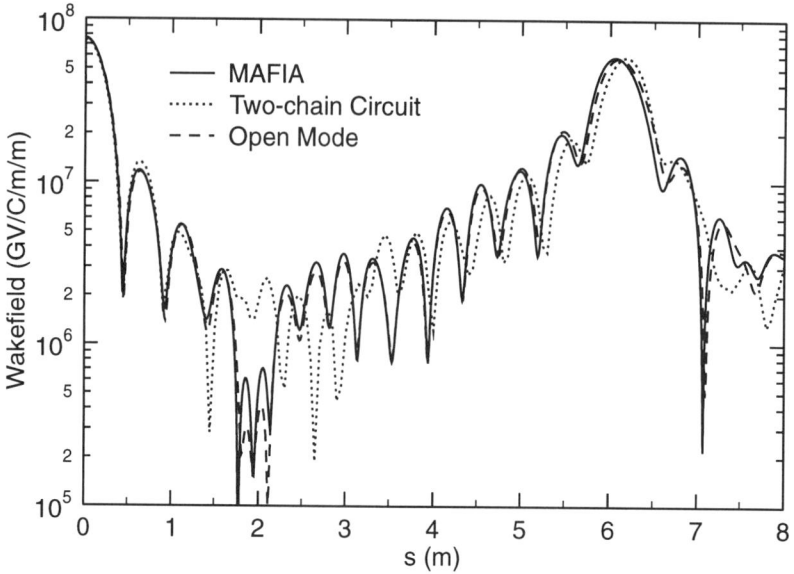

Figure 4: Wakefield envelopes for a 21-cell structure.

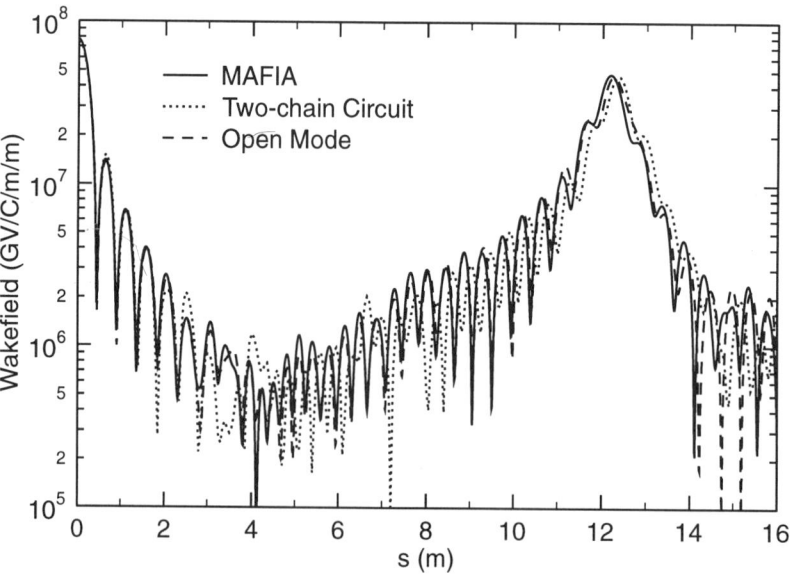

Figure 5: Wakefield envelopes for a 41-cell structure.

Simulation of High-Brightness Electron Photoinjectors

Eric R. Colby

Fermi National Accelerator Laboratory *
Mail Stop 308, P. O. Box 500, Batavia, IL 60510-0500

Abstract

The problem of accurately modeling RF photoinjector electron sources in three dimensions with reasonable computational power remains a challenge. Development work on extensions to the Los Alamos particle tracking code PARMELA (Phase and Radial Motion in Electron Linear Accelerators) to handle fully 3-D problems including non-rotationally symmetric RF cavity and space-charge fields is discussed. Algorithms for computing the space charge forces in a point-to-point fashion including the effects of image charges on the cathode and retardation effects will be presented. A SIMD implementation of the code using the PVM (Parallel Virtual Machine) routines to perform far-reaching optimization studies is also discussed.

INTRODUCTION

The optimization of high charge low emittance RF photocathode driven electron injectors poses an interesting numerical modeling challenge for several reasons. The intense RF fields give rise to initial electron accelerations of order $10^{19} m/s^2$, causing electron velocities to range from thermal to relativistic within a few tens of picoseconds. Bunch densities (generally $\sim 10^{12}/cm^3$) are high enough initially to have plasma periods in the hundreds of picoseconds and high enough to develop space charge fields that are comparable in magnitude to the RF accelerating field. Wakefield effects are of particular

---

*Operated by the University Research Association, Inc. for the U. S. Department of Energy.

interest when the cathode is conductive, as the transverse component of the space charge field will be reduced by the presence of image charges, and are of interest in the design of very high brightness injectors where wakefield effects may be a dominant source of emittance dilution if unremedied. Bunch dimensions typically are small fractions of the overall accelerating structure dimensions (generally a hundredth the size) requiring spatial modeling to encompass features spanning three orders of magnitude in size. Photoinjectors of even the azimuthally symmetric type are manifestly 3-D problems owing to such asymmetry-inducing perturbations as RF power input couplers, magnet misalignments, and quadrupole focussing.

PROBLEMS OF INTEREST

Progress in understanding photoinjector dynamics has brought photoinjector beam quality within striking distance of the demands placed on electron sources by the next generation of linear colliders. The Trillion Electron-Volt Superconducting Linear Accelerator (TESLA) will require interaction point (IP) emittances of $1 \times 20\pi$ mm-mr normalized transverse emittances, comparable to the brightest sources today[1]. Traditionally, the low emittances required at the IP were produced in a damping ring, which for the extremely long bunch trains planned for TESLA (800 bunches spaced 1 μS apart) must be many kilometers in circumference. Given the size and complexity of the TESLA damping ring, designing a source capable of eliminating the ring becomes extremely attractive.

For the near term, however, a source capable of emitting a pulse train of the correct structure but with relaxed emittances ($20 \times 20\pi$ mm-mr) is needed to address more immediate problems of wakefield effects in the superconducting accelerator sections of TESLA. Consequently, two different electron injectors have been examined: the first, a symmetric emittance injector, is intended for immediate use at the TESLA Test Facility (TTF) [2]; the second, an asymmetric emittance photoinjector, is intended for future use as the injector to the TESLA-500 accelerator, with the goal of providing emittances low enough to eliminate the costly electron damping ring. The design of the symmetric emittance photoinjector is virtually complete [3], with a prototype having been under test at Argonne National Laboratory for eight months. Figure 1 shows a wireframe sketch of the asymmetric emittance RF gun structure, and a drawing of the symmetric emittance gun, including the focussing solenoids. Table 1 summarizes the most significant parameters which affect the beam qualities of interest (primarily the three emittances), together with the parameter ranges considered, expressed as dimensionless quantities where possible.

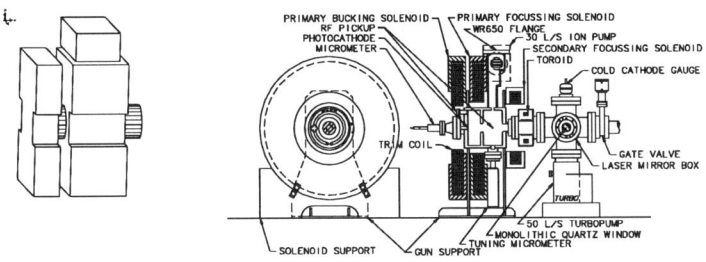

Figure 1: Wireframe drawings of the asymmetric (left) and symmetric (right) emittance gun structures

Parameter	Definition	Range Considered
Initial bunch density	$N/(\sigma_x \sigma_y \gamma \sigma_z)$	$1 - 2 \times 10^{12} cm^{-3}$
Bunch aspect ratio	$\sqrt{\sigma_x \sigma_y}/\gamma \sigma_z$	3 – 40
Beam launch phase	(after $E_z = 0$)	30-90 Degrees
Gun accelerating gradient	$E_z/(2m_e c k_{rf})$	1.1-1.8
Gun cavity geometry	N.A.	Pillbox and Reentrant
Solenoid focussing strength: primary	f_L/λ_{rf}	1-30
Solenoid focussing strength: secondary	f_L/λ_{rf}	5-∞
Drift distance to linac	l_d/λ_{rf}	0-10
Linac phase	$\phi - \phi_{sync}$	−30 to 30 Degrees
Linac gradient	$E_z/(2m_e c k_{rf})$	0.36-0.90
Linac cavity geometry	N.A.	TESLA geometry only
Pulse compression ratio	σ_{z0}/σ_{zf}	1-4
Pre-compression optics configuration	N.A.	doublet, triplet
Post-compression optics configuration	N.A.	doublet, triplet

Table 1: Principal parameters influencing photoinjector beam quality

ENHANCEMENTS TO PARMELA

Modelling of the asymmetric emittance photoinjector required that a number of additions be made to PARMELA. RF cavity fields are represented in two dimensional polar coordinates, unacceptable for the manifestly three-dimensional fields of the asymmetric RF structures considered (see figure 1 above). Space charge fields are calculable in three dimensions using a direct implementation of Coulomb's law, but require large numbers of macroparticles ($\geq 10^3$) to model small emittance ($1\pi mm - mr$) beams without excessive numerical noise.

RF, Wakefield and Space Charge Modelling

Two new fully three-dimensional methods for representing RF cavity fields have been implemented, via quadruple Fourier series expansion in TM-like space harmonics, and through direct importation of both **E** and **H** components (calculated by Superfish (2-D) or HFSS (3-D)) and subsequent interpolation using Legendre polynomials.

Wakefield kicks from the RF structure have been modeled in a non-self-consistent fashion by evaluating kicks to each of the particles at the end of the structure from a Green function developed by an external wakefield code (ABCI and Xwake, for instance).

Several new space charge calculating algorithms have been developed, and several more are on the way. Already developed is an ellipsoidal macroparticle generalization of the standard spherically screened point-by-point calculation. For the asymmetric emittance injector, the beams large aspect ratio ($\sigma_x/\sigma_y \sim 40$) causes fields perpendicular to the small dimension to be underestimated owing to the small number of nearby macroparticles. Modelling the macroparticles as charged ellipsoids with aspect ratios matched to the beam increases the number of sources contributing significantly at each point within the bunch. As pointed out by H. Liu [4], screening the macroparticles applies a spatial filter to the fields; with independent control of the macroparticle size in each coordinate direction, it is possible to control the filtering in each coordinate direction separately.

Still under development are 3-D methods exploiting the nearly 2-D nature of the asymmetric emittance injector problem which model the beam as a set of infinite charged rods oriented parallel to the beams large dimension (in this case, the horizontal). Also under development is a causally correct algorithm including the radiation field contributions due to bending magnets. The method is simple but time-consuming: second order Taylor series expansions are developed and periodically updated for each particle's coordinates and momenta. Retarded times and relative velocities are then evaluated and used to compute the relativistically correct fields acting on each macroparticle in a point-by-point fashion.

Diagnostics and Particle Loading

Diagnostics to examine the RF kicks, wakefield kicks, space charge kicks and magnetic element kicks have been added to allow validation of the calculations. Numerous phase space diagnostics have been added to aid in understanding the emittance correlation removal process [5] including "slice" emittance computation, representative particle tracing, Twiss parameter computation, and others. Modelling of the "slit" mask emittance measurement technique has been included for more direct comparison with measurement.

Bunch distributions with unusual spatial distributions arise and require modeling in a number of instances: Gaussian pulse "stacking" to achieve longer laser pulses yields a fringe pattern on the cathode when the pulses cross at a small angle [6]. Additionally, non-uniformities in the laser transport system give rise to spatial structure within the beam. Modelling of arbitrary charge distributions has been implemented using standard Monte Carlo rejection techniques to generate electron distributions from video image data (or, more generally from a matrix of $I(x, y)$) to permit modeling of these deviations from the ideal.

SIMD PARALLELIZATION OF PARMELA

Optimization of the injector involved guided parameter scanning over a very limited region of the total parameter space shown in table 1. Typical execution times for a modeling of the full symmetric emittance injector are on the order of 4-6 hours on the faster workstations, (e.g. SGI's 150 MHz R4010), making the time required for a two parameter 10×10 search two to three weeks. Typical execution times for modeling of the full asymmetric emittance injector are on the order of 12-16 hours, making the same 10×10 search last more than two months. To expedite searching of the 16-dimensional parameter space described above, a single-instance multiple-data (SIMD) parallelization of PARMELA using the Parallel Virtual Machine (PVM) library [7] was implemented. More than 60 workstations comprised the virtual machine used in the computations, with job scheduling, load leveling, input parameter generation, output file analysis and archiving being accomplished from a single "master" node in a fully automated fashion.

CONCLUSIONS

Several improvements to the widely used electron injector code PARMELA have been made in an effort to reproduce the three dimensional dynamics of an asymmetric emittance injector. Provisions for modeling wakefield effects, arbitrary field geometries, and space charge effects have been added, and the diagnostics enhanced to aid understanding of the beam dynamics. A parallel version of PARMELA has been implemented to permit more wide-ranging searches of the huge design space.

ACKNOWLEDGMENTS

I thank Bruce Carlsten, Richard Sheffield and Jean-François Ostiguy for many helpful discussions leading both to improvements in the algorithms used in PARMELA, and to implementing the PVM version. I thank Jerome Gonichon for contributing his pseudo random sequence generation routines.

References

[1] C. Travier, "An Introduction to Photo-Injector Design", LAL/RT 93-08, (1993).

[2] D. A. Edwards (Ed.),"TESLA Test Facility Linac - Design Report",v.1.0,(1995).

[3] E. R. Colby,"High Brightness Symmetric Emittance RF Photoinjector Preliminary Design Report",Fermilab TM-1900, (1994).

[4] H. Liu, "Analysis of Space Charge Calculations in PARMELA and its Application to the CEBAF FEL Injector Design",CEBAF-PR-93-008, (1993).

[5] B. E. Carlsten, NIM **A285**, p.313-9, (1989).

[6] I. Will, *et al*,"A Laser System for the TESLA Photo-Injector",MBI internal document, (1994).

[7] Al Geist, *et al*, "PVM 3 User's Guide and Reference Manual", ORNL/TM-12187, (1994).

Methods Used in WARP3d, a Three-Dimensional PIC/Accelerator Code

D. P. Grote, A. Friedman
Lawrence Livermore National Laboratory
L-440, P.O. Box 5508, Livermore CA 94550

I. Haber
Naval Research Laboratory
Code 6790, Washington DC 20375

Abstract. WARP3d(1,2), a three-dimensional PIC/accelerator code, has been developed over several years and has played a major role in the design and analysis of space-charge dominated beam experiments being carried out by the heavy-ion fusion programs at LLNL and LBNL. Major features of the code will be reviewed, including: residence corrections which allow large timesteps to be taken, electrostatic field solution with subgrid scale resolution of internal conductor boundaries, and a bent beam algorithm. Emphasis will be placed on new features and capabilities of the code, which include: a port to parallel processing environments, space-charge limited injection, and the linking of runs covering different sections of an accelerator. Representative applications in which the new features and capabilities are used will be presented along with the important results.

INTRODUCTION

Heavy-ion-fusion (HIF) requires low temperature, or emittance, beams so that they can be focused on to a small fusion target. The presence of non-linear self-fields and applied fields can lead to an increase in emittance and make a self-consistent model necessary. Since the space-charge-dominated beams in an induction accelerator are effectively non-neutral plasmas, analysis of the beams can be carried out with computational modeling techniques related to those used in plasma physics. Particle-in-cell (PIC) simulation techniques are especially effective and have proved valuable in the design and analysis of both ongoing experiments and future machines, as well as in the study of basic physics issues of intense beams.

The three-dimensional electrostatic PIC simulation code, WARP3d, was developed for HIF to study beam behavior from first principles in realistic geometries. Here, we present an overview of the WARP3d code, describing in detail some of the major features which in combination make WARP3d a versatile beam simulation tool. We also present applications of the code that demonstrate the use of these features of WARP3d and highlight important results.

OVERVIEW OF WARP3D

WARP3d is part of a larger code, WARP, which includes different levels of high-current beam modeling, including WARPrz, a two-dimensional, axisymmetric PIC code, and a simple envelope equation solver which is primarily used to obtain an initial beam state well-matched to the accelerator. The code is built atop the Basis code development and run-time system, which affords a flexible and powerful interpretive user interface.

The WARP3d code combines PIC with a description of the "lattice" of accelerator elements. The fields affecting the beam particles include both the self-fields of the beam and the external fields from the lattice. The self-fields are assumed to be electrostatic and are calculated in the beam frame by solving Poisson's equation on a Cartesian grid which moves with the beam. The electrostatic solution is valid since, with the high mass of the beam ions, the beam is nonrelativistic throughout most of the accelerator. The Poisson solvers can optionally include any conductor geometry self-consistently in the field solution.

The lattice is a general set of finite-length accelerator elements, including focusing, bending, accelerating gaps, elements with arbitrary transverse multipole components, and curved accelerator sections (described in reference 1). The field from the lattice elements can be specified at one of several levels of detail which are described below.

METHODS USED IN WARP3D

A number of features and capabilities have been implemented in WARP3d to make possible simulation of a broader class of accelerators, both by way of inclusion of the relevant physics, and by optimization of the code to allow simulation of larger scale problems. The most important and novel of these are described.

Lattice Description

The lowest level of description of lattice elements is a hard-edged, axially uniform field. As the particles enter and exit such an element, "residence corrections" are used; the force applied on the particles is scaled by the fraction of the time-step spent inside the element. These residence corrections allow use of larger time-steps by ensuring that the particles receive approximately the correct impulse from the element independent of the time-step size. The applied field can be a sum of arbitrary multipole components.

At a higher level of detail, fields which vary axially can be applied. The coefficients of the multipole components are stored as tabulated data along the axial direction and are interpolated to the particles. This allows application of realistic fields, including for example fringe fields and more rapidly varying higher order multipole moments associated with a quadrupole focusing element.

The highest level of detail is specification of the field via a three dimensional grid. This allows application of fields more efficiently than a multipole description for a element with many multipole components. The field can be calculated in a different code, for example using TOSCA(3) to calculate the magnetic fields.

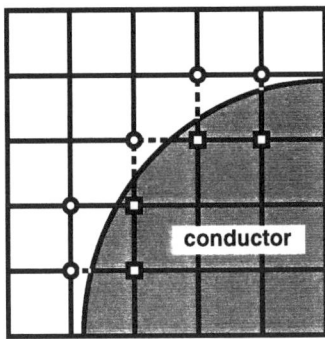

FIGURE 1. A curved conductor is shown on a grid. Poisson's equation at the circled points outside of the conductor is modified to explicitly include the location of the conductor surface by replacing the term which would use the squared points inside.

With electric elements, the field can be calculated self-consistently with WARP3d by inclusion of the electrodes in the field solution, thereby obtaining the realistic fields and image affects from the electrodes on the beam.

Subgrid-Scale Conductor Boundary Resolution

To allow more accurate modeling of electrodes for a given grid cell size and to avoid problems associated with the jaggedness of representing tilted or curved electrodes on a Cartesian grid, the iterative SOR field solver allows subgrid-scale resolution of the conducting boundary locations. Referring to Figure 1, this can be explained as follows. For grid points just outside the surface of a conductor, the form of Poisson's equation is changed to explicitly include the location of the surface. The potential at points inside the conductor, which is normally used in the evaluation of Poisson's equation at the points outside, is replaced by a value which is extrapolated from the potential at the conductor surface and at the grid point outside the surface. The method is similar to one used in EGUN(4), extended to three dimensions and time dependence.

Beam Injection

Several models of particle beam injection are used, all of which can model emission from planar or curved surfaces. Beside injection at a preset current, two space-charge limited injection methods have been implemented where the injected current is calculated self-consistently from the conditions near the emitting surface. To calculate the emitted current, the first model uses the Child-Langmuir relation in a region in front of the emitting surface, the length of which is small compared to the length of the diode. In the second model, the surface charge required to produce the desired zero normal electric field on the surface is calculated assuming an infinite conducting plane. The amount of charge injected into a grid cell is that surface charge integrated over the grid cell, subtracting the charge already in the grid cell from earlier times.

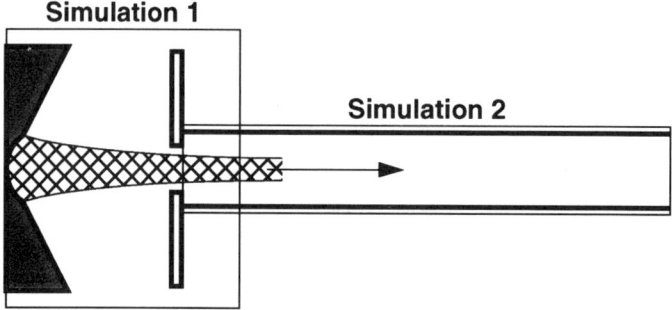

FIGURE 2. A simulation of an ion gun is linked to the following transport channel. With the simulation regions overlapping, data is saved in simulation 1 at the interface to simulation 2 and used as initial conditions in simulation 2.

For both self-consistent models, the normal electric field is obtained by calculating the potential at a point far enough in front of the emitting surface so that the grid points from which interpolation is done are all in front of the emitting surface. This avoids the problem of estimating a potential behind the emitting surface. The normal electric field is used both in the calculation of the injected current and as the field applied to particles near the emitting surface.

With space-charge limited injection, WARP3d can be run in an efficient iterative manner which captures only the steady state behavior. The algorithm is similar to that used in the two-dimensional electron gun design code EGUN(4). It should be pointed out that this capability was added to the code entirely through the Basis interpreter and required no modification of the existing code.

Linking of Simulations

For efficiency, sections of the accelerator often need to be simulated separately, linking from one to the next. For example, in simulating an ion gun followed by a transport channel, a much larger transverse grid is needed in the diode, to include the Pierce electrode, than is needed in the channel with a small pipe radius as shown in Figure 2. Other examples are the transition from a straight to a bent lattice where different symmetries should be used, and the transition from electric focusing to magnetic focusing, which have different boundary conditions and usually need different field solvers.

To link the simulations, the region covered by the first overlaps into that covered by the second, and data is saved from the first and is used as the initial conditions in the second. At each time step in the first simulation, the position and velocity of particles which pass the transition plane are saved, as well as the potential at the transition plane and the plane one grid cell to the left. In the second simulation, at each time step the particles previously saved are "injected" and the two planes of potential are loaded onto the field grid as a Dirichlet boundary condition. Two planes of potential are needed for the correct evaluation of the finite difference to obtain the electric field for particles near the transition plane. The overlap must be large enough so the fields in the region of the transition are calculated accurately.

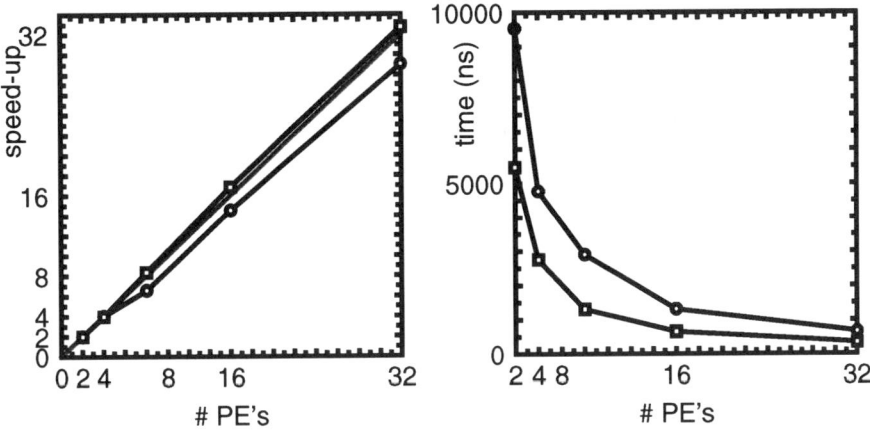

FIGURE 3. The timings of WARP3d on the massively parallel Cray T3D show nearly linear speed-up with increasing number of processors. The squares give the timings of the field solver and the circles the timing of the particles. The gray line is the ideal linear speed-up. On the right are the actual timings, nanoseconds per grid cell per field solve and nanoseconds per particle per timestep.

Parallel Processing

WARP3d was originally designed to run on vector supercomputers and on workstations. With the present trend toward massively parallel processing, WARP3d was ported to a parallel environment. WARP3d now can run on the CRAY T3D, using both the PVM(5) message passing library and the native shared memory library, and on multi-workstation clusters using PVM.

Domain decomposition is done only along the axial dimension, which is generally longer than the transverse dimensions. With the FFT field solver, the field grid is divided evenly among the processors who share one grid cell with adjacent processors. A transpose is done for the axial FFT. The number of processors and number of grid cells both must be powers of two. With the SOR field solver, an overlap of 2 grid cells is needed so that a plane which is calculated by one processor can be passed to its neighbor to be used as a Dirichlet boundary condition. The data is passed on each iteration in the field solve. The particles are divided into same regions as the field grid. This is not ideal since the beam may not fill entire axial extent of the grid leaving some processors with fewer or no particles.

An effort was made to keep the same user interface for the parallel code as for the serial code. The effort was complicated by the fact that the Basis interpreter does not run on the parallel machine. The parallel code is run through the Basis interpreter on a serial machine, allowing same structured input decks as the serial code. The processes on the parallel machine are then spawned via PVM and the initial data is passed down to them. The user can then issue commands, which are passed to the parallel code, allowing the user to run the simulation, set or access variables, and retrieve grid, particle, diagnostic, or time history data.

Timings were made, shown in Figure 3, with a model simulation on a Cray T3D

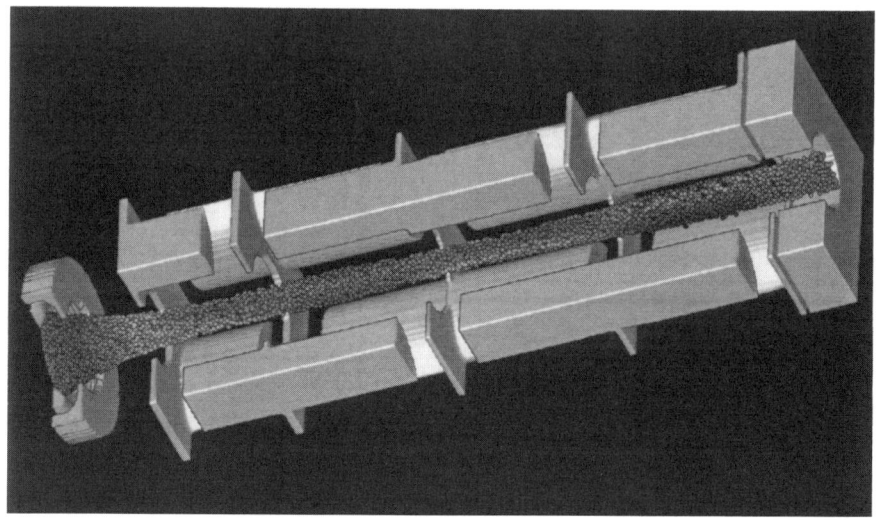

FIGURE 4. The electrostatic quadrupole injector is shown filled with beam. The shading of the beam particles is proportional to the particle's axial velocity relative to the velocity along the axis, lighter being faster. One quadrant is cut away to show the beam.

with up to 32 processors, following a beam through five lattice periods. The field solver showed the slightly superlinear speed-up due to better use of cache with smaller problem size in each processor. The particles showed slightly less than linear speed-up, likely due to small load imbalances.

APPLICATIONS

WARP3d has been used in a wide variety of applications, making use of the methods described above. Several recent applications are described, showing the use of the methods and highlighting important results.

Electrostatic Quadrupole Injector

A HIF driver-scale injector and matching section is being built at LBNL, in collaboration with LLNL, to address beam physics and technology issues(6). The injector uses a sequence of electrostatic quadrupole lenses arranged to obtain a voltage gradient along the axis. The net effect is to both confine the beam transversely and accelerate it. The beam is emitted from a hot-plate source and is accelerated through a diode section into the set of quadrupole lenses. Figure 4 shows an image from a detailed injector simulation. WARP3d was used to understand a major issue in the injector design, namely the increase in emittance caused by the high order multipole components associated with the electrode geometry.

The methods used in the simulations allowed realistic beam injection and modelling of the electric quadrupoles. The beam was formed by space-charge limited injection, using both the EGUN-like iterative mode and time-dependent injection. The subgrid-scale resolution of quadrupole electrode surfaces was used to obtain accurate fields. Four-fold transverse symmetry was used, simulating only one quadrant. Typically, the mesh size used was 40x40x348 with 300,000 particles. The runtime was approximately 30 minutes on a Cray C90, single processor. Code predictions influenced the design geometry, mitigating emittance growth, and good agreement was obtained between the simulation and experimental results.

Bending Experiments

A recirculating heavy-ion induction accelerator, or "recirculator," offers the prospect of reduced cost relative to a conventional linear accelerator because the accelerating and focusing elements are re-used many times per shot(7). A small recirculator is being developed at LLNL to explore the beam dynamics of a HIF driver in a scaled manner; the key dimensionless parameters that characterize the beam are similar to those of a driver-scale ring but the physical scale is much smaller(8).

As a precursor to full recirculator experiments, the beam is being examined in a lattice with an 45 degree bend. The bend experiment will be the first detailed examination of a space-charge dominated heavy-ion beam in a bent, alternating gradient lattice. The experiment consists of a series of electric and then magnetic quadrupoles with 9 degree bends and electric dipoles in the last five half lattice periods. The experiment has been examined with WARP3d. The simulations predict an acceptable emittance growth of about a factor of two, due entirely to the matching of the beam from a straight lattice to a bent lattice.

The beam is initially launched in a simulation of the diode which is then linked to a simulation of the following transport channel. The field from the electric quadrupoles was calculated with WARP3d using the electrode geometry and decomposed into multipole components. The multipole components from the magnetic quadrupoles was calculated analytically. The fields are applied as elements with axially varying multipole components. The electric dipole plates are also modeled in detail using the WARP3d field solver, and the fields are applied to the particles as elements with data on a 3-D grid. The simulation of the bends required the use of the bent lattice algorithm.

Many of the same methods which are used in the simulations of the bent beam experiment are also used in the simulations of the full recirculator experiment. The major difference is that the simulations are carried out over a much longer time, following the beam for hundreds of lattice periods.

Over 15 laps, the simulations show little emittance growth after the initial growth due to the matching of the beam from a straight lattice to a bent lattice. The overall degradation of the beam is acceptable.

AVLIS

An electron gun is used in atomic vapor laser isotope separation (AVLIS) to melt and vaporize the material to be separated. The electron gun has shown anomalous asymmetric beam behavior that cannot be studied with axisymmetric codes. We have begun to use WARP3d to examine the gun. The methods used are space-charge limited injection of the electrons off the emitter surface and use of magnetic fields calculated by TOSCA and specified on a 3-D grid. The full conductor geometry is simulated using subgrid-scale conductor resolution.

CONCLUSIONS AND FUTURE WORK

Many future code enhancements and applications are planned for WARP3d. Some example enhancements that are envisioned are an interface with a CAD system to input general 3-D conductor shapes, a more advanced field solver which uses an unstructured mesh outside of the beam, and inclusion of beam self-magnetic fields. Also, a number of methods in WARP3d will be further improved. For example, the mechanism for linking simulations can be made more flexible by allowing different grid cell and time step sizes in the linked run.

WARP3d makes use of many different features, all built into one package, which allows it to be applied on a wide variety of beam problems. This makes it a versatile tool is the analysis of the dynamics of space-charge-dominated beams.

ACKNOWLEDGMENTS

This work was performed under the auspices of the U.S. Department of Energy by LLNL under contract W-7405-ENG-48 and at NRL under contracts DE-AI02-93ER40799 and DE-AI02-94ER54232

REFERENCES

1. A. Friedman, D. P. Grote, and I. Haber, "Three-Dimensional Particle Simulation of Heavy-Ion Fusion Beams," *Phys. Fluids B* **4**, 2203 (1992).
2. D. P. Grote, A. Friedman, I. Haber, and S. S. Yu, "Three-Dimensional Simulation of High-Current Beams in Induction Accelerators with WARP3d," in *Proc. Int. Sympos. on Heavy Ion Inertial Fusion*, Princeton, Sept. 6-9, 1995; to be publ. in *Journal of Fusion Engineering Design*, 1996.
3. "TOSCA Reference Manual", technical paper VF-11-92-14, Vector Fields Limited, 24 Bankside, Kidlington, Oxford OX5 1JE, England.
4. W. B. Herrmannsfeldt, "EGUN-An Electron Optics and Gun Design Program", technical report 331, SLAC, 1988.
5. Al Giest, *et. al.*, "PVM3 User's Guide and Reference Manual", technical report ORNL/TM-12187.
6. S. Yu, "Heavy Ion Fusion Injector Program", in *Proceedings of the 1993 Particle Accelerator Conference*, Washington D.C., pp 703-705, IEEE, 1993
7. J. J Barnard, *et. al.*, "Recirculating Induction Accelerators as Drivers for Heavy-Ion Fusion," *Physics of Fluids B* **5**, no. 7 (July 1993): 2698.
8. A. Friedman, *et. al.*, "Progress toward a prototype recirculating induction accelerator for heavy-ion fusion", in *Proc. of the 1995 Particle Accelerator Conf*, Dallas, Texas, pp 828-830, IEEE (1996). (LLNL Report UCRL-JC-119538, 1995).
9. J. J. Barnard, *et. al.*, "Emittance Growth in Heavy Ion Recirculators," in *Proceedings of the 1992 Linear Accelerator Conference*, AECL-10728 (1992): 229.

MAGY: Time Dependent, Multifrequency, Self-Consistent Code for Modeling Electron Beam Devices

M. Botton*, T. M. Antonsen
Institute for Plasma Research, University of Maryland
College Park, MD 20742

and

B. Levush
Naval Research Laboratory
Washington D.C. 20375

A new MAGY code is being developed for three dimensional modeling of electron beam devices. The code includes a time dependent multifrequency description of the electromagnetic fields and a self consistent analysis of the electrons. The equations of motion are solved with the electromagnetic fields as driving forces and the resulting trajectories are used as current sources for the fields. The calculations of the electromagnetic fields are based on the waveguide modal representation, which allows the solution of relatively small number of coupled one dimensional partial differential equations for the amplitudes of the modes, instead of the full solution of Maxwell's equations. Moreover, the basic time scale for updating the electromagnetic fields is the cavity fill time and not the high frequency of the fields. In MAGY, the coupling among the various modes is determined by the waveguide non-uniformity, finite conductivity of the walls, and the sources due to the electron beam. The equations of motion of the electrons are solved assuming that all the electrons traverse the cavity in less than the cavity fill time. Therefore, at each time step, a set of trajectories are calculated with the high frequency and other external fields as the driving forces. The code includes a verity of diagnostics for both electromagnetic fields and particles trajectories. It is simple to operate and requires modest computing resources, thus expected to serve as a design tool.

The operation of electron beam devices is, in principle, described by Maxwell's equations for the electromagnetic fields, along with the relativistic equations of motion of the electrons. The sources of the electromagnetic fields are the electronic currents and possibly external sources (e.g. amplifier), while the driving forces of the electrons are both static fields (e.g. an axial magneto-static field or focusing lenses) and the electromagnetic fields generated by the beam. One would like to simulate these complex systems in order to gain a better understanding and possibly improve designs of existing set-ups. The most straight-forward method is to solve simultaneously Maxwell's equations and the equations of motion using the Particle In Cell (PIC) scheme, in which fields components are assigned on a spatio-temporal grid, calculated for successive time steps and used to "push" the particles. Unfortunately, three-dimensional PIC simulations usually requires extensive computing resources, and therefore are not yet at a stage where they can be used effectively.

In this work we present the development of a new MAGY code which is designed to be an effective tool for modeling electron beam devices, using a reduced description of the system. The key step in the derivation of the reduced description is the description of the fields as a superposition of (relatively few) eigen-modes of the waveguide where the interaction takes place. The complex amplitudes of the modes evolve slowly in time on the scale of the frequency, and the basic time scale for updating the radiation field is a fraction of the cavity fill time and not a fraction of the high frequency period. Accordingly, Maxwell's equations are reduced to a set of coupled partial differential equation(time and axial coordinates), and the fields are locally computed as a superposition of the eigen functions with the calculated amplitudes. As the wall radius of the waveguide is, in most cases, a function of the axial coordinate (tapering or discontinuities), the modal representation makes use of "local eigen modes" which are axially dependent, and coupling terms among the various modes are introduced. This coupling is added to the one introduced by the finite conductivity of the walls, and the coupling due to the electronic current source. Besides speeding up the electromagnetic calculations, the reduced description also facilitates the computation of the electronic trajectories. Assuming that all the injected electrons traverse the interaction region in fraction of the cavity feel time, and that the Larmor radius is small compared to the cavity scale length one can obtain a significant savings of computation time by periodically launching ensembles of beam particles each time step, calculate the trajectories and deduce the currents sources for the electromagnetic fields. Both assumption can be relaxed within the framework of the reduced description, but with an increased cost of computation. The reduced description was successfully applied in a series of MAGY codes (see Refs. (1)-(3)). The current version of MAGY is intended to generalize the analysis and include additional features. The result is a relatively simple and compact code which can be used for modeling of wide verity of electron beam systems utilizing modest computing resources. In what follows we describe the code and bring examples of its operation.

Our derivation of the transmission line equations follows Reiter (4), and is based on the representation of the transverse field components as a superposition of the waveguide modes, $E_T = \sum_n V_n(z,t)\vec{e}_n(r_T,z)\exp\{-i\omega t\}$, $B_T = \sum_n I_n(z,t)\vec{b}_n(r_T,z) \cdot \exp\{-i\omega t\}$, including both transverse electric (TE) and transverse magnetic (TM) modes. Compared to the uniform waveguide, the eigen functions are axially dependent, as the boundaries are changing along the structure. Substituting these expressions in Maxwell's equations and assuming a slow time variation of the amplitudes, we get for the TM amplitudes I_k^I, V_k^I:

$$\frac{2}{c}\frac{\partial I_k^I}{\partial t} = \Gamma_k^I I_k^I - \frac{\partial V_k^I}{\partial z} + K_{lk}^I V_l^I + M_{lk} V_l^{II} - \alpha_{lk}^{I/I} I_l^I - \beta_{lk} I_l^{II} - J_{zk}^I$$

$$ik_0 V_k^I = \frac{\partial I_k^I}{\partial z} + K_{kl}^I I_l^I + J_{Tk}^I$$

and for the TE amplitudes I_k^{II}, V_k^{II}:

$$\frac{2}{c}\frac{\partial V_k^{II}}{\partial t} = \Gamma_k^{II} V_k^{II} - \frac{\partial I_k^{II}}{\partial z} - K_{kl}^{II} I_l^{II} - M_{kl} I_l^I - \alpha_{lk}^{V/II} V_l^{II} - J_{Tk}^{II}$$

$$ik_0 I_k^{II} = \frac{\partial V_k^{II}}{\partial z} - K_{lk}^{II} V_l^{II} + \alpha_{KL}^{i/II} i_L^I + \beta_{lk} I_l^I$$

where $k_0 = \omega/c$, $\Gamma_k^I = i\omega/c(1 - k_{mk}^2 c^2/\omega^2)$, $\Gamma_k^{II} = i\omega/c(1 - k_{ek}^2 c^2/\omega^2)$, the coupling coefficients due to wall radius variations are given by:

$$K_{lk}^I = \int \vec{e}_l^{I} \cdot \frac{\partial \vec{e}_k^{I*}}{\partial z} da, \quad K_{lk}^{II} = \int \vec{e}_l^{II} \cdot \frac{\partial \vec{e}_k^{II*}}{\partial z} da, \quad M_{lk} = \int \vec{e}_l^{II} \cdot \frac{\partial \vec{e}_k^{I*}}{\partial z} da,$$

the coupling due to finite conductivity of the walls is:

$$\alpha_{lk}^{I/I} = Z_s \sqrt{1+r_w'^2} \oint (\hat{n}\cdot\vec{e}_k^{I*})(\hat{n}\cdot\vec{e}_l^I)dl, \quad \alpha_{lk}^{I/II} = Z_s \sqrt{1+r_w'^2} \oint (\hat{n}\cdot\vec{e}_k^{II*})(\hat{n}\cdot\vec{e}_l^{II})dl,$$

$$\beta_{lk} = Z_s \sqrt{1+r_w'^2} \oint (\hat{n}\cdot\vec{e}_k^{II*})(\hat{n}\cdot\vec{e}_l^I)dl, \quad \alpha_{lk}^{V/II} = \frac{Z_s}{k_0^2}\sqrt{1+r_w'^2} \oint (\nabla\cdot\vec{b}_k^{II*})(\nabla\cdot\vec{b}_l^{II})dl$$

and $J_{Tk}^I, J_{zk}^I, J_{Tk}^{II}$ are the projection of the electronic current on the electromagnetic modes. The transmission line equations are solved on a staggered spatial grid as an implicit finite difference time domain, utilizing the predictor-corrector scheme. The convergence operator is chosen so as to increase stability, but under the restriction that the resulting equations can be cast as a tri-diagonal matrix form. As an example, we analyze the modes content in a tapered waveguide (radius changes from 1mm to 3mm over a length of 5mm). The excitation is a

TE_{11} mode (frequency 90Ghz), and we include 20 modes in the calculations ($TE_{1,1} - TE_{1,10}$, and $TM_{1,1} - TM_{1,10}$). The results of the calculation are shown in Fig. 1a for the $TE_{1,1}$ mode, Fig. 1b for the higher TE modes, and Fig. 1c, for the TM modes (amplitudes normalized to excited $TE_{1,1}$). The tapering induces energy

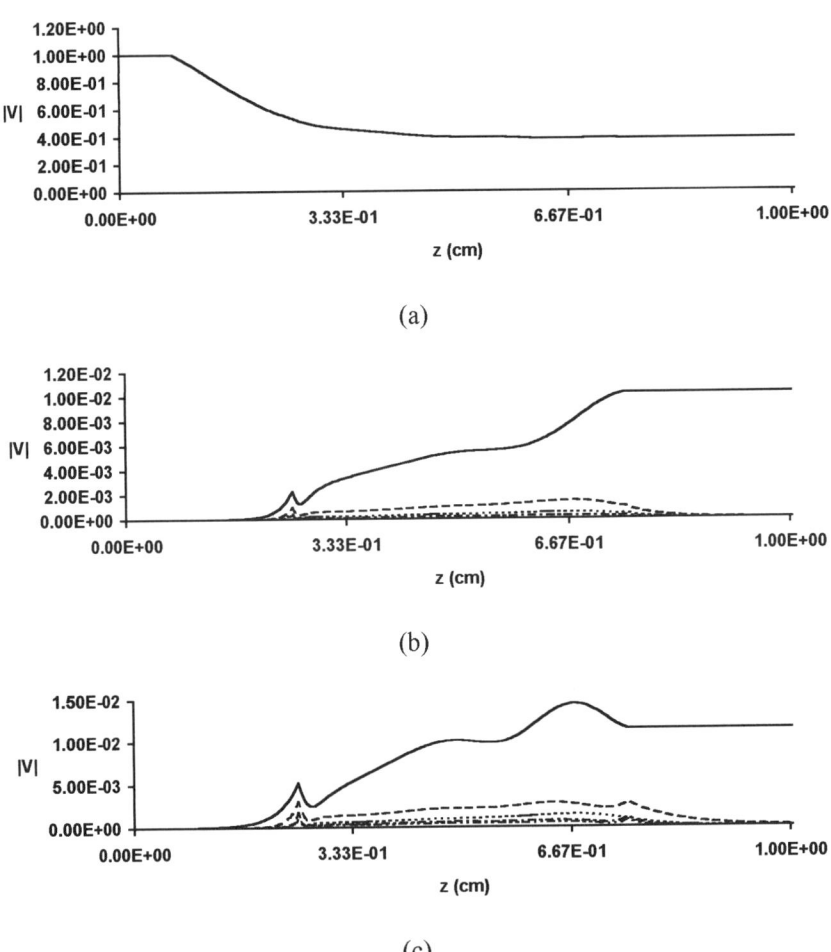

Figure 1: Mode content in a tapered waveguide ($r_1 = 1mm, r_2 = 3mm, \Delta z = 5mm$) (a) Excited $TE_{1,1}$, (b) $TE_{1,2} - TE_{1,5}$ modes, (c) $TM_{1,1} - TM_{1,5}$ modes.

transfer from the excited $TE_{1,1}$ mode to higher TE modes and to the TM modes. The $TE_{1,2}$ and $TM_{1,1}$ are above cutoff at the 3mm waveguide, and shown to propagated. The higher modes are below cutoff and thus contain energy only at the tapered section. The mode content naturally depends on the exact geometrical form of the cavity, however, these results demonstrate that an accurate description of the electromagnetic fields can be obtained even with relatively few modes.

The electron dynamics is described by the relativistic equations of motion. In view of the axial magnetic field which is applied in many applications, we use the guiding-center formulation, namely, $\vec{r} = \vec{X}_0 - 1/\Omega_0 \, \vec{u} \times \vec{1}_z$, $\vec{u} = u_\perp \vec{1}_\perp + u_z \vec{1}_z$ where $\vec{u} = \gamma \vec{v}$, (γ is the relativistic energy factor), $\vec{1}_\perp = \cos\xi \, \vec{1}_r + \sin\xi \, \vec{1}_y$, ξ is the gyrophase, $\vec{X}_0$ is the guiding center coordinate, and $\Omega_0 = eB_0/mc$ is the gyrofrequency. The equations of motion are then:

$$\dot{\vec{X}}_0 = \frac{1}{\Omega_0} \vec{a} \times \vec{1}_z \quad , \quad \dot{u}_z = \vec{a} \cdot \vec{1}_z \quad , \quad \dot{u}_\perp = \vec{a} \cdot \vec{1}_\perp \quad , \quad \dot{\xi} = \frac{1}{u_\perp} \vec{a} \cdot (\vec{1}_z \times \vec{1}_\perp) - \frac{\Omega_0}{\gamma}$$

where $\vec{a} = \frac{e}{m}\left(\vec{E} + \frac{1}{c}\vec{v} \times \vec{B} - \frac{1}{c}\frac{\partial B_0}{\partial z}\frac{r}{2}\vec{v} \times \vec{1}_r\right)$ is the driving force. Substituting the modal representation of the electromagnetic fields into the equations of motion we get:

$$\frac{\partial}{\partial t}\vec{X}_0 = -\frac{c}{B_0}\Re\left\{\sum_n \left[(V_n - \beta_z I_n)\vec{b}_n - \frac{1}{ik_0}V_n \nabla_T \cdot \vec{b}_n \vec{\beta}\right]e^{-i\omega t}\right\} - c\beta_z \frac{r}{2}\frac{\partial \ln B_0}{\partial z}\vec{1}_r$$

$$\frac{\partial}{\partial t}(\gamma\beta_z) = \frac{e}{mc}\Re\left\{\sum_n I_n\left(\vec{\beta}_\perp \cdot \vec{e}_n - \frac{1}{ik_0}\nabla_T \cdot \vec{e}_n\right)e^{-i\omega t} + \frac{4\pi}{i\omega}J_z\right\} + \Omega_0 \frac{r}{2}\frac{\partial \ln B_0}{\partial z}\vec{\beta}_\perp \cdot \vec{1}_\varphi$$

$$\frac{\partial}{\partial t}(\gamma\beta_\perp) = \frac{e}{mc}\Re\left\{\sum_n (V_n - \beta_z I_n)(\vec{1}_\perp \cdot \vec{e}_n)e^{-i\omega t}\right\} - \Omega_0 \frac{r}{2}\frac{\partial \ln B_0}{\partial z}\beta_z \vec{1}_\perp \cdot \vec{1}_\varphi$$

$$\frac{\partial}{\partial t}\xi = -\frac{\Omega_0}{\gamma} - \frac{e}{mc\gamma}\Re\left\{\sum_n \left[\frac{1}{\beta_\perp}(V_n - \beta_z I_n)(\vec{1}_\perp \cdot \vec{b}_n) + V_n \frac{1}{ik_0}\nabla_T \cdot \vec{b}_n\right]e^{-i\omega t}\right\}$$
$$- \frac{\Omega_0}{\gamma}\frac{\beta_z}{\beta_\perp}\frac{r}{2}\frac{\partial \ln B_0}{\partial z}\vec{1}_\perp \cdot \vec{1}_r$$

These equations include the effects due to nonuniform axial magnetic field, and the space charge of the beam. A significant simplification of this set can be obtained if we assume that all the particles traverse the cavity in time shorter than the cavity fill time, and the Larmor radius is small compared to the characteristic cavity length. When these conditions are met, we can avoid solving the equation for the guiding center coordinate, and approximate its value using the conservation of magnetic flux. Furthermore, with these assumptions, we change the free variable from t to z in the other three equations and solve them each time step for the ensemble of particles using a fourth order Runge-Kutta integration method with a constant step. The obtained trajectories are then used to compute the electronic currents which in turn affect the electromagnetic field, hence the description is a self-consistent one.

The method of solution outlined above enables a detailed level of diagnostics for both the electromagnetic fields and the particle dynamics. With the computed amplitudes of the various modes, we can generate spatial sections or time samples, power distribution in the cavity (mode content) and calculate the field components. For the particles we can obtain plots of the trajectories and sections of the phase space.

In conclusion, the new MAGY code is expected to be an effective and simple tool for time dependent three dimensional modeling of electron beam systems.

Acknowledgments: Work supported by AFOSR and NRL.
We wish to thank Dr. S. Cooke for valuable discussions during the preparations of this work.

* On leave from Rafael, Haifa, Israel.

References:

1. Kleva R. G., Antonsen T. M., Jr., and Levush B., *Physics of Fluids,* **31**, 375-386, 1988.
2. Levush B, and Antonsen T. M., Jr, *IEEE Transaction of Plasma Science,* **18**, 260, 1990.
3. Cai S. C., Antonsen T. M., Jr, Saraph G., and Levush B., *Int. J. Electronics,* **72**, 759-777, 1992.
4. Reiter G., *Convention on Long Distance Transmission by WaveGuide,* 54-57, 1959.

MAFIA Version 4

T.Weiland[1], M.Bartsch[2], U.Becker[1], M.Bihn[1], U.Blell[3], M.Clemens[1], M.Dehler[1], M.Dohlus[1,4], M.Drevlak[4], X.Du[1], R.Ehmann[2], A.Eufinger[2], S.Gutschling[1], P.Hahne[2], R.Klatt[1], B.Krietenstein[1], A.Langstrof[1], P.Pinder[1], O.Podebrad[1], T.Pröpper[1], U. van Rienen[1], D.Schmidt[1], R.Schuhmann[1], A.Schulz[2], S.Schupp[1], P.Schütt[1], P.Thoma[1], M.Timm[1], B.Wagner[5], R.Weber[6], S.Wipf[4], H.Wolter[1], Z.Min[1,4]

1: TH-Darmstadt, Fachbereich 18, Schloßgartenstr.8, D64289 Darmstadt, Germany
2: CST Computer Simulations Technology GmbH, Lauteschlägerstr. 38, D64289 Darmstadt
3: GSI mbH, Planckstr. 1, 64291 Darmstadt
4: Deutsches Elektronen-Sychrotron DESY, Notkestr. 85, D22607, Germany
5: CSS Computer Simulation Services GbRmbH, Lauteschlägerstr. 38, D64289 Darmstadt
6: AET Inc. , 1-1-7 Mukaibara, Asaoku, Kawasaki City, Japan

Abstract. MAFIA Version 4.0 is an almost completely new version of the general purpose electromagnetic simulator known since 13 years. The major improvements concern the new graphical user interface based on state of the art technology as well as a series of new solvers for new physics problems. MAFIA now covers heat distribution, electro-quasistatics, S-parameters in frequency domain, particle beam tracking in linear accelerators, acoustics and even elastodynamcis. The solvers that were available in earlier versions have also been improved and/or extended, as for example the complex eigenmode solver, the 2D-3D coupled PIC solvers. Time domain solvers have new waveguide boundary conditions with an extremely low reflection even near cutoff frequency, concentrated elements are available as well as a variety of signal processing options. Probably the most valuable addition are recursive sub-grid capabilities that enable modeling of very small details in large structures.

INTRODUCTION

Since its first appearance in the literature in 1993(1) MAFIA has become a worldwide known and widely distributed tool for designing and analyzing a vast variety of accelerator components. Starting with the pioneering wake field modules for cylindrical structures and fully three dimensional structures it was soon extended to treat frequency domain problems such as cavity resonator design as well. The fundamental equations that gave the solid mathematical basis for the unique solution in three dimensions were the so called Maxwell's Grid Equations(2). Meanwhile MAFIA has grown with the advance of computational power to a fully graphical user interface driven up-to-date em-simulator with an unsurpassed broadness of possible applications. Now that virtually all electromagnetic problems can be solved, it was logical to apply the FIT (Finite Integration Theory) also to similar physical problems, such as heat distribution and acoustic waves. We will briefly describe the new features and solvers but refer to more extensive publications for detail.

MAFIA'S GRAPHICAL USER INTERFACE

Based on a platform independent language, MAFIA's new GUI provides over 140 new graphical menus and online context sensitive help in HTML style. Over 500kLOC have been necessary to create a consistent and flexible interface that is organized as supervisor process with the ordinary MAFIA running in the former command line modus. All solvers and the optimizer are composed to one single program. Fig. 1 shows the main MAFIA icon bar and as an example the interactive drawing window for modeling two and three dimensional structures.

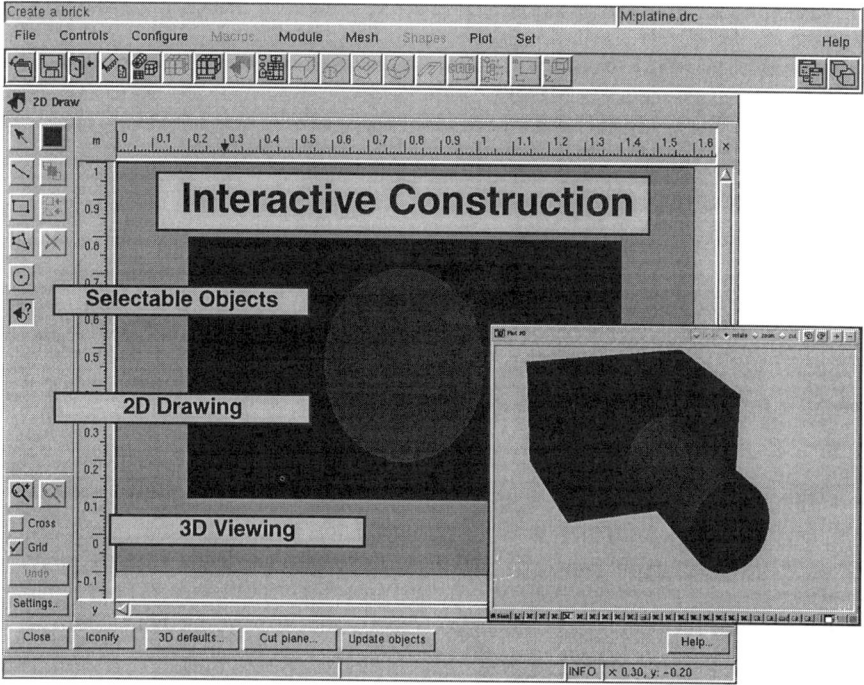

Figure 1. MAFIA's main icon bar (top) and the 2D interactive modeling window for geometric modeling. Overlaid is the 3D view window.

The most important key features of the new GUI for MAFIA 4.0 are
- interactive modeling of 3D structures made of primitives
- transformations and boolean operations on all primitives
- logical tree of all geometrical objects with hypertext links to their properties
- macro recorder for generating user defined macros and associated icons
- configurable icon bar with predefined and self made icons
- CAD data import and export, various data formats are supported
- wizard for frequently used calculations (e.g. S-parameters, antenna problems)

NEW PHYSICS SOLVERS

The generality of the FIT method using matrices replacing integrals along lines, over surfaces and volumes offers a simple way to extend the application to a variety of physics problems with similar differential or integral equations. It is straight forward to include heat distribution problems that occur in rf cavities due to wall currents. It is thus possible to consistently solve the electromagnetic and thermal problem of e.g. an accelerating cavity within one computer program(3). A simulation of a cavity welding process with eddy currents is shown in Fig. 2.

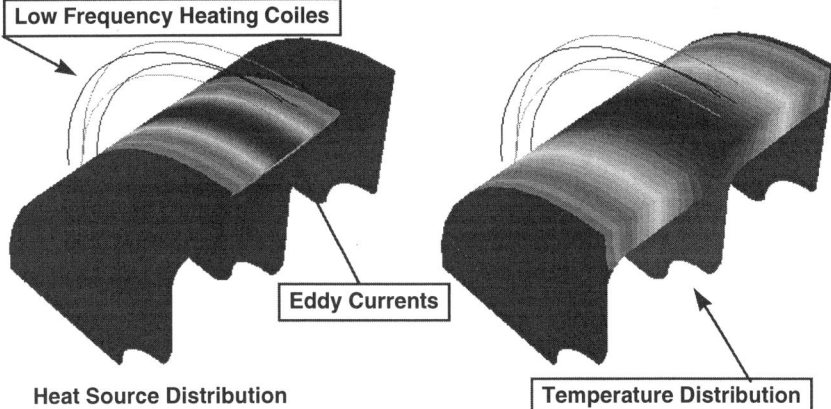

Figure 2. Combined rf and temperature analysis of rf cavity welding by externally excited eddy currents. The left plot shows a quarter of a two-cell rf cavity and low frequency coils surrounding the structure. The coil current excites eddy currents in the copper structure. The gray scale shows the density of heat sources created by the eddy current. The right hand side plot shows the resulting temperature distribution with its maximum temperature at the weld location.

In the electrodynamics area new solvers are provided for electro-quasistatics. In this low frequency regime the displacement current is an important quantity while the time derivative of the magnetic flux is not. Thus one can derive a Poisson type of equation with complex matrix elements and a complex scalar potential(4). The field of an wetted insulator computed by this new module is shown in Fig. 3.

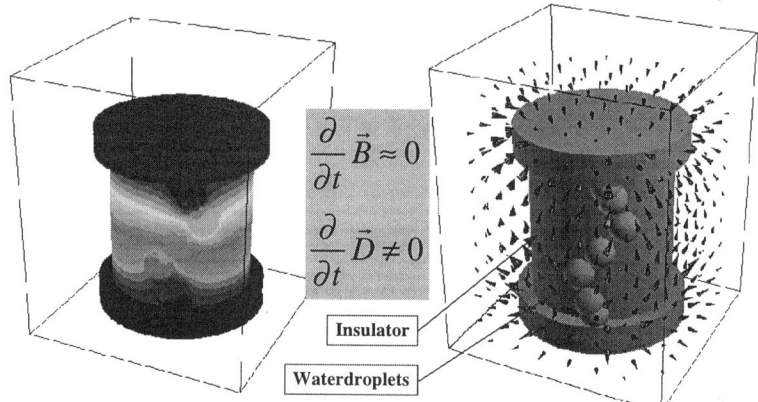

Figure 3. A high voltage insulator wetted by water droplets (right) and the absolute value of the resulting electric field strength on the insulator surface (left).

A new linac tracking module L was added for simulating beams in linear accelerators. This module includes collective effects such as cumulative beam breakup due to bands of higher modes in cavities as well as single bunch wake field effects. The linac is constructed recursively from basic elements and groups thereof. Ground motion is simulated using the ATL law and a choice of all currently used orbit correction schemes is included as well. Thus MAFIA-L provides a unique tool to simulate realistic operation of large linear accelerators.

The computation of S-parameters of rf structures is probably the most frequently used application of MAFIA. Traditionally, MAFIA uses broadband time domain analysis and associated FFT/DFT to obtain the results as function of frequency. This well known technique is very fast and usually requires only about the same computational effort for the entire frequency range of interest as does a frequency analysis for one single frequency. However, in the rather rare case of very low frequencies the time step limitation may hurt and a frequency domain analysis can be advantageous. For theses cases MAFIA now provides a frequency domain S-parameter solver. Of course, both methods yield exactly the same results (within rounding error limits) as both use the basically identical discretization and the same basic Maxwell's Grid Equations.

IMPROVEMENTS AND EXTENSIONS

Besides the new GUI and new solvers many improvements and extensions have been worked out for the existing older modules of MAFIA 3.2. Only the most important ones shall be mentioned here.

Virtually any rf devices is driven by some kind of power source which is connected to the object of interest by a coaxial line or hollow waveguide. Thus a simulation needs to offer a waveguide boundary condition that has a little as possible parasitic reflections. MAFIA 4.0 possesses a new boundary condition that has a reflection of less than -130db. Moreover, unlike most absorbing boundary versions this is true even at and below cutoff frequency (see Fig. 4).

Three dimensional structures are often operated in connection with concentrated elements such as resistors and capacitors. In version 4 it is now possible to locate concentrated elements into the mesh made of R, L, C and parameterizable source. Thus it is possible to simulate a truly wave type of problem of e.g. a micro strip line terminated by a resistor.

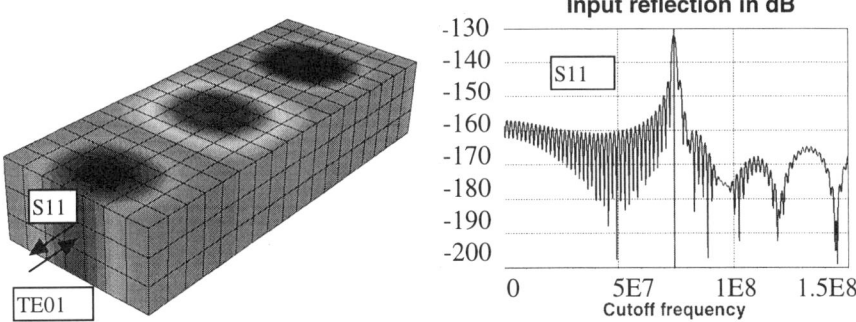

Figure 4. A section of a hollow rectangular waveguide as object for a transmission calculation. The residual reflection at the entrance is below -130dB even at and below cutoff frequency.

Often, three dimensional structures have very detailed localized sub-structures, such as drift tubes in a large Interdigital H-Mode Ion accelerator (IH structure) or near coupling antennas in large resonators. MAFIA 4.0 can treat such problems using a new recursive sub-grid technology(4). This new method is the first one published so far that is stable and consistent. The combined grid and sub-grid fields hold the same fundamental properties of the grid fields as known from MAFIA. Examples are shown in Fig. 5 and 6.

SUMMARY

With MAFIA Version 4 electromagnetic simulation of accelerator components has become easier than ever. More and more different problems including non linear permanent magnets at the low frequency end up to full three dimensional klystron simulation including 3D waveguide output circuits can be attacked(6). With the new parameterized optimizer, MAFIA 4.0 can actually design components for you including geometric layout.

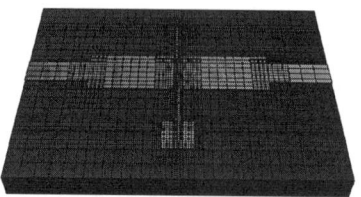
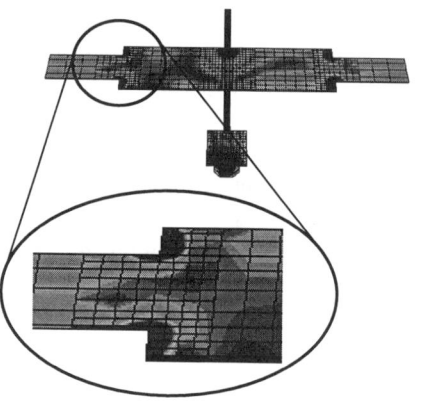

Phase of S21 in degrees

Frequency/GHz	base mesh	subgrids	measured
6	-64.33	-63.82	-63.54
9	-127.09	-125.57	-124.28
12	-184.17	-181.48	-179.76
15	-241.55	-237.49	-237.29
18	-312.79	-301.35	-301.22

Figure 5. Sub-grid application in a micro strip line with stubs. The sub grids near the sharp edges greatly improve accuracy as can be seen from the comparison with the measured data and those computed with a coarse mesh only.

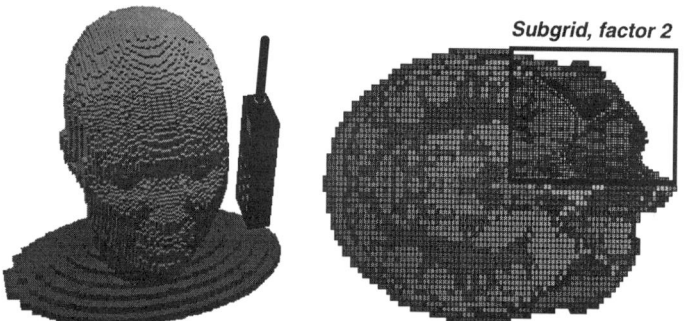

Figure 6. A three dimensional high resolution model of the head of an accelerator physicists using a mobile telephone(7). In order to compute accurately the heat distribution in the eye region MAFIA can use local mesh refinement, shown here for the front left quarter of the head. Inside the eye further sub-grids are used.

REFERENCES

1. T.Weiland, *Lecture Notes in Physics*, Vol.215(1983), pp.21-32, Computing in Accelerator Design and Operation, Proceedings of Europhysics Conference, Berlin, September 1983
2. T.Weiland, *Particle Accelerators*, Vol.15(1984), pp. 245-291
3. U.van Rienen, P.Pinder, T.Weiland, *The Seventh Biennial IEEE Conference on Electromagnetic Field Computation* (CEFC), Okayama 1996, to be published
4. U.van Rienen, M.Clemens, T.Weiland, *10th Conference on the Computation of Electromagnetic Fields*, Berlin, 1995, pp.318-319
5. P.Thoma, T.Weiland, *Proceedings of the 25th European Microwave Conference*, Vol.2, 1995, pp.770-774
6. U.Becker, M.Dohlus, T.Weiland, *Particle Accelerators*, Vol.51 (1995), pp.135-154
7. S.Gutschling, T.Weiland, *EMC95, 11th Int. Zürich Symp. On EMC Problems*, p.291-296

QUICKSILVER — A General Tool for Electromagnetic PIC Simulation

David B. Seidel, Rebecca S. Coats, William A. Johnson,
Mark L. Kiefer, L. Paul Mix, Michael F. Pasik, Timothy D. Pointon,
Jeffrey P. Quintenz, Douglas J. Riley, and C. David Turner

*Information & Pulsed Power Research & Development Division,
Sandia National Laboratories, Albuquerque, New Mexico 87185*

Abstract. The dramatic increase in computational capability that has occurred over the last ten years has allowed fully electromagnetic simulations of large, complex, three-dimensional systems to move progressively from impractical, to expensive, and recently, to routine and widespread. This is particularly true for systems that require the motion of free charge to be self-consistently treated. The QUICKSILVER electromagnetic Particle-In-Cell (EM-PIC) code has been developed at Sandia National Laboratories to provide a general tool to simulate a wide variety of such systems. This tool has found widespread use for many diverse applications, including high-current electron and ion diodes, magnetically insulated power transmission systems, high-power microwave oscillators, high-frequency digital and analog integrated circuit packages, microwave integrated circuit components, antenna systems, radar cross-section applications, and electromagnetic interaction with biological material. This paper will give a brief overview of QUICKSILVER and provide some thoughts on its future development.

OVERVIEW OF QUICKSILVER

Charged-particle simulations in three dimensions are performed routinely in the Pulsed Power Sciences Center at Sandia with the QUICKSILVER (1) suite of codes. QUICKSILVER is a three-dimensional, relativistic, finite-difference, electromagnetic, particle-in-cell (PIC) code developed at Sandia. It was originally targeted for vector supercomputers, such as the Cray Y-MP, which are characterized by large, shared memory and multiple processors. It now runs on a wide variety of platforms, including most UNIX workstations, and is presently being ported to the Intel Paragon, a massively-parallel distributed memory supercomputer.

QUICKSILVER is actually a suite of codes; in addition to the main simulation code there are several support codes. The problem geometry is generated using various preprocessors and the simulation results are examined with one or more postprocessors. The original MERCURY command-driven preprocessor assists the user in defining the mesh, boundary conditions, and other input parameters. Recently, a set of widget-based tools has been developed to further simplify the process of mesh generation. These widget-based tools, built upon the IDL data analysis and visualization tool[1] have been incorporated in PFIDL (2), QUICKSILVER's primary simulation data postprocessor. In its role as postprocessor, PFIDL provides the capability to manipulate and examine 3D scalar and vector field data as well as 6D particle phasespace data. Additionally, PFIDL can be used to examine and manipulate time histories of various simulation quantities. AVS[2] is used for the visualization and/or animation of field and particle distributions as well as the 3D model geometry. These pre- and postprocessing tools are available on a wide variety of platforms. The potentially vast amount of simulation data is shared between the simulation code and the postprocessors via the Portable File Format (PFF) (3), a portable, compact, machine-independent binary file format developed expressly for the QUICKSILVER suite but widely used for many other applications.

QUICKSILVER's Preprocessors

Generating input data for three-dimensional simulations can be difficult, time-consuming, and error-prone. MERCURY is a command-driven preprocessor that is used in defining the finite-difference grid, the problem geometry, the boundary conditions, and other input parameters. MERCURY allows free-format input and provides menus for guiding simulation setup and on-line help. It processes all input for a QUICKSILVER simulation and checks for errors and inconsistencies. QUICKSILVER uses a nonuniform, multiple-block, rectilinear grid with staggered grids. The MERCURY grid generator provides straightforward tools to facilitate the generation of these multi-block, nonuniform grids, automatically ensuring that the grid is both continuous and smoothly varying.

A nonuniform grid defined by the relationship between the grid index i and the physical grid location x,

$$x(i) = x_0 + ai + bi^2 + ci^3,$$

is supported by QUICKSILVER. Different regions of the grid can have different descriptions of the relationship between i and x. Across any interface between such grid regions, $x(i)$ and $dx(i)/di$ must be continuous in order to retain accuracy in the field solution. The MERCURY grid generator ensures that this condition is met as the mesh is produced. Cartesian, cylindrical, and spherical coordinate system multiple-block grids can be generated. By multiple block, we mean that the grid is composed of logically connected blocks, each of which is a coordinate-system-conformal region of space with its own local grid.

---

[1] IDL is a product of Research Systems, Inc., 2995 Wilderness Place, Suite 203, Boulder, Colorado 80301.

[2] AVS is a product of Advanced Visual Systems, 300 Fifth Ave., Waltham, MA 02154.

Conducting and dielectric volumes are easily generated with MERCURY by combining (sequentially adding or removing) objects selected from a provided set of simple solid-object primitives. MERCURY then fits the resulting compound volume description to the simulation's underlying finite-difference grid. MERCURY also computes the memory requirements for arrays in QUICKSILVER so that only the minimum memory required for a simulation is used.

In addition to MERCURY's capabilities, widget-based tools have been added to PFIDL to ease some of the more difficult facets of simulation setup. For example, since often a description of the problem geometry is available from solid modeling or CAD tools, PFIDL currently has a tool for editing DXF[1] files and converting them to MERCURY format, and we are working on extending that capability to allow the use of ACIS[2] files. Also, one of the difficulties of constructing grids for complex system simulations is that often the locations of structure surfaces are critical to the fidelity of the simulation, and since material interfaces must coincide with grid locations it is thus difficult and time consuming to design a grid conforming to these constraints. To address this problem we have added an interactive "point-and-click" graphical tool to PFIDL that allows the user to tie grid locations to various surfaces of the model structure while providing continuous visual feedback on the size and quality of the resulting mesh.

The QUICKSILVER 3D Physics Simulation Code

QUICKSILVER, the member of the suite for performing 3D physics simulations, can be divided into two distinct parts, the field solver and the particle handler. In the following sections, the features of each will be described.

The QUICKSILVER field solver utilizes explicit (4) and implicit (5) finite-difference, leap-frog algorithms. Multiple lossy, non-dispersive dielectrics are allowed for regions without particles. Available boundary conditions include conductors, inlet and outlet boundaries, mirror symmetry, and periodic symmetry. Simulations can be performed in Cartesian, cylindrical or spherical coordinate systems. Currently, inlet wave boundaries can be driven either with multiple, independent TEM modes or a 1D, multi-line Telegraphers' model. In both cases, outgoing waves are treated with a 1st-order Mur (6) radiation-absorbing boundary condition. The code also supports a variety of outlet boundary conditions, including 1st-order Mur, 2nd-order dispersive (7), and the Perfectly Matched Layer (PML) (8). QUICKSILVER also has models for embedded current source excitation and surface impedance.

Recently, the capabilities of the QUICKSILVER's field solver have been extended to use unstructured grid methods. Although unstructured grids provide consider-

---

[1] DXF is a registered trademark of Autodesk, Inc., 111 McInnis Parkway, San Rafael, California 94903.

[2] ACIS is a registered trademark of Spatial Technology, Inc., 2425 55th St., Bldg. A, Boulder, Colorado 80301.

able modeling flexibility for complex geometries, they can require an extremely large number of cells; in addition, their cost of computation (and memory) per cell is significantly higher than that of QUICKSILVER's structured-grid finite-difference algorithm. The finite-volume hybrid-grid (FVHG) field algorithm (9) enables unstructured grids to be combined with rectangular-cell structured grids, thus combining the modeling flexibility of unstructured grids with the efficiency of standard finite-difference methods. This approach was first incorporated in the VOLMAX solver (10) and, with its inclusion in QUICKSILVER's field solver, allows its modeling flexibility to be combined with QUICKSILVER's extensive boundary condition and diagnostic features. Currently, we use I-DEAS[1] to build the solid models and to subsequently generate the grid for the unstructured region of the simulation space.

The second major portion of the QUICKSILVER code is its particle handler, whose job is to advance particle positions with 3D, fully-relativistic kinematics and to subsequently allocate each particle's contribution to the current back to the finite-difference grid for use by the field solver. QUICKSILVER's particle handler allows multiple particle species with particle creation via preloading, beam injection and space-charge-limited field emission. It supports the same boundary conditions and coordinate systems as the field solver. Currently the code uses a current/charge density allocation algorithm that locally conserves charge exactly. A pseudo-current algorithm (11) which diffuses errors in the charge to the simulation boundaries has recently been added to PML regions to allow that boundary condition to function properly with a low density particle flux at the boundary. It will also be used to ensure charge conservation on both the structured and unstructured grid regions of the hybrid QUICKSILVER/VOLMAX code when it is extended to treat particles.

Diagnostics

The QUICKSILVER code has a wide variety of diagnostics available to the user which can be divided into two basic types: snapshots and time histories. Snapshot diagnostics provide detailed spatial information about some simulation quantity at specified instants of time (or averaged over specified intervals of time). On the other hand, time histories provide, as a function of time, a simulation quantity at a fixed spatial location or integrated over some spatial region of the simulation.

QUICKSILVER can provide snapshots of both vector and scalar field quantities: electric (**E**) and magnetic (**B**) fields, current density (**J**), and charge density (ρ). Snapshots of simulation particles in 6D phasespace (x, y, z, p_x, p_y, p_z), or a subset of that phasespace, can also be obtained.

Time histories can be requested for ρ or any component of **E**, **B**, or **J** at any spatial location in the simulation. In addition, line integrals $\int \boldsymbol{E} \cdot d\boldsymbol{l}$ and $\int \boldsymbol{B} \cdot d\boldsymbol{l}$ are available, each along a complex path composed of one or more coordinate-confor-

---

[1] I-DEAS is a product of Structural Dynamics Research Corporation, Milford, Ohio.

mal subpaths. Similarly, the area and volume integrals $\int \boldsymbol{E} \cdot d\boldsymbol{A}$, $\int \boldsymbol{B} \cdot d\boldsymbol{A}$, $\int \boldsymbol{J} \cdot d\boldsymbol{A}$, $\int \boldsymbol{S} \cdot d\boldsymbol{A}$, $\int \rho dV$, $\int W_E dV$, $\int W_M dV$, and $\int W dV$ can be obtained, each over one or more coordinate-conformal subareas or subvolumes, respectively. Here, S is the Poynting vector ($\boldsymbol{E} \times \boldsymbol{H}$), and W_E, W_M, and W are the electric ($\varepsilon E^2/2$), magnetic ($B^2/2\mu$), and total ($W_E + W_M$) field energy densities, respectively. Time histories are also available for several particle-related items, including count, energy, or charge of surviving, created, or killed particles, by species. Also, time-history data for the current, energy, and momentum of particles killed on specified conductor surfaces is available. Furthermore, selected subsets of such killed particles can be saved, either for postprocessing in IDL, or as input to other codes, e.g., beam transport codes. To examine simulation charge conservation, maximum and RMS values of the error in charge density ($\nabla \cdot D - \rho$) are also available as time histories.

A LOOK TO QUICKSILVER'S FUTURE

As the demands on the modeling capabilities of QUICKSILVER expand, it has become increasingly difficult, and consequently more expensive, to enhance the code to meet these new demands. A substantial portion of this difficulty is connected with the limitations inherent in current versions of FORTRAN-77, in which almost all of the QUICKSILVER suite is written. However, the modern object-oriented design software methods and their embodiment in programming languages such as C++ offer the prospect of developing an integrated capability that will be portable, extendable, reusable, and will be well suited to next-generation high performance computing platforms. It is our goal to develop an EM-PIC toolset based upon these methods that will both integrate our present capabilities as well as provide for a cost-effective route for extensions in those capabilities that will be needed to meet our simulation requirements for at least the next decade.

New physics modeling capability will be needed to provide the high degree of confidence that will be required as our ability to perform key experiments decreases and consequently our reliance upon simulation increases. This includes adding multi-scale physics (e.g., wires, slots, thin films, etc.) on hybrid mesh structures, non-linear devices (e.g., coupled solid-state components), fluid descriptions of high density plasmas, the desorption and ionization of contaminants from electrodes, secondary electron emission models, and beam transport and scattering through low density gas. These physics packages will require greatly increased computational capacity as well as a highly-modular object-based code framework.

To meet our modeling needs in the next decade, even with the expected advances in computing hardware, we will need the capability to partition a problem into coupled regions, each of which could treat a potentially different subset of available physics models, using different algorithms, on different types of meshes.

The QUICKSILVER/VOLMAX integration is only a first small step toward this capability; more generalized modeling/meshing of this sort can only be achieved using modern object-oriented design techniques.

The need for higher resolution for increasingly complex systems, coupled with more realistic physics modeling, clearly drives us to take full advantage of the new generation of massively-parallel supercomputers. QUICKSILVER is now running (fields only) on Sandia's 1800+-processor Intel Paragon; in one year it will be running in full PIC mode on that machine as well as on the Sandia/Intel Teraflop machine that is scheduled for availability in early 1997. To take full advantage of the capacity afforded by these machines, it is also clear that we will need to significantly alter the way in which we set up and mesh our simulations as well as the way in which we retrieve, store, and analyze their complex and voluminous data. We anticipate that our pre- and postprocessing tools will require significant, if not radical, new development in order to meet this challenge.

ACKNOWLEDGMENTS

This work supported by the U.S. Department of Energy under Contract No. DE-AC04-94AL85000.

REFERENCES

1. D. B. Seidel, M. L. Kiefer, R. S. Coats, T. D. Pointon, J. P. Quintenz, and W. A. Johnson, "The 3-D, Electromagnetic Particle-In-Cell Code, QUICKSILVER." in *The CP90 Europhysics Conference on Computational Physics*, Armin Tenner, Ed., World Scientific, Amsterdam, 1991, pp. 475–482; J. P. Quintenz, D. B. Seidel, M. L. Kiefer, T. D. Pointon, R. S. Coats, S. E. Rosenthal, T. A. Mehlhorn, M. P. Desjarlais, and N. A. Krall, *Laser and Particle Beams* **12**, 283-324 (1994).
2. L. P. Mix, R. S. Coats, and D. B. Seidel, "PFIDL: Procedures for the Analysis and Visualization of Data Arrays," presented at the 1st Biennial Tri-Laboratory Engineering Conference on Computational Modeling, Pleasanton, California, Oct. 31–Nov. 2, 1995.
3. D. B. Seidel, R. S. Coats, M. L. Kiefer, T. D. Pointon, and L. P. Mix, "PFF — A Compact, Machine-Independent File Format for Simulation Data," presented at the 9th Biennial CUBE Symposium, Santa Fe, New Mexico, Nov. 27–30, 1990.
4. K. S. Yee, *IEEE Trans. Antennas Propagat.* **14**, 2155–2163 (1966); O. Buneman, "Fast Numerical Procedures for Computer Experiments on Relativistic Plasmas," in *Relativistic Plasmas*, O. Buneman and W. Pardo, Eds., New York: Benjamin, 1968, pp. 205–219.
5. B. B. Godfrey, presented at the 9th Conference on Numerical Simulation of Plasmas, Evanston, Illinois, June 30–July 2, 1980.
6. G. Mur, *IEEE Trans. Electromagnetic Compatibility* **23**, 1191–1196 (1982).
7. Z. Bi, K. Wu, C. Wu, and J. Litva, *IEEE Trans. Microwave Theory Tech.* **40**, 774–777 (1992).
8. J.-P. Berenger, *J. Comp. Physics* **114**, 185–200 (1994).
9. D. J. Riley and C. D. Turner, *IEEE Microwave and Guided Wave Letters* **5**, 284–286 (1995).
10. D. J. Riley and C. D. Turner, "VOLMAX: A Solid-Model Based, Transient *Vo*lumetric *Max*well Solver Using Hybrid Grids," to appear in *IEEE Antennas & Propagation Magazine*.
11. B. Marder, *J. Comp. Phys.* **68**, 48–55 (1987).

High Accuracy 3D Electromagnetic Finite Element Analysis

Eric M. Nelson

Los Alamos National Laboratory,
Los Alamos, New Mexico 87545

Abstract. A high accuracy 3D electromagnetic finite element field solver employing quadratic hexahedral elements and quadratic mixed-order one-form basis functions will be described. The solver is based on an object-oriented C++ class library. Test cases demonstrate that frequency errors less than 10 ppm can be achieved using modest workstations, and that the solutions have no contamination from spurious modes. The role of differential geometry and geometrical physics in finite element analysis will also be discussed.

INTRODUCTION

Electromagnetic finite element analysis (FEA) is becoming more popular as the accuracy and reliability of FEA codes improve. Accurate models of complicated structures have long been sought, and FEA with warped and/or unstructured meshes has been perceived as one path to achieve this accuracy. Thermal and mechanical FEA has met great success, but electromagnetic FEA has traditionally been plagued with reliability problems. Usually these reliability problems take the form of spurious modes (see (1) for many references). Fortunately, numerous workers have made steady progress in the past decade to eliminate these problems.

Differential geometry has been mentioned a few times (1, 2) in the electromagnetic FEA literature. It is an excellent tool for understanding electromagnetic FEA. Unfortunately, most of the current literature continues to use vector calculus notation, thus obscuring the simple nature of electromagnetic FEA.

Some reliable 3D electromagnetic FEA codes have existed for some time (see (3) and (4) for example). I have written an electromagnetic FEA code which should be equivalent to the code described in (4). In this paper I would like to (1) describe what I have learned about electromagnetic FEA from studying some differential geometry, and (2) demonstrate the accuracy and reliability of this FEA code.

---
* Work supported by DOE, contract W-7405-ENG-36.

FINITE ELEMENT FORMULATION

In vector calculus notation, this FEA code is based on the following weak formulation of the eigenmode problem for electric fields: find eigenvalues ω^2/c^2 and the corresponding eigenmode fields $\mathbf{E} \in \mathcal{U}_E$ such that $\forall \mathbf{F} \in \mathcal{U}_E$,

$$\int_\Omega (\nabla \times \mathbf{F}) \cdot \mu^{-1}(\nabla \times \mathbf{E}) - \frac{\omega^2}{c^2} \mathbf{F} \cdot \epsilon \mathbf{E} \, d\Omega = 0, \tag{1}$$

where Ω is the cavity interior and the space $\mathcal{U}_E$ of test functions $\mathbf{F}$ and trial functions $\mathbf{E}$ is

$$\mathcal{U}_E = \{ \mathbf{E} \in \mathcal{H}_{\text{curl}}(\Omega) : \hat{\mathbf{n}} \times \mathbf{E} = 0 \text{ on } d\Omega \}, \tag{2}$$

and $\mathcal{H}_{\text{curl}}(\Omega)$ is the space of vector fields on Ω which are square integrable in the following sense,

$$\mathcal{H}_{\text{curl}}(\Omega) = \{ \mathbf{E} : \int_\Omega |\nabla \times \mathbf{E}|^2 + |\mathbf{E}|^2 \, d\Omega \text{ exists} \}. \tag{3}$$

A similar formulation is based on the magnetic fields: find eigenvalues ω^2/c^2 and the corresponding eigenmode fields $\mathbf{H} \in \mathcal{U}_H$ such that $\forall \mathbf{G} \in \mathcal{U}_H$,

$$\int_\Omega (\nabla \times \mathbf{G}) \cdot \epsilon^{-1}(\nabla \times \mathbf{H}) - \frac{\omega^2}{c^2} \mathbf{G} \cdot \mu \mathbf{H} \, d\Omega = 0, \tag{4}$$

where the space $\mathcal{U}_H = \mathcal{H}_{\text{curl}}(\Omega)$.

The cavity interior Ω is partitioned into quadratic hexahedral (27-node) elements. Curved edges and faces allows these elements to closely follow curved boundaries. On each element there are 54 quadratic mixed-order 1-form basis functions, which are described in more detail below.

Numerical integration is used to compute the matrix components, and a simple subspace iteration scheme with a conjugate gradient solve is used to solve the sparse algebraic eigenvalue problem. A C++ class library handles matrices and bookkeeping of elements, faces, edges, nodes and basis functions.

LESSONS FROM GEOMETRICAL PHYSICS

In the finite element method, the problem domain Ω is divided into elements with simple shapes like tetrahedra and hexahedra. Each element Ω_e has a local coordinate system (i.e., a master element) and a map x_e from local to global coordinates. This map is typically only used to define basis functions and to change variables to numerically integrate equation (1) over Ω_e.

There is a close parallel between FEM and differential geometry. In differential geometry one considers a manifold (the problem domain Ω) which is covered by coordinate patches (the elements). Where the coordinate patches overlap, the coordinate system of one patch is a differentiable function of the coordinates in the other patch. Differential geometry does not demand that a global coordinate system exist, but it accomodates one very well.

What does one learn from this comparison? First, the local coordinates of an element are a valid coordinate system. The physical equations can be expressed in local coordinates just like they are expressed in global coordinates.

Consider a vector basis function. There is no conceptual difference between a vector in global coordinates and a vector in local coordinates. The two are related by the transformation rule for vectors,

$$v^i = \sum_{j=1}^{3} \frac{\partial x_e^i}{\partial u^j} \bar{v}^j, \qquad (5)$$

where v^i and $\bar{v}^j$ are the components of the vector in the global and local coordinate basis, respectively, and $x_e^i(u^j)$ is the map from local coordinates u^j ($j = 1, 2, 3$) to global coordinates x^i ($i = 1, 2, 3$). Note that early attempts at 3D electromagnetic FEA violated this transformation rule. They would use scalar basis functions for the vector components, and simply map these components from local to global coordinates, $v^i = \bar{v}^i$. This procedure is geometrically incorrect, and thus the results of these FEA codes were usually flawed.

Now consider which basis functions are appropriate for electromagnetic FEA. In reference (5), Maxwell's equations are described in terms of differential geometry and geometrical physics. In particular, it is pointed out that the electric field is most naturally expressed as a 1-form, or covariant vector. Hence the appropriate basis functions for electric fields is most easily expressed as 1-forms, not vectors. Recent literature typically shows complicated constructions for these basis functions (see references (1) and (6) for example), but if one writes the basis functions as 1-forms in local coordinates (u, v and w), they are simple polynomials. This is the spirit of FEM—the field is a linear combination of simple basis functions on simple elements.

The choice of basis functions is important, as described in (7) and (8). The basis functions should be mixed-order, with the basis for the field along a coordinate direction, say E_u, being complete to order p in v and w, but only $p-1$ in u. The basis functions employed in this FEA are listed in table 1. The local coordinates of the hexahedra are $0 \leq u \leq 1$, $0 \leq v \leq 1$ and $0 \leq w \leq 1$, and the coordinate basis for 1-forms is du, dv and dw.

The basis functions are assembled so that the tangential component of the field at an interface between elements is continuous. Thinking of the basis functions as 1-forms makes it easy to verify that the assembly process works. The basis functions are characterized by their non-zero tangential field on an edge or face. The first 12 basis functions have constant tangential field along one edge. These basis functions, by themselves, are appropriate for a linear mixed-order approximation to the fields. The next 12 basis functions (a_{13}

Table 1. Quadratic 1-form basis functions a_i for hexahedral elements.

$a_1 = (1-v)(1-w)\,du$	$a_5 = (1-u)(1-v)\,dw$	$a_9 = (1-v)w\,du$
$a_2 = (1-u)(1-w)\,dv$	$a_6 = u(1-v)\,dw$	$a_{10} = (1-u)w\,dv$
$a_3 = u(1-w)\,dv$	$a_7 = (1-u)v\,dw$	$a_{11} = uw\,dv$
$a_4 = v(1-w)\,du$	$a_8 = uv\,dw$	$a_{12} = vw\,du$

$a_{13} = (2u-1)a_1$	$a_{17} = (2w-1)a_5$	$a_{21} = (2u-1)a_9$
$a_{14} = (2v-1)a_2$	$a_{18} = (2w-1)a_6$	$a_{22} = (2v-1)a_{10}$
$a_{15} = (2v-1)a_3$	$a_{19} = (2w-1)a_7$	$a_{23} = (2v-1)a_{11}$
$a_{16} = (2u-1)a_4$	$a_{20} = (2w-1)a_8$	$a_{24} = (2u-1)a_{12}$

$a_{25} = 4u(1-u)(1-w)\,dv$	$a_{31} = u\,4w(1-w)\,dv$
$a_{26} = 4v(1-v)(1-w)\,du$	$a_{32} = u\,4v(1-v)\,dw$
$a_{27} = (1-v)\,4w(1-w)\,du$	$a_{33} = 4u(1-u)\,v\,dw$
$a_{28} = 4u(1-u)(1-v)\,dw$	$a_{34} = v\,4w(1-w)\,du$
$a_{29} = (1-u)\,4v(1-v)\,dw$	$a_{35} = 4v(1-v)\,w\,du$
$a_{30} = (1-u)\,4w(1-w)\,dv$	$a_{36} = 4u(1-u)\,w\,dv$

$a_{37} = (2v-1)a_{25}$	$a_{41} = (2w-1)a_{29}$	$a_{45} = (2w-1)a_{33}$
$a_{38} = (2u-1)a_{26}$	$a_{42} = (2v-1)a_{30}$	$a_{46} = (2u-1)a_{34}$
$a_{39} = (2u-1)a_{27}$	$a_{43} = (2v-1)a_{31}$	$a_{47} = (2u-1)a_{35}$
$a_{40} = (2w-1)a_{28}$	$a_{44} = (2w-1)a_{32}$	$a_{48} = (2v-1)a_{36}$

$a_{49} = 4v(1-v)\,4w(1-w)\,du$	$a_{52} = (2u-1)a_{49}$
$a_{50} = 4u(1-u)\,4w(1-w)\,dv$	$a_{53} = (2v-1)a_{50}$
$a_{51} = 4u(1-u)\,4v(1-v)\,dw$	$a_{54} = (2w-1)a_{51}$

to a_{24}) have linear tangential field along one edge. The next 24 basis functions (a_{25} to a_{48}) have non-zero tangential field on one face but no edges. Finally, the last 6 basis functions (a_{49} to a_{54}) have no tangential field on the element boundary.

To assemble global basis functions, the first 24 basis functions must coordinate with all elements which share their one edge. The second 24 basis functions must coordinate with the element which shares their one face. The last 6 basis functions are valid global basis functions by themselves, and need not coordinate with any neighboring elements.

In terms of exterior products ($\wedge$) and exterior derivatives (d), equation (1) can be expressed as

$$\int_\Omega dF \wedge \mu^{-1} dE - \frac{\omega^2}{c^2} \mathbf{F} \wedge \epsilon E = 0, \tag{6}$$

where μ and ϵ are now Hodge-star operators which convert 1-forms (e.g., E and H) to 2-forms (e.g., D and B) using some tensor (i.e., the permittivity and

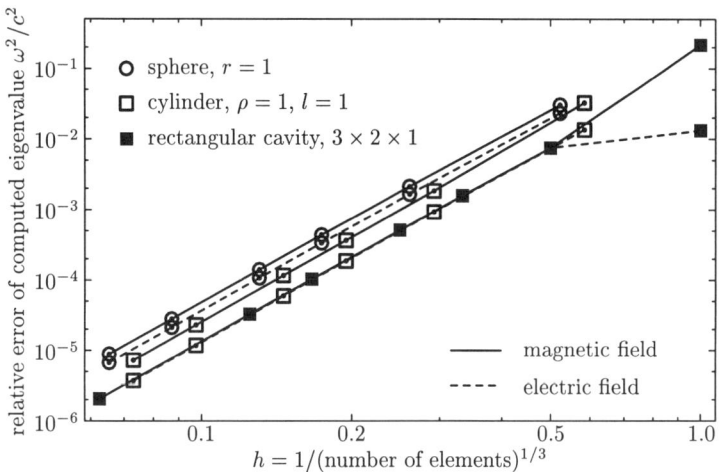

Figure 1. Relative eigenvalue error versus element size h for three test cavities.

permeability tensors). Differential geometry reminds us that we can differentiate in local coordinates (where the basis and their derivatives are simple) and transform the result to global coordinates (where the permittivity, permeability and metric tensors are usually simple). This avoids messy applications of the chain rule and extra calculation. In some cases it may be cost-effective to transform the tensors to local coordinates and perform all of the computations in local coordinates.

TESTS OF THE FEA CODE

This FEA code was tested on (1) a $3 \times 2 \times 1$ rectangular cavity, (2) a pillbox cavity with radius $\rho = 1$ and height $l = 1$ and (3) a $\rho = 1$ spherical cavity. The mesh was refined in a regular manner from a coarse mesh to a fine mesh. The relative error of the computed eigenvalues for both electric and magnetic field calculations is shown in figure 1. Excellent accuracies (less than 10ppm error) are achieved on a modest workstation. The error is proportional to h^4, where h is the element size. The numerical eigenvalue of the spurious modes is less than 10^{-12}, so spurious modes are well separated from the physical modes.

Test cases with inhomogeneously filled cavities show similar results. The spurious modes still have zero eigenvalue, even when the dielectric properties change within an element. The accuracy is excellent, with the caveat that sharp corners cause a significant reduction in accuracy, so the mesh needs to be refined in these locations.

The execution times are reasonable when compared with MAFIA. Figure 2 compares this FEA code (YAP) with MAFIA on a pillbox test cavity. A uniform grid was employed in the MAFIA calculations in order to produce the typical accuracy achieved by MAFIA in a more complicated structure. This

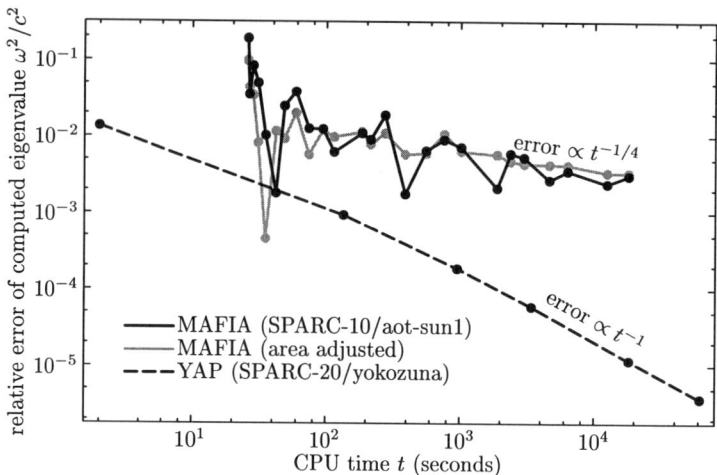

Figure 2. Relative eigenvalue error versus CPU time for MAFIA and YAP. The test structure is a pillbox cavity with radius $\rho = 1$ and height $l = 1$. The YAP results are electric field calculations. The MAFIA results employed a uniform grid.

FEA excels when good accuracy is desired since the FEA error scales like t^{-1} instead of $t^{-1/4}$. This difference is due to the discretization error (proportional to h^4 for this FEA, and h for MAFIA).

CONCLUSION

A 3D electromagnetic finite element analyis code with high accuracy and with no contamination from spurious modes has been demonstrated. Differential geometry and geometrical physics is useful for understanding the features of these codes and why they work.

References

1. Cendes, Z. J., *IEEE Trans. Magnetics* **27**, 3958–3966 (1991).
2. Bossavit, A., *IEE Proc.* **135**, 493–500 (1988).
3. Barton, M. L. and Cendes, Z. J., *J. Appl. Phys.* **61**, 3919–3921 (1987).
4. Crowley, C. W. et. al., *IEEE Trans. Magnetics* **24**, 397–400 (1988).
5. Deschamps, G. A., *Proc. IEEE* **69**, 676–696 (1981).
6. Nelson, E. M., Ph.D. thesis, Stanford University (1993); also SLAC-431.
7. Nédélec, J. C., *Numer. Math.* **35**, 315–341 (1980).
8. Nédélec, J. C., *Numer. Math.* **50**, 57–81 (1986).

Electromagnetic PIC Modeling with a Background Gas

J. P. Verboncoeur and D. Cooperberg
Electronics Research Laboratory
University of California
Berkeley, CA 94720-1774
johnv@eecs.berkeley.edu

Abstract

Modeling the interaction of relativistic electromagnetic plasmas with a background gas is described. The timescales range over many orders of magnitude, from the electromagnetic Courant condition ($\sim 10^{-12}$ sec) to electron-neutral collision times ($\sim 10^{-7}$ sec), to ion transit times ($\sim 10^{-5}$ sec). For this work, the traditional Monte Carlo algorithm [1] is described for relativistic electrons. Subcycling is employed to improve efficiency, and smoothing is employed to reduce particle noise. Applications include plasma-focused electron guns, gas-filled microwave tubes, surface wave discharges driven at microwave frequencies, and electron-cyclotron resonance discharges. The method is implemented in the OOPIC code [2].

The Monte Carlo collision (MCC) model statistically describes the collision processes, using cross sections for each reaction of interest. An electrostatic MCC model [1] is adapted for electromagnetic simulations. Techniques for electron-neutral and ion-neutral collisions are described, as well as methods for improving performance in simulations with widely varying timescales.

Consider a set of particles incident on a second set of targets. For the ith incident particle of energy $\mathcal{E}_i = \frac{1}{2}mv_i^2$, the probability, P_i, of a collision event can be written

$$P_i = 1 - \exp\left[-n_g(\mathbf{x})\,\sigma_T(\mathcal{E}_i)v_i\Delta t\right], \tag{1}$$

where the total cross section is the sum over all processes, $\sigma_T(\mathcal{E}_i) = \sum_j \sigma_j(\mathcal{E}_i)$. Here $n_g(\mathbf{x})$ is the spatially varying target density, v_i is the incident speed, and Δt is the time interval.

For a pure Monte Carlo method [3], the probability is inverted to solve for the time interval between collisions for the ith particle:

$$\Delta t_i = -\frac{\ln(1-R)}{n_g(\mathbf{x})\,\sigma_T(\mathcal{E}_i)v_i}, \tag{2}$$

where $0 < R < 1$ is a uniformly distributed random number. The equations of motion can then be integrated for Δt_i before applying the collision behavior. While this technique has the benefit of taking the longest possible timestep, resulting in maximum computational efficiency, it is evident that the particles are no longer synchronized in time. This method can only be applied when space charge and self-field effects can be neglected. Nonetheless, once a collision event does occur, the energy and angular distributions are computed as described below.

There is a finite probability that the ith particle will undergo more than one collision in the time interval Δt_i. For nearly lossless collisions, such as elastic scattering with a massive

© 1997 American Institute of Physics

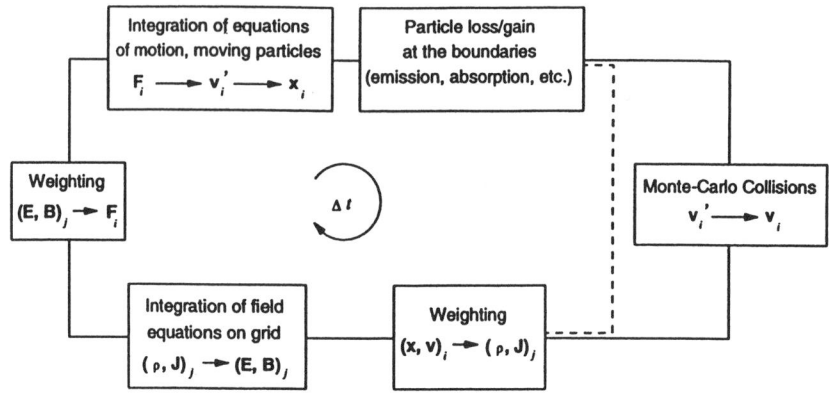

Figure 1: Flow chart for an explicit PIC-MCC code.

target, the probability of n collisions in Δt_i is P_i^n, so the total number of missed collisions can be written

$$r_i \approx \sum_{k=2}^{\infty} P_i^k = \frac{P_i^2}{1-P_i}. \qquad (3)$$

Since $P_i \ll 1$, $r_i \approx P_i^2$ provides a measure of the under-representation of the collision operator. Hence, we are constrained to choose $\nu_{T,\max}\Delta t \ll 1$ for accuracy, where $\nu_{T,\max} = \max[n_g(\mathbf{x})\sigma_T(\mathcal{E}_i)v_i]$ is the maximum collision frequency in space and energy.

Computing the collision probability for each particle each timestep is computationally expensive, since it involves computing the particle energy, a square root to obtain the speed, and either interpolation of tabled cross sections or computation of a curve fit for each process. The cost of the MCC can exceed the cost of integrating the equations of motion by an order of magnitude, so we seek a more efficient method.

Defining a maximum collision frequency in space and energy,

$$\nu_{\max} = \max_{\mathbf{x}}\left(n_g(\mathbf{x})\right) \max_{\mathcal{E}}\left(\sigma_T(\mathcal{E})v\right), \qquad (4)$$

we can write a total collision probability independent of particle energy and position, $P_T = 1 - \exp\left(-\nu_{\max}\Delta t\right)$. The physical interpretation is shown in Fig. ??. The collision frequencies are incrementally summed for electron impact on neon, with ν_1 representing elastic scattering, ν_2 representing all excitations summed, and ν_3 representing electron impact ionization. The area between $\nu_{\max}$ and $\nu_1 + \nu_2 + \nu_3$ represents the null collision event.

The fraction of particles undergoing a collision each time step is now given by P_T, and the particles can be chosen at random from the particle list. Depending upon the implementation, duplicates may be discarded, resulting in the error described in Eq. 3. A more accurate method is to apply multiple collisions to a single particle sequencially. Once the particles undergoing collisions have been selected, the type of collision for each particle is determined by choosing a random number, $0 \leq R \leq \nu_{\max}$. R is mapped onto the collision frequencies shown in Fig. ??.

First, we consider electron-neutral scattering. The differential cross section is required to compute the final velocity of the incident electron. For elastic scattering of electrons in

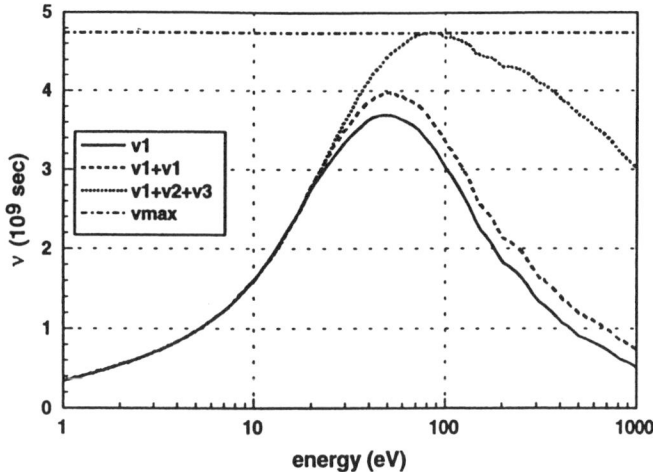

Figure 2: Summed collision frequencies for the null collision method.

argon, one possible cross section is [4]:

$$\frac{\sigma(\mathcal{E}_i, \chi)}{\sigma(\mathcal{E}_i)} = \frac{\mathcal{E}_i}{4\pi \left[1 + \mathcal{E}_i \sin^2(\chi/2)\right] \ln(1 + \mathcal{E}_i)} \quad (5)$$

with incident electron energy $\mathcal{E}_i$ in eV, χ the scattering angle. It should be noted that many other choices are possible; for example, see [5] for differential ionization cross sections in a number of gases. The cumulative distribution function is:

$$R = \frac{\int_0^\chi \sigma(\mathcal{E}, \chi) \sin \chi \, d\chi}{\int_0^\pi \sigma(\mathcal{E}, \chi) \sin \chi \, d\chi} \quad (6)$$

If R is a random number $0 \leq R \leq 1$, the scattering angle becomes:

$$\cos \chi = \frac{2 + \mathcal{E}_i - 2(1 + \mathcal{E}_i)^R}{\mathcal{E}_i} \quad (7)$$

The scattering angles for electrons incident on argon are shown in Fig. 3. The angular distribution varies from isotropic at 10 mV to small-angle at 10 kV. The azimuthal angle is uniformly distributed, $0 \leq \theta \leq 2\pi$.

Once the scattering angle is specified, the fractional energy loss in the scattering event can be computed by classical collision mechanics [6]:

$$\Delta \mathcal{E} = \frac{2m}{M}(1 - \cos \chi)\mathcal{E}_i. \quad (8)$$

Next, we consider electron-neutral inelastic collisions, such as excitation and ionization. For ionization, the energy balance is $\mathcal{E}_f = \mathcal{E}_i - \mathcal{E}_2 + \mathcal{E}_N - \mathcal{E}_+ - \mathcal{E}_{iz}$, where $\mathcal{E}_i$ and $\mathcal{E}_f$ are the initial and final primary electron energies, $\mathcal{E}_2$ is the energy of the secondary electron, $\mathcal{E}_N$ is the neutral energy, $\mathcal{E}_+$ is the ion energy, and $\mathcal{E}_{iz}$ is the ionization threshold. For excitation, the energy balance becomes $\mathcal{E}_f = \mathcal{E}_i - \mathcal{E}_{ex}$, where $\mathcal{E}_{ex}$ is the excitation threshold. Since the

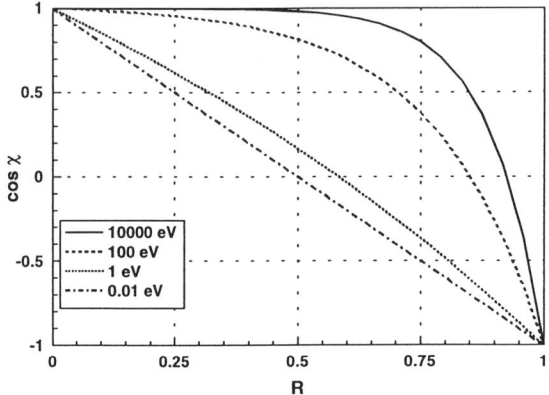

Figure 3: Scattering angle distribution for a range of energies.

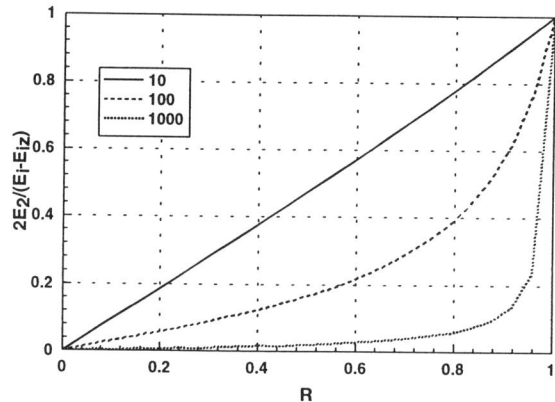

Figure 4: Normalized scattered electron energy distribution for $10 \leq \mathcal{E} \leq 1000$ eV.

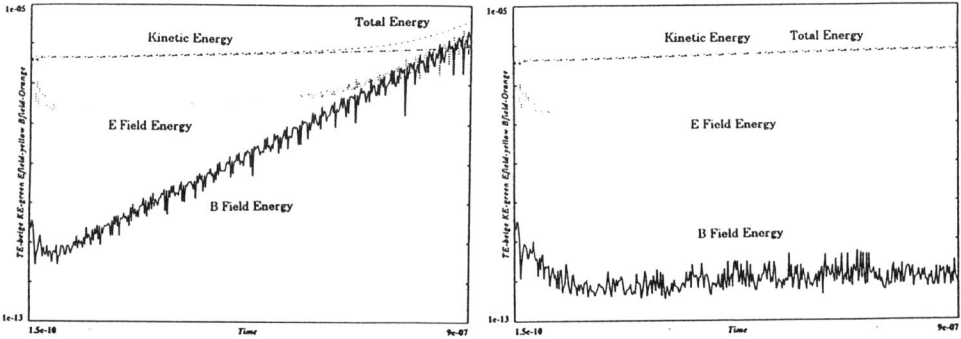

Figure 5: Temporal evolution of field energy in a surface wave plasma for undamped (left) and damped (right) cases.

mass of the electron is small compared to the mass of the neutral, $m \ll M$, we neglect the momentum change of the neutral, so that $\mathcal{E}_N = \mathcal{E}_+$.

For ionization at low $\mathcal{E}_i$, a differential cross section may be chosen of the form [7]

$$S(\mathcal{E}_i, \mathcal{E}_2) = \frac{\sigma_{iz}(\mathcal{E}_i) B(\mathcal{E}_i)}{\arctan\{[\mathcal{E}_i - \mathcal{E}_{iz}]/[2B(\mathcal{E}_i)]\} [\mathcal{E}_2^2 + B^2(\mathcal{E}_i)]}. \tag{9}$$

Here, $B(\mathcal{E}_i)$ is a known function for many gases. Similar differential ionization cross sections for a series of gases are given in [5]. Inverting the distribution,

$$R = \frac{\int_0^{\mathcal{E}_2} S(\mathcal{E}_i, \mathcal{E}_2') d\mathcal{E}_2'}{\int_0^{(\mathcal{E}_i - \mathcal{E}_{iz})/2} S(\mathcal{E}_i, \mathcal{E}_2') d\mathcal{E}_2'} \tag{10}$$

gives the energy of the secondary electron:

$$\mathcal{E}_2 = B(\mathcal{E}_i) \tan\left[R \arctan\left(\frac{\mathcal{E}_i - \mathcal{E}_{iz}}{2B(\mathcal{E}_i)}\right)\right]. \tag{11}$$

Although the primary and secondary electrons are indistinguishable, we have chosen the convention that the more energetic electron is the primary. In Fig. 4, the energy of the secondary electron normalized to the maximum secondary energy, $\mathcal{E}_{2,\max} = \frac{1}{2}(\mathcal{E}_i - \mathcal{E}_{iz})$, is plotted for a number of incident electron energies. The energy is distributed uniformly at low energies, and the normalized secondary energy is reduced significantly at higher energies.

In the high-energy regime, the first Born approximation accurately predicts the differential cross section [9]:

$$f^{(1)}(\chi) = -\frac{2m}{\hbar^2} \frac{1}{2k_i \sin(\chi/2)} \int_0^\infty rV(r) \sin(2k_i \sin(\chi/2)) \, dr \tag{12}$$

where k_i is the incident electron wavenumber, $V(r)$ is the interaction potential. Using a potential appropriate to the collision of interest, one can obtain the energy and angular distributions as discussed above. For the Coulomb potential, $V(r) = Z_1 Z_2 e^2/r$, the differential cross section becomes the classical Rutherford scattering cross section:

$$\frac{d\sigma}{d\Omega} = \frac{1}{16}\left(\frac{Z_1 Z_2 e^2}{\mathcal{E}_i}\right)^2 \frac{1}{\sin^4(\chi/2)}. \tag{13}$$

Ion-neutral collisions are similar to electron-neutral collisions, except the collision mechanics must be performed in the rest frame of the neutral since the momenta are similar, $v_i \sim v_N$. For scattering, the scattered ion energy can be written:

$$\mathcal{E}_f = \left(1 - \frac{2M_i M_N}{(M_i + M_N)^2}(1 - \cos\Theta)\right) \mathcal{E}_i = \mathcal{E}_i \cos^2\chi \tag{14}$$

where $\cos\chi = \sqrt{1-R}$ for isotropic scattering, and Θ is the scattering angle in the center of mass frame. The subscripts i and N refer to the incident ion and neutral respectively. The azimuthal angle is chosen randomly. Charge exchange uses the same process for computing the probability, but the neutral velocity is chosen from an analytic distribution. The neutral identity is then exchanged with the ion.

The wide variation of timescales requires use of a number of techniques to accelerate the computation. The cost can be reduced to that of fastest species (electrons). The maximum subcycling ratio is given by:

$$\frac{\Delta t_i}{\Delta t_e} = \frac{\omega_{pe}}{\omega_{pi}} = \sqrt{\frac{M}{m}}. \tag{15}$$

The efficiency becomes $\eta = (N_e + N_i \Delta t_e / \Delta t_i)^{-1}$. For $N_e = N_i$, $50\% \leq \eta \leq 99.6\%$ for argon. A narrow band instability occurs for $\omega_{pe} \Delta t_i \approx l\pi$, so it is important in discharge-type problems avoid the instability in the areas where particles are trapped [10][11].

Subcycling fields leads to repeated application of source terms containing statistical noise. This can cause high frequency modes to grow, resulting in particle heating. A technique for temporal filtering of the fields can reduce the heating [12]:

$$\bar{E}^{n-1} = (1-\zeta)E^n + \zeta \bar{E}^{n-2}, \ 0 \leq \zeta \leq 1/2 \tag{16}$$

$$E^{n+1} = E^n + c\Delta t \nabla \times B^{n+1/2} - \varepsilon \Delta t J^{n+1/2} \tag{17}$$

$$B^{n+3/2} = B^{n+1/2} - c\Delta t \nabla \times \tag{18}$$

$$\left[(1 + \zeta/2)E^{n+1} - E^n/2 + (1-\zeta)\bar{E}^{n-1}/2\right]. \tag{19}$$

Here $\bar{E}$ defines a lag-averaged electric field, and ζ is the damping parameter. This method allows a reduction of the number of particles per Debye sphere, $N_{D,0}/N_{D,f} \sim 10$, for equivalent numerical heating. A comparison of energy growth rates is shown in Fig. 5.

The electromagnetic modeling of devices which include background gases requires careful attention to timescales to achieve acceptable performance. Methods such as subcycling can improve performance, while methods such as temporal filtering can reduce noise. These techniques have been used with success in many applications, including large area surface wave discharges, series resonance discharges, plasma-focused electron beams, and microwave tubes with plasma background. A number of additional techniques can also be considered, including implicit solution of field and particle equations and numerical splitting/coalescing of particles.

This work was supported in part by AFOSR-MURI grant F49620-95-1-0253, U.S. Enrichment Corporation, and ONR AASERT N100014-93.

References

[1] V. Vahedi and M. Surendra, *Comp. Phys. Comm.* **87**, 179 (1995).

[2] J. P. Verboncoeur, A. B. Langdon and N. T. Gladd, *Comp. Phys. Comm.* **87**, 199 (1995). Codes available via http://ptsg.eecs.Berkeley.EDU.

[3] B. M. Penetrante and J. N. Bardsley, *J. Appl. Phys.* **54**, 6150 (1983).

[4] M. Surendra, D. B. Graves I. J. Morey, *Appl. Phys. Lett.* **56**, 1022 (1990).

[5] A. E. S. Green and T. Sawada, *J. Atm. Terr. Phys.* **34**, 1719 (1972).

[6] E. W. McDaniel, *Atomic Collisions*, Wiley N.Y. (1989).

[7] C. B. Opal, W. K. Peterson and E. C. Beaty, *J. Chem. Phys.* **55**, 4100 (1971).

[8] D. Vender and R. W. Boswell, *IEEE Trans. Plasma Sci.* **18**, 725 (1990).

[9] J. J. Sakurai, *Modern Quantum Mechanics*, Addison-Wesley, N. Y. (1985).

[10] J. C. Adam, A. Gourdin-Serveniere and A. B. Langdon, *J. Comp. Phys.* **47**, 229 (1982).

[11] B. I. Cohen, "Orbit Averaging and Subcycling in Particle Simulation of Plasmas", *Multiple Timescales*, Ed. by J. U. Brackbill and B. I. Cohen, Academic Press, Orlando FL (1985).

[12] P. Rambo, J. Ambrosiano, A. Friedman and D. E. Nielson, *Proc. 13th Conf. Num. Sim. Plasmas*, Santa Fe, NM (1989).

Higher Order Modes in Tapered Disc-Loaded Structures

Ursula van Rienen

*Technische Hochschule Darmstadt, Fachbereich 18, FG TEMF,
Schloßgartenstr. 8, 64289 Darmstadt, Germany*

Abstract. Several designs for linear colliders for electron-positron-collisions in the TeV range are currently under discussion. Most of them will use long trains of bunches. Therefore the study of higher order modes excited by previous bunches in a train became even more important for the optimal design of the accelerator components. Many of the designs use long tapered disc-loaded waveguides for acceleration. Numerical reasons hinder the mode analysis of these long structures by discretization methods. In this paper the electromagnetic waves in tapered multi-cell structures with circular cross section are calculated by a modal field matching method in which the scattering matrix formulation is used to calculate the amplitudes of the field expansion. Some remarks on convergence aspects are made. The field distribution of deflecting dipole modes is analyzed for a 180-cell tapered disc-loaded S-Band structure as well as for a 36-cell test structure. For the 36-cell test structure comparisons with a grid-oriented numerical method, an equivalent circuit model and measurements are also presented.

INTRODUCTION

Deterioration of the intense particle beams in a multibunch collider scheme is caused by the long range wake fields. These deflecting fields build up by parasitic higher order modes (HOMs) which are excited by preceeding bunches. The loss factor (1) is a measure of the interaction between the particle beam and some mode in the accelerating structure:

$$k_{mn}(r) = \frac{|V|^2}{4U} = \frac{|\int_{z=0}^{L} E_z(r_0, \varphi, z) e^{j\beta z} dz|^2}{4U}, \qquad (1)$$

with m the azimuthal dependance of the n-th mode of frequency ω_{mn}, V the voltage seen by a particle which travels along the structure at radius r and U

the stored energy. The wake potential in a cylindrically symmetric structure can be calculated by the transverse wake functions (1)

$$w_m(s) = 2q \sum_n k_{mn} \sin(\omega_{mn} s/c). \qquad (2)$$

Detuning and damping are two measures to overcome the HOM problem. Single and multibunch dynamics of the beam can be studied by computer simulations. For these simulations the loss factors, determining the wake potential, and the quality factors of the HOMs have to be taken into account.

NUMERICAL METHOD

The application of grid-oriented methods for tapered disc-loaded structures is limited by the realizable number of mesh points and the accuracy of the eigenvalue solver in case of a large cluster of eigenvalues. Semi-analytical methods do not know this limitation but most structures have in general to be simplified in order to apply the methods. The Modal Field Matching Technique can be applied for structures which can be split into subregions where analytic solutions of Maxwell's equations can be given as an expansion of an orthogonal series over discrete modes. The field solutions for the whole structure are then obtained by field matching at the interfaces between the subregions. The placement of these interfaces is problem-dependent.

The derivation of the Mode Matching Technique has been explained in more detail in (2). Regarding an iris-loaded cylindrical waveguide, the transverse field of an E_z- or H_z-wave in each subregion is given by a Bessel-Fourier Series:

$$\vec{E}_t(r,\varphi,z) = \sum_{n=1}^{\infty} U_n(z)\, \mathbf{e}_n(r,\varphi), \quad \vec{H}_t(r,\varphi,z) = \sum_{n=1}^{\infty} I_n(z)\, \mathbf{h}_n(r,\varphi). \qquad (3)$$

with the field eigenfunctions $\mathbf{e}_n(r,\varphi)$ and $\mathbf{h}_n(r,\varphi)$. They are differential functions of the sectional eigenfunctions which are composed of a trigonometric and a Bessel function and form a complete orthonormal system.

The next step is to match the solutions at the interfaces of neighbouring subregions. Voltage and current amplitudes $U_n^i(z)$ and $I_n^i(z)$ of region $i = I, II$ are connected to the amplitudes of the eigenwaves in the longitudinal direction:

$$U_n^i = \pm \sqrt{Z_n^i}\, (a_n^i\, e^{-jk_{z,n}^i z} + b_n^i\, e^{jk_{z,n}^i z}) \qquad (4)$$

$$I_n^i = \pm \sqrt{Y_n^i}\, (a_n^i\, e^{-jk_{z,n}^i z} - b_n^i\, e^{jk_{z,n}^i z}). \qquad (5)$$

The transverse fields at the cross-sectional area between regions I and II are given by Equation (3) with appropriate local coordinate z. Expanding in the reciprocal direction assures equal power flow from both directions onto the common cross-sectional area. Exploiting the orthonormality of the eigenfunctions finally leads to the scattering matrix expressing the connection between the wave amplitudes of an incident wave **a** with that of the reflected or transmitted waves **b** (compare figure 1):

$$\begin{pmatrix} \mathbf{b}^I \\ \mathbf{b}^{II} \end{pmatrix} = \begin{pmatrix} \underline{S}_{I,I} & \underline{S}_{I,II} \\ \underline{S}_{II,I} & \underline{S}_{II,II} \end{pmatrix} \begin{pmatrix} \mathbf{a}^I \\ \mathbf{a}^{II} \end{pmatrix}. \qquad (6)$$

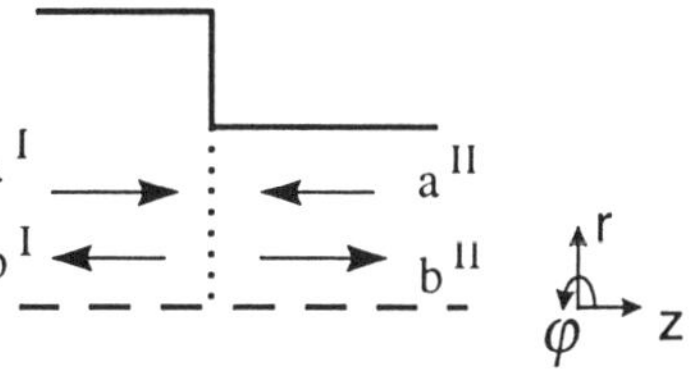

FIGURE 1: Step in cylindrical waveguide with amplitudes of incident waves $\mathbf{a}^I, \mathbf{a}^{II}$ and reflected or transmitted waves $\mathbf{b}^I, \mathbf{b}^{II}$.

In a structure with multiple changes in diameter the scattering matrix is constructed for each subregion, first. These matrices are concatenated to obtain the scattering matrix for the complete structure or parts of it. The outgoing amplitudes $\mathbf{b}^I$ and $\mathbf{b}^{II}$ are calculated from the scattering matrix for the complete structure. The inner amplitudes in each subregion needed to determine fields, voltages and loss factors are computed from the scattering matrices for the parts left and right of their location. In case of a trapped mode which does not touch any end cell of the constant-gradient structure it is necessary to split up the complete structure at a convenient iris inside the structure. Resonant frequencies are obtained by the code RESO (3). For each resonant mode fields, voltages and loss factors are calculated separately with the program ORTHO (2).

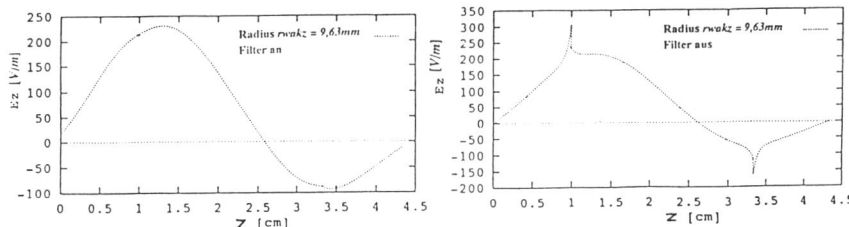

FIGURE 2: E_z over z. Left: Filter *and* intermediate steps. Right: Filter alone.

Convergence criteria and studies have been treated in (2). Recently, some improvements of ORTHO have been made: Using "artificial intermediate steps" of nearly zero length in addition to a cosine filter it is possible to improve the smoothness of the longitudinal field across a matching plane as shown in Figure 2. Besides this, ORTHO was speeded up by a factor of 2-3 by storing more of the intermediate parameters that are needed more than once, but this changes only the coefficient in the quadratic dependance of cpu-time on number of sections and modes for the expansion. This quadratic dependance is one of the major drawbacks of the Mode Matching Technique.

RESULTS AND COMPARISONS

S-Band Quasi-Constant Gradient Structure

To make the problem tractable the 180 cell constant gradient structure of the SBLC Design (3) was approximated by a 180 cell model which is tapered in groups of six cells with rectangular instead of rounded corners (2). The curve displayed in Figure 4 shows a nearly flat maximum extended over 2/3 of the first dipole passband. Therefore, not only the first π-like dipole modes influence the beam dynamics but about 120 modes, in fact. A major part of these modes is trapped *inside* the structure, that is without contact to the end cells (compare Figure 3). The first and second dipole band overlap. The local periodicity of the structure is reflected by a somewhat oscillatory behaviour of the loss factor curve. Averaging results in the curve for a linear taper. Many further numerical investigations have been carried out which can't be reported here because of lack of space. It became obvious that the form of the loss factor curve depends strongly on the degree of tapering.

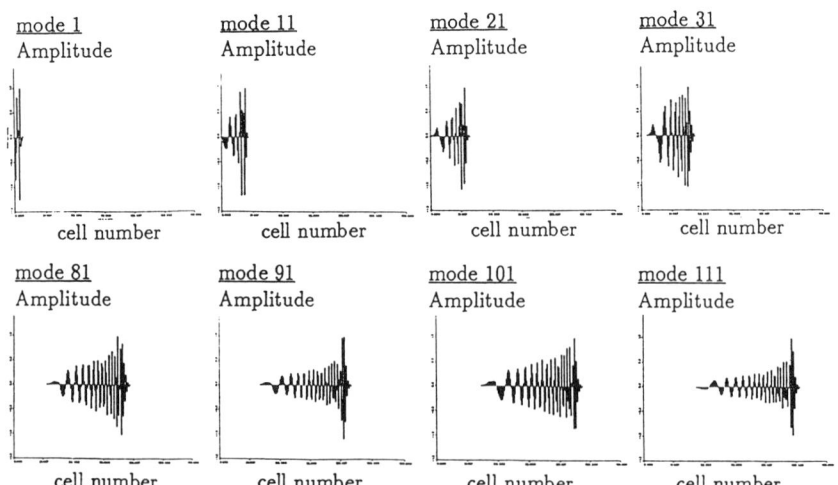

FIGURE 3: Normalized voltage amplitudes of some trapped HOMs in a 180-cell quasi-constant gradient structure as function of cell number.

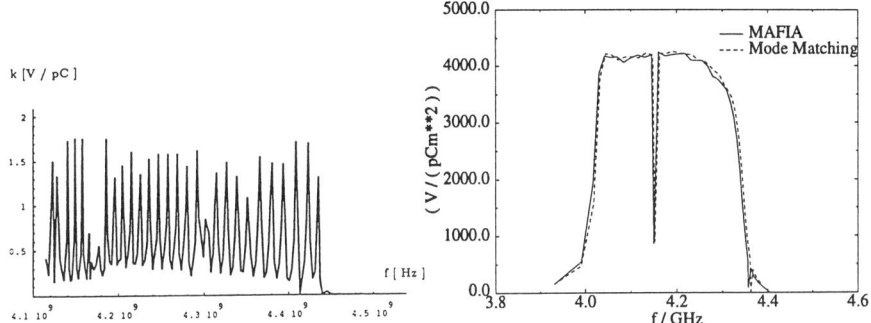

FIGURE 4: Loss-parameter as function of frequency for the 180-cell structure (left) and the 36-cell structure (right).

In order to further study the phenomenon of trapped modes a 36-cell structure has been designed (5) and a double-band coupled oscillator model (COM) (6) has been developed. As a direct consequence of the simulations the damping system for the SBLC was completely redesigned. A global damping with Kanthal-coated irisses is foreseen now (8) and two local HOM-dampers, which will mainly be used for beam diagnostics purposes (9).

36-Cell-Experiment

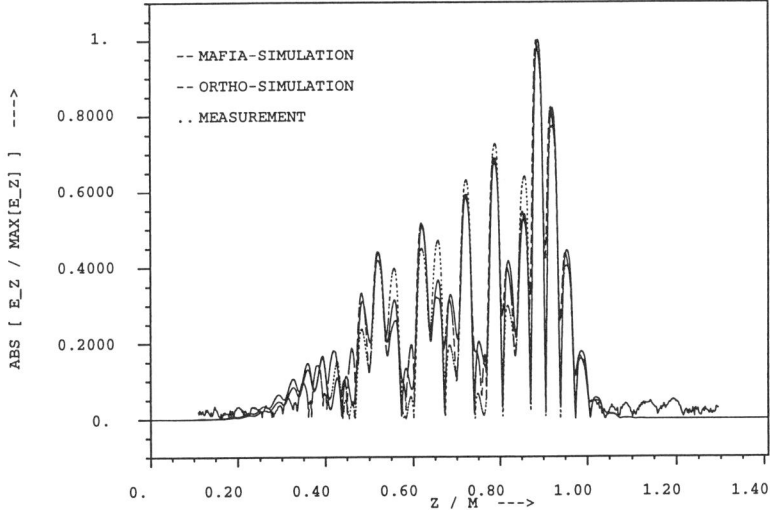

FIGURE 5: Plot of $|\vec{E}_z|/\|\vec{E}_z\|_\infty$ at $r_{\text{offset}} = 6$ mm for the 19-th dipole mode of the 36-cell structure.

The design goals for the 36-cell structure were to get a structure which is easy to measure, easy to manufacture, computable without geometric approximations by several numerical methods (MAFIA, URMEL–T (10), ORTHO,

COM) and which shows a clear appearance of trapped modes. A relatively short structure was chosen in order to avoid the appearance of mode overlap limiting the performance of RF-measurements. The simple geometry has numerous practical and numerical advantages (11). The strong linear taper ensures the appearance of trapped modes. Figure 4 shows the loss factors of the final design. In this structure, all modes with high loss factor are trapped modes. Figure 5 displays a comparison between different simulation methods and some bead-pull measurement carried out at the University of Frankfurt (11) for one of the trapped modes. The agreement between simulation and measurement is good.

FIGURE 6: Electric field of the 15-th dipole mode in the 36-cell structure.

CONCLUSION

The mode matching method to calculate modes in tapered disc-loaded structures was described. It is an appropriate tool for *long tapered* structures. *Trapped modes* and their loss factors are shown for a S-Band quasi-constant gradient structure of 180 cells. For further studies of the phenomenon of trapped HOMs a 36-cell structure has been designed, simulated with proven discretization methods and a coupled oscillator model as well as measured by the bead pull technique. The calculations and the measurements agreed well.

The author would like to thank the many people that have been involved in the theoretical and experimental aspects of this work.

REFERENCES

1. Bane, K.L.F.; Wilson, P.B. and Weiland, T., *AIP* **127** 875–928 (1983)
2. van Rienen, U., *Part.Acc.* **41** 173–201 (1993)
3. Steinigke, K., PhD thesis, TH Darmstadt, Germany
4. Weiland, T. et al., "Design Study for a 500 GeV Linear Collider", in *Proceedings of the Int. Linear Accelerator Conf.*, 1990
5. Krietenstein, B. et al., "The S-Band 36-cell Experiment", in *Proceedings of the Particle Accelerator Conf.*, 1995, pp. 695–697
6. Dohlus, M., private communication
7. Weiland, T., *Part.Acc.* **17** 227–242 (1985)
8. Dohlus, M. et al., "Higher Order Mode Damping by Artificially Increased Surface Losses", DESY-Report, in preparation
9. Hülsmann, P. et al., "Beam Position Monitoring for SBLC Using HOM-Coupler Signals", in *Proceedings of the European Part. Acc. Conf.*, Sitges, Spain, 1996
10. van Rienen, U. and Weiland, T., *Part.Acc.* **20** 239–267 (1986/87)
11. Kurz, M. et al., "Higher-Order Modes in a 36-Cell Test Structure", in *Proceedings of the European Particle Accelerator Conf.*, Sitges, Spain, June 10-14, 1996

EFFECTS OF SPACE CHARGE ON THE CURRENT-VOLTAGE CHARACTERISTICS OF FIELD EMITTER ARRAYS

K. L. Jensen, E. G. Zaidman, M. A. Kodis

Naval Research Laboratory, Washington, D.C. 20375

Abstract. Field emitter arrays are microfabricated very high electron current density sources. For rf amplifier applications, typical current densities are typically on the order of 100 Amps/cm^2. Unlike thermionic emitters, the current densities at the emission sites on field emitters can approach 10^8 Amps/cm^2 at high fields. Consequently, the high current from the array can affect the I(V) characterization of the emitters. In this manuscript, we use a simple model of a field emitter to calculate the one dimensional space charge effects on the current versus gate voltage characteristics. Two effects are treated: charge between the gate and anode, and charge within the FEA unit cell, which gives rise to a new space charge component. It is shown that space charge effects can taint the Fowler Nordheim parametrization of field emitters and consequently affect the estimates of their transconductance.

INTRODUCTION

Gated electron emitters, such as the field emitter array (FEA) hold the potential for significantly impacting next-generation RF amplifiers (1,2) especially linear beam tubes (3). Given the high current densities which are required for rf devices, it becomes crucial to estimate under what conditions space charge effects occur in the I(V) characterization of these emitters, as field emitter arrays are not typically intended to operate in a space charge limited regime (under which conditions Child's law (4) relates the gate to anode field to the current density over the array). Space charge can therefore in principle can render FEA characterizations using a collecting anode irrelevant in the prediction of IOA performance characteristics. In order to assess the effects of space charge, we will postulate a simple model of a field emitter and investigate under what conditions space charge effects are important.

ANALYTICAL MODEL WITHOUT SPACE CHARGE

Consider a hyperbolic field emitter whose surface is specified by the cylindrical coordinates $\rho = a_s \cot\beta \sinh\alpha$ and $z = a_s \cot^2\beta (1 - \cosh\alpha)$, where the apex of the tip lies at the origin, a_s is the radius of curvature of the emitter tip, and β is the cone half-angle. Define the quantity $z_o = a_s \cot^2\beta$; in the absence of a gate plane (*i.e.*, a diode), z_o would specify the tip to anode separation. The field at the apex of a hyperbolic emitter in a diode geometry has an analytical solution, as does the tip field in the simple "Saturn Model" of a triode. As shown elsewhere (5), the two analytical solutions can be united in the "hybrid approximation" to the tip field as

© 1997 American Institute of Physics

$$F_{tip} = \frac{\pi}{a_s \ln\left(k a_g / a_s\right)} V(z_o) \qquad (1)$$

where $V(z_o)$ remains to be specified, a_s is the tip radius, a_g is the gate hole radius, and k depends on all other geometrical parameters associated with the unit cell (and weakly on gate and tip radius; it may be determined using Boundary Element simulations.

The potential on axis, $V(z)$, may be approximated by

$$V(z) = \frac{F_{tip} z}{V_{gate} + F_{tip} z}\left(1 + \frac{F_o z}{V_{gate}}\right) V_{gate} \qquad (2)$$

where the term in parentheses is an *ad hoc* term to ensure proper behavior as $z \to D$, but which otherwise has a negligible effect near the emitter. Inserting this expression into Eq. (1) allows F_{tip} to be found

$$F_{tip} \approx \left(\frac{\pi}{\ln\left(k \frac{a_g}{a_s}\right)} - \tan^2\beta\right) \frac{V_{gate}}{a_s} \qquad (3)$$

where terms proportional to F_o have been ignored. Once k is determined, Eq. (3) gives good agreement with Boundary Element simulations for variations in tip and gate radius.

Using the Fowler Nordheim approximation for current density, $J_{FN}(F) = a_{fn} F^2 \exp(-b_{fn}/F)$, where a_{fn} and b_{fn} are constants which depend on the work function ϕ (6), the total current is obtained by integrating the charge density over the surface of the emitter. To a very good approximation, it is given by

$$I_{tip}(V_{gate}) = 2\pi a_s^2 \left[\frac{F_{tip} \cos^2\beta}{b_{fn} + F_{tip}\sin^2\beta}\right] J_{FN}(F_{tip}) \qquad (4)$$

where F_{tip} implicitly depends upon the gate potential. For small fields, I is proportional to $a_{fn} F_{tip}^3 \exp(-b_{fn}/F_{tip})$; the presence of an extra F_{tip} will cause positive convexity in a Fowler Nordheim plot of $I(V_{gate})$ (7). The agreement between Eq. (4) and Boundary Element (BE) simulations is very good if F_{tip} is calculated using BE methods, and is within a factor of 2 if calculated using Eq. (3).

ANALYTICAL MODEL INCLUDING SPACE CHARGE EFFECTS

The presence of electrons between the emitter tip and the collection anode will serve to depress the field at the surface of the emitter. To assess their effects, we make the following assumptions: *(i)* electrons are emitted along the symmetry axis of a conical emitter every $\tau = 1/I_{tip}$ femtoseconds (8), *(ii)* all of the emitters in an array emit at the same time, and *(iii)* at a point z_g, where $V(z_g) = V_{gate}$, the electrons have a velocity equal to $v_g = (2V_{gate}/m)^{1/2}$. For a unit cell, equipotential lines at z_g resemble a boss (*i.e.*, a hemisphere) on a plane. Outside of that boss, the electrons over the array resemble a sheet of charge traveling from the gate plane to the anode for which the transit time is designated by Δt. The number of sheets between the gate and anode is given by $\Delta t/\tau$. Each sheet gives rise to a force component (9) equal to $2\pi\alpha_{fs}hc\eta$, where α_{fs} is the fine structure constant and η is the (surface) charge density of the sheet and is equal to the packing density of the emitters. The field enhancement factor on the sphere approximating the unit cell emitter is simply a factor of 3 and is independent of radius.

Near the emitter, the distance an electron travels in a time τ is given by $z(\tau) \equiv L = F_{tip} \tau^2 / 2m$. The departing electron generates an image charge on the tip equivalent sphere). Assuming that an electron is emitted every τ femtoseconds, and adding up the field components of all emitted electrons plus their image charges gives rise to the unit cell component of space charge. Unlike the gate-anode space charge component, this component is independent of the presence of other emitters, and will exist even for a single emitter: *this effect is not present for thermionic emitters and is peculiar to field emitters.*

In the absence of charges between the gate and anode, the transit time Δt is simply given by the *source-limited* expression (10) $\Delta t_o = 2(D - z_g)/(v_a - v_g)$, where $v_a = \sqrt{(2V_{anode}/m)}$, and likewise for v_g, which is simply the time of flight of a particle falling in a uniform field with an initial velocity v_g. When charge is present between the gate and anode, the transit time increases due to the effects of the retarding fields produced by the sheets of charge. We shall use the $\Delta t = \Delta t_o$ as given, even though in space charge limited flow, a potential maximum exists outside the gate region, and electrons will consequently be delayed much longer than Δt_o would suggest. To reiterate, we are concerned with the effects on the field emitter prior to the space charge limited regime, in which the transit time is in fact closer to the approximation we use. As the gate-anode portion of δF dominates, making this approximation actually *underestimates* the influence of space charge on the $I(V)$ characteristics.

The sum of the unit cell and gate-anode (inner- and outer-boss, respectively) field components on-axis is thus

$$\delta F = \pi \alpha_{fs} \hbar c \left(6 \frac{\eta \Delta t}{\tau} + \frac{\pi}{6 a_c L} + \frac{\pi^3}{45 L^2} \right) \tag{8}$$

The field at the surface of the emitter is thus given by $F_{tip} - \delta F$. As δF implicitly depends on I_{tip}, Eq. (4) and Eq. (3) must be solved self-consistently. However, two approximations are convenient, and can be found under the assumption that δF is small: for the contribution to δF from outside the emitter, we have

$$\delta F_{outer} \approx \frac{6 F_{tip}^2 \pi \alpha_{fs} \hbar c \eta \Delta t I_{tip}}{6 (b_{fn} + 3F_{tip}) \pi \alpha_{fs} \hbar c \eta \Delta t I_{tip} + F_{tip}^2} \quad (9)$$

while from inside

$$\delta F_{inner} \approx \frac{(4m\pi^2 a_c I_{tip}^2 + 15 F_{tip}) F_{tip}^2 m \pi \alpha_{fs} \hbar c I_{tip}^2}{45 a_c F_{tip}^4 + (2b_{fn} + 5F_{tip})(8m\pi^2 a_c I_{tip}^2 + 15 F_{tip}) m \pi \alpha_{fs} \hbar c I_{tip}^2} \quad (10)$$

For low tip current, Eqs. (9) and (10) are good approximations to the effect of electrons upon the field at the tip. In both of the cases considered herein, Eq. (9) typically dominates Eq. (10) in magnitude.

SIMULATION

Let us consider field emitters typical of MIT Lincoln Laboratory (11) and SRI (12), for which the parameters are shown in Table 1. The deviation of Eqs. (9) and (10) from their exact values is shown in Fig. 1. As can be seen, the approximations are adequate for low fields, and qualitatively describe the behavior at higher fields. The effects of the field suppression at the emitter surface are shown in Fig. 2: as can be seen, at high fields, enough current is drawn from the emitters to make δF significant, thereby decreasing the magnitude of I_{tip}. Such deviations are clearly visible on a Fowler Nordheim plot of $\ln(I(V_g)/V_g^2)$ vs. $1/V_g$.

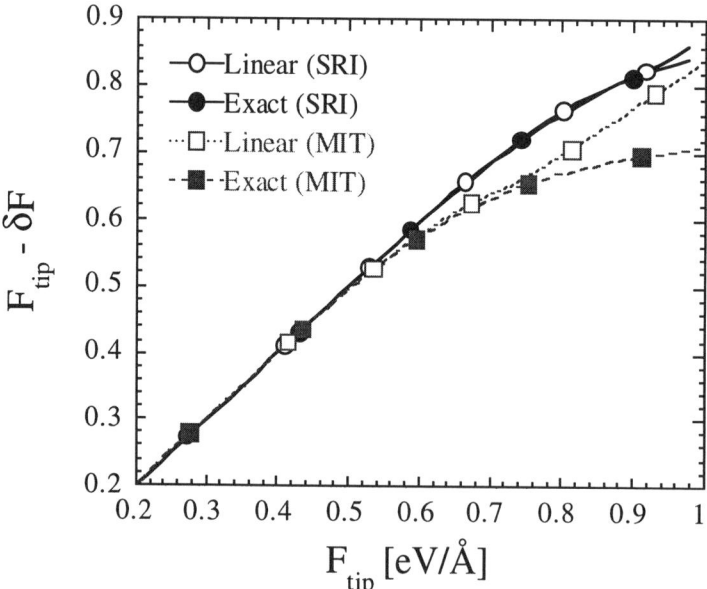

Fig. 1: Effects of space charge on the field at the tip of an emitter for both MIT-like and SRI-like FEA geometries and compared to the approximations in Eqs. (9) and (10).

The Linear Least Squares estimations ([13]) of the slope ($\ln(A_{fn})$) and intercept ($-B_{fn}$) characterize the $I(V)$ data; the Fowler Nordheim A_{fn} and B_{fn} parameters are given in Table 1. A figure of merit associated with FEA performance (2) is given by the ratio of the transconductance over the capacitance (g_m/C_{tip}), where $g_m = \partial I_{tip} / \partial V_g$. As the transconductance depends exponentially on B_{fn}, one may conclude, as shown in Table 1, that space charge effects can strongly affect the figure of merit, and that, if the anode extraction field is insufficient, the Fowler Nordheim parametrization of a field emitter can be tainted.

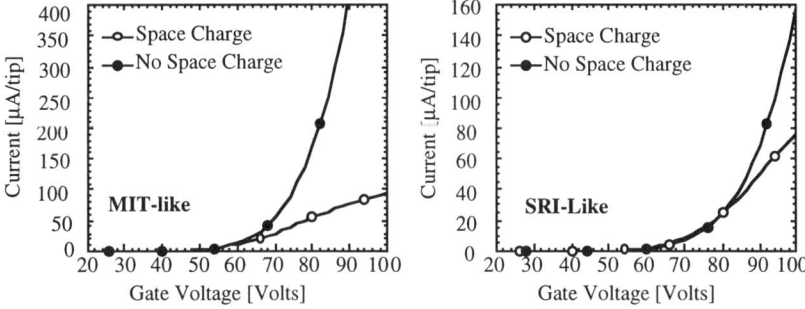

Fig. 2: Effects of space charge on the I(V) characteristics of FEAs with MIT-like and SRI-like geometries.

TABLE 1

TERM	UNITS	MIT.like	SRI.like
Work function	eV	3.70	4.07
Tip Radius	Å	57.20	43.00
Gate Radius	µm	0.08	0.25
Gate to anode	µm	457.00	800.00
Cone Angle	deg	23.00	15.00
k	–	4.79	10.20
% Emitting	%	16.41	70.00
tip-to-tip	µm	0.32	1.00
Anode Voltage	Volts	100.00*	300.00†
A_{fn} (SC) §	µA/V²	0.95	2.52
A_{fn} (No SC)	µA/V²	11.40	4.54
B_{fn} (SC)	Volts	378.27	529.46
B_{fn} (No SC)	Volts	485.47	564.76
g_m (SC) **	µA/V²	0.24	4.00x10⁻²
g_m (No SC)	µA/V²	0.41	3.75x10⁻²

* Anode measured with respect to gate potential.
† Anode measured with respect to ground potential.
§ Linear Least Squares fit was performed on I(V) for which $I_{tip} \geq 1$ nA.
** Transconductance was calculated at $V_g = 50$ V.

CONCLUSION

We have compared a one-dimensional theory of a field emitter to estimate the effects of space charge. The analytical model was applied to show how the onset of curvature at high gate voltages appears in $I(V)$ data, and shows that unless care is taken to insure that the extraction field is adequate, the Fowler Nordheim characterization of field emitters (in terms of A_{fn} and B_{fn}) may be in error.

ACKNOWLEDGEMENTS

We gratefully acknowledge funding provided by the *Office of Naval Research*.

REFERENCES

1. J. P. Calame, H. F. Gray, and J. L. Shaw, J. Appl. Phys. 73 (1993) 1485.
2. M. A. Kodis, K. L. Jensen, E. G. Zaidman, B. Goplen, and D. N. Smithe, NRL Technical Memorandum NRL/FR/6840--95-9783 (1995).
3. R. K. Parker, IEEE Int. Conf. on Plasma Science (ICOPS), 5-8 June 1995, Madison, WI.
4. C. K. Birdsall and W. B. Bridges, *Electron Dynamics of Diode Regions*, (Academic Press, New York, 1966) p36.
5. "Electron Emission from a single Spindt-type Field Emitter: Comparison of Theory with Experiment," K. L.Jensen, P. Mukhopadhyay-Phillips, E. G. Zaidman, K. Nguyen, M. A. Kodis, L. Malsawma, C. Hor *(accepted for publication in Applied Surface Science)*
6. A. Modinos, Field, Thermionic, and Secondary Electron Spectroscopy (Plenum, New York, 1984).
7. K. L. Jensen, E. G. Zaidman M. A. Kodis, B. Goplen and D. N. Smithe, J. Vac. Sci. Tech. B14 (1996) 1942 and 1947.
8. Units natural to this discussion are electron volts, femtoseconds, angstroms, and electron charge. In these units, current density is given by $e/(fs\ Å^2)$ and current by (e/fs).
9. L. Eyges, *The Classical Electromagnetic Field* (Dover, New York, 1972).
10. C. K. Birdsall and W. B. Bridges, Electron Dynamics of Diode Regions, (Academic Press, New York, 1966) p36.
11. R. A. Murphy and M. Hollis *(private communication)*; MIT-LL array is "Wafer 11-16-B2b" which contained 9760 nickel-coated molybdenum emitters.
12. C. A. Spindt and P. Schwoebel *(private communication)*; SRI array is "Quadrode #106L-1-3W"which contained 7304 molybdenum emitters.
13. J. R. Taylor, *An Introduction to Error Analysis* (University Science Books, Mill Valley, CA, 1982).

Enhancements to the Opera-3d Suite

Christopher P. Riley

Vector Fields Ltd., 24 Bankside, Kidlington, Oxford OX5 1JE, UK

Abstract. The OPERA-3D suite of programs has been enhanced to include 2 additional 3 dimensional finite element based solvers, with complimentary features in the pre- and postprocessing. SOPRANO computes electromagnetic fields at high frequency including displacement current effects. It has 2 modules – a deterministic solution at a user defined frequency and an eigenvalue solution for modal analysis. It is suitable for designing microwave structures and cavities found in particle accelerators. SCALA computes electrostatic fields in the presence of space charge from charged particle beams. The user may define the emission characteristics of electrodes or plasma surfaces and compute the resultant space charge limited beams, including the presence of magnetic fields. Typical applications in particle accelerators are electron guns and ion sources. Other enhancements to the suite include additional capabilities in TOSCA and ELEKTRA, the static and dynamic solvers.

INTRODUCTION

OPERA-3d is a suite of programs for the design of electromagnetic devices using finite element methods. It includes interactive graphical pre- and postprocessing with a menu driven graphical user interface (GUI), and a comprehensive command language. The latter allows creation of "macro" files of instructions to perform often repeated functions and supports parameterized modeling, as well as providing facilities to communicate with other software.

The preprocessor creates data for finite element based analysis programs in the suite. In the accelerator community, the TOSCA analysis (1) for non-linear magneto- and electrostatic analysis is widely used within laboratories and industrial manufacturers for the design of both conventional and superconducting magnets. More recently, many groups have also acquired ELEKTRA (2) which allows modeling of eddy current effects in magnets under steady state a.c. or transient behaviour, for example in (3). The analysis programs produce a direct access binary database containing the mesh, geometry and solution, which is interrogated directly by the postprocessing module (4).

TOSCA and ELEKTRA are joined by 2 new analysis programs – SOPRANO, for high frequency components, such as microwave structures and cavities,

© 1997 American Institute of Physics

and SCALA, for space charge limited particle beams. This paper describes these modules, including the additional facilities incorporated into the pre- and postprocessors to support them, and also some recent enhancements to TOSCA and ELEKTRA.

SOPRANO

SOPRANO has been developed to solve problems in microwave applications, especially cavity problems in accelerators. There are 2 modules – SOPRANO/SS which solves deterministic problems with user defined modes at input ports and known frequencies, and SOPRANO/EV which calculates the modes of a structure using an eigenvalue analysis. Both modules are based on the finite element method, giving the advantage of irregular meshing and correct representation of geometric shapes, but there are important differences in their implementations.

Deterministic Analysis Using SOPRANO/SS

The SOPRANO/SS solver (5) solves the wave equation in terms of the magnetic vector potential, $\mathbf{A}$, and the electric scalar potential, V. The Lorentz gauge

$$\nabla \cdot \mathbf{A} = -\mu\sigma V - \mu\epsilon\frac{\partial V}{\partial t} \qquad (1)$$

allows the magnetic vector potential solution to be decoupled giving the new governing equation

$$\nabla \times \frac{1}{\mu}\nabla \times \mathbf{A} - \nabla\frac{1}{\mu}\nabla \cdot \mathbf{A} + \sigma\frac{\partial \mathbf{A}}{\partial t} + \epsilon\frac{\partial^2 \mathbf{A}}{\partial t^2} = 0 \qquad (2)$$

Making the substitution $\sigma = \omega\epsilon''$ and using a complex permeability, $\epsilon = \epsilon' - j\epsilon''$, the electric field is determined from the solution of the continuity equation

$$\nabla \cdot \epsilon\nabla V + \nabla \cdot \epsilon\frac{\partial \mathbf{A}}{\partial t} = 0 \qquad (3)$$

The algorithm is implemented using a Galerkin weighted residual method based on 1st or 2nd order hexahedral finite elements. Decoupling has the advantage that equation (2) can be solved independently, reducing the matrix size. Equation (3) is then solved using $\frac{\partial \mathbf{A}}{\partial t}$ values obtained from the solution to (2). SOPRANO/SS uses a complex representation of the vector and scalar potentials (e.g. $\mathbf{A} = \mathbf{A}_c e^{j\omega t}$). Problems are excited from user specified modes at the ports. Defined currents in coils may also be specified as a method of excitation.

Eigenvalue Analysis Using SOPRANO/EV

The eigenvalue analysis module solves for the modes of a cavity or other structure using a formulation of the wave equation in terms of the electric field.

$$\nabla \times \frac{1}{\mu} \nabla \times \mathbf{E} - \epsilon \omega^2 \mathbf{E} = 0 \qquad (4)$$

The finite element implementation of the equation is also different to SOPRANO/SS with the use of tetrahedral elements and edge variables. The reasons for doing this are fully explained in (6), but simply result from the need to reliably separate the "spurious" from the true modes. The inexact representation from nodal based finite elements cause spurious modes to no longer be at zero frequency, giving a spectrum of intermingled real and spurious modes. Edge variable formulations overcome this problem as the spurious modes remain identically at 0 Hz., and can be eliminated.

The preprocessor has been enhanced to include the boundary conditions at the cavity wall, which are assumed perfectly conducting for the analysis but are assigned finite conductivity for determining losses in the postprocessor. Standard macro files are used to compute the total loss in the cavity wall for determination of Q-factor and shunt impedance.

Modal frequency results for the program have been compared with measurements on a cavity at Daresbury Laboratory, UK (7) showing good agreement

TABLE 1. Resonant Frequency Calculations

Mode*	Soprano (MHz)	Experiment (MHz)
Zero	125.54	125.12
L1	546.82	541.79
L2	730.90	728.48

* All longitudinal modes

A calculation on a more complex 3-dimensional cavity has been made to investigate the effect of connecting ports (8). Results for resonant frequency,

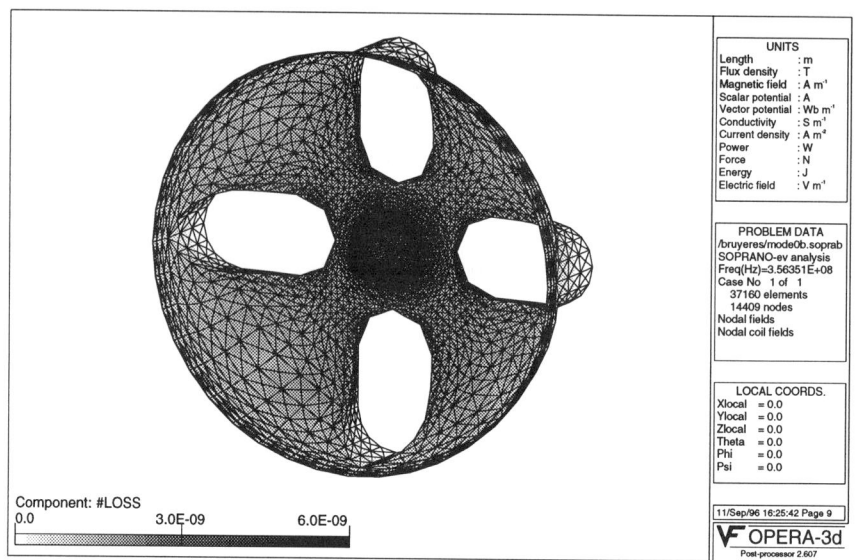

Figure 1: Loss distribution in cavity wall

Q-factor and geometric impedance compare well with other codes. For example, in the zero mode with the ports connected

TABLE 2. Zero mode parameters

Code	F(MHz)	Q	$g(=\frac{R}{Q})$ (Ω)
Mafia	356.72	40373	149.9
Antigone E	355.05	44678	152.8
Antigone H	356.33	42517	147.5
SOPRANO	356.35	44687	150.5

Figure 1 shows the computed losses in the wall of one half of the cavity for the zero mode. The model consisted of only a 45^0 section, but the postprocessor has been used to replicate the model for clarity.

SPACE CHARGE LIMITED BEAMS IN SCALA

SCALA is based on developments made by Vector Fields, University of Genova, Italy and Philips, Netherlands (9). The algorithm is an extension of the TOSCA electrostatic solution, solving the Poisson equation

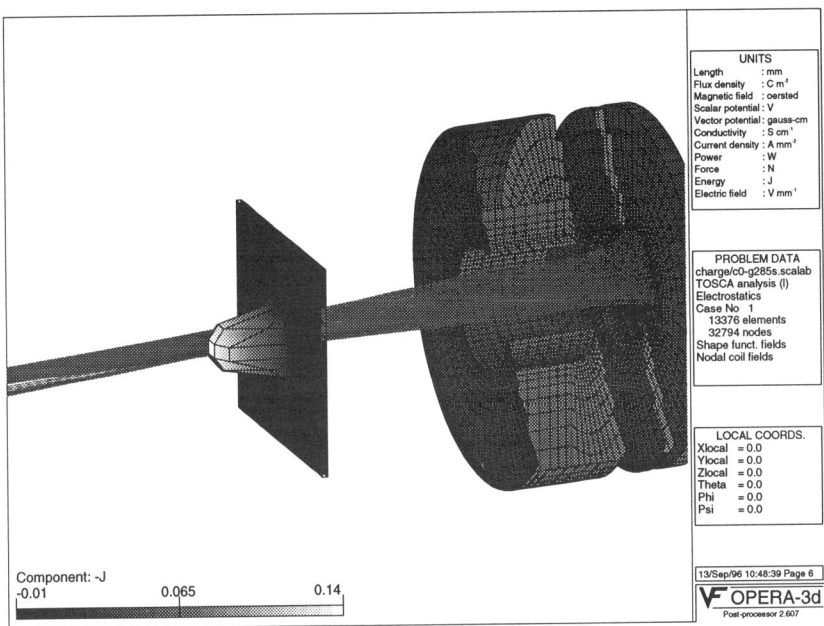

Figure 2: Representative particle tracks and current distribution

$$-\nabla \cdot \epsilon \nabla V = \rho \qquad (5)$$

to include the self-consistent space charge from charged particle beams. Equation (5) is solved iteratively until convergence is obtained. At the end of each iteration, the tracks from a set of representative particles launched from electrodes or plasma surfaces are computed. The current associated with each particle is used to provide an updated distribution of the space charge on the mesh nodes. The user defines the emission surfaces within the model and gives the characteristics to determine the emission current. These include a user defined current, Child's law and Langmuir-Fry limit, thermal saturation limit, field effect emission and a plasma free surface model.

Magnetic fields from a TOSCA (or other) analysis may be added to the problem, although the beam current does not alter their distribution. The particle tracking will take into account the combined electric and magnetic fields. Figure 2 shows the tracks from the converged solution of an electron gun and the beam current density distribution on an intersection plane.

ENHANCEMENTS TO TOSCA AND ELEKTRA

TOSCA now allows the solution of the d.c. current continuity equation in an arbitrary shaped geometry. The postprocessor has been upgraded to allow the magnetic fields from the resultant current distribution to be computed. In ELEKTRA, the steady state a.c. solver now includes non-linear materials. The transient solver has been enhanced to allow all coils (or families of coils) to vary independently in time, allowing different rise or decay times from different electrical circuit sources.

CONCLUSIONS

OPERA-3d has been enhanced by new modules to provide facilities for microwave and cavity structures. Eigenvalue and deterministic analysis is available. A new module is also included to model electric fields subject to space charge from particle beams. TOSCA and ELEKTRA also provide new features.

REFERENCES

1. Simkin, J. & Trowbridge, C.W., "3 dimensional nonlinear electromagnetic field calculations using scalar potentials", Proc. IEE, vol. 127, part B, Nov. 1980
2. Bryant, C.F. et al, "A general purpose 3d formulation for eddy currents using the Lorentz gauge", IEEE Trans. Mag., vol. 28, no. 5, Sep. 1990
3. Barnes, M.J. & Wait, G.D., "Comparison of measured and predicted inductance per cell for a travelling wave kicker magnet", presented at EPAC96, Sitges, Spain, Jun. 1996
4. Biddlecombe, C.S. & Riley, C.P., "Postprocessing of 3 dimensional electromagnetic fields", IEEE Trans. Mag., vol. 24, no. 1, Jan. 1988
5. Emson, C.R.I. & Trowbridge, C.W., "Finite element solutions to 3d waveguide problems using the magnetic vector potential", presented at IEE Conference on Computational Electromagnetics, London, Nov. 1991
6. Walsh, D.A. et al, "Resonant cavity design using the finite element method", presented at EPAC96, Sitges, Spain, Jun. 1996
7. MacIntosh, P.A., "An investigation of model performance of low frequency RF cavities", proceedings of EPAC94, London, Jun. 1994
8. Balleyguier, P., "Coupling slots without shunt impedance drop", presented at Linac 96 Conference
9. Girdinio, P. et al, "Finite element modelling of charged beams", IEEE Trans. Mag., vol. 30, no. 5, Sep. 1994

Code Update: MicroWaveLab

John F. DeFord
Ansoft Corporation, 4300 W. Brown Deer Rd., Suite 300, Milwaukee, WI 53223
deford@ansoft.com

Abstract. MicroWaveLab is a commercial software package that is used for design of RF and microwave components. The finite-element method is used to solve the frequency-domain Maxwell equations in 3-D volumes containing isotropic and/or anisotropic media. Edge elements are employed to avoid spurious modes, and several different element topologies have been implemented. The finite-element matrix is solved via either LU decomposition or a conjugate-gradient iterative method. Another approach is available wherein the total fields due to an excitation are expanded in terms of the 3-D eigenmodes of the structure. MicroWaveLab has geometry creation capabilities based on the ACIS solid geometry kernel, and pre- and postprocessing support for network parameters and antennas.

SOLVER

The solver takes an ASCII file description of the problem, performs the analysis, then outputs a results database file that is accessed by the user-interface. Supported analysis types include frequency-domain analysis (one matrix solution at each frequency point), eigenmodes for the 3-D model, and eigenmodes for the 2-D ports.

In the 3-D volume the following equation is solved using the finite-element method:

$$\nabla \times \frac{1}{\mu} \nabla \times \vec{E} = j\sigma\omega\vec{E} - \epsilon\omega^2\vec{E},$$

where $\vec{E}$ is the electric field. Using the finite-element method, this equation is reduced to a matrix equation of the form:

$$\left[-\omega^2 M + i\omega B + K\right]u = F,$$

where u is the vector of unknowns. MicroWaveLab uses H_0-curl and H_1-curl edge elements[1] in the formulation of the finite element method, which support unknowns (electric fields) which are projections of the field along, and normal to, the element edges and faces (see Fig. 1). The H-curl elements eliminate spurious modes that plague so-called nodal elements by providing an adequate representation of gradient subspace of the curl operator [2]. Three different element topologies are supported: tetrahedrons, triangular prisms, and hexahedrons or bricks. The second order elements (H_1-curl) are isoparametric, which means that their edges can conform to curved geometry.

The matrix equation obtained in frequency-domain analysis may be solved either via LU decomposition (using logic and storage schemes appropriate for sparse matrices), or using a pre-conditioned conjugate gradient iterative solver. Although several pre-conditioning schemes may be used, the default is a block-incomplete Cholesky technique.

© 1997 American Institute of Physics

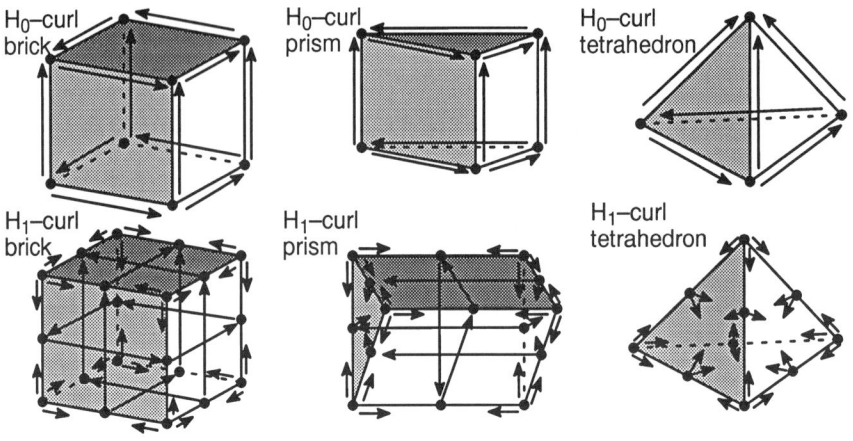

Figure 1: Element types supported in MicroWaveLab. Electric field projection unknowns are indicated by arrows.

Another approach to frequency-domain analysis is available in which the fields in the 3-D model that result from a port excitation are expanded in terms of the 3-D eigenmodes. This approach incurs an initial expense associated with the computation of the eigenmodes, but once the modes are computed S-parameters for individual frequency points are essentially free. A limitation of the technique as implemented is that a *real* eigenvalue analysis is performed on the 3-D geometry, and thus only no loss (or, using perturbation, low loss) structures may be analyzed. The eigenvalue analysis uses the Lanczos method with Sturm sequencing, so one is guaranteed to get all the modes in a specified range.

USER INTERFACE

The MicroWaveLab user interface is built on top of the ACIS solid geometry kernel from Spatial Technologies. ACIS provides support for constructive solid geometry in MicroWaveLab, including creation of solid primitives such as boxes, cylinders, and spheres. Construction of complex solids from wireframes is also supported, as are boolean operations between solids such as union, intersection, subtraction, and regioning.

The complete solid construction history may be accessed and modified to change the final geometry. Dimensions such as the length of an edge or the radius of a hole may be parameterized using a variable, and a new geometry created by changing the variables. A user may also modify geometry by picking a displayed dimension and changing it in an edit field (see Fig. 2).

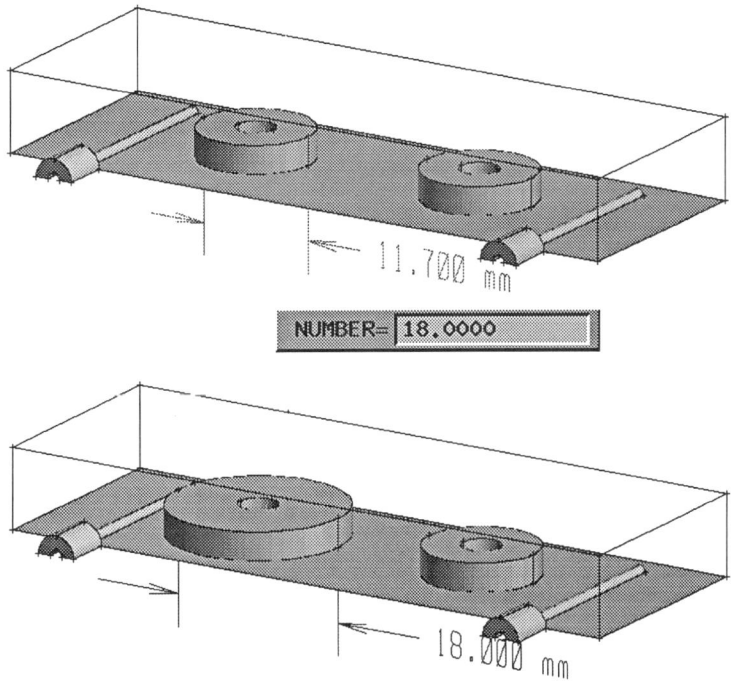

Figure 2: Geometry dimension editing feature in MicroWaveLab.

Boundary conditions are assigned by selecting the desired condition from a menu and then picking the appropriate face(s) of the solid model using the mouse. Boundary conditions that are supported include perfect electric and magnetic conductors, symmetry planes, ground plane, and radiation boundary. Boundary conditions are displayed on the associated surfaces using icons (Fig. 3).

Meshing the model is accomplished using one of three distinct meshers: automatic, mapped, and extrusion. The automatic mesher generates tetrahedrons (or triangles/quadrilaterals in 2-D) with a minimum of user intervention. Mesh density is controlled by placing size constraints at points, or along edges or faces, as well as by setting the target element size. There are also options which allow control over the mesh density along/near curved surfaces.

The map mesher is used to generate brick and prism elements in regions that have the appropriate topology. For example, brick elements may be generated in solids, or in regions within solids, that are logically hexagonal. A logically hexagonal volume does not have to appear brick-shaped, but it must have six faces that have the topology of a brick. Mesh density is determined by setting the number of elements along one edge in each logically perpendicular direction, and constraints can be used to vary the distribution of elements along an edge and within the volume.

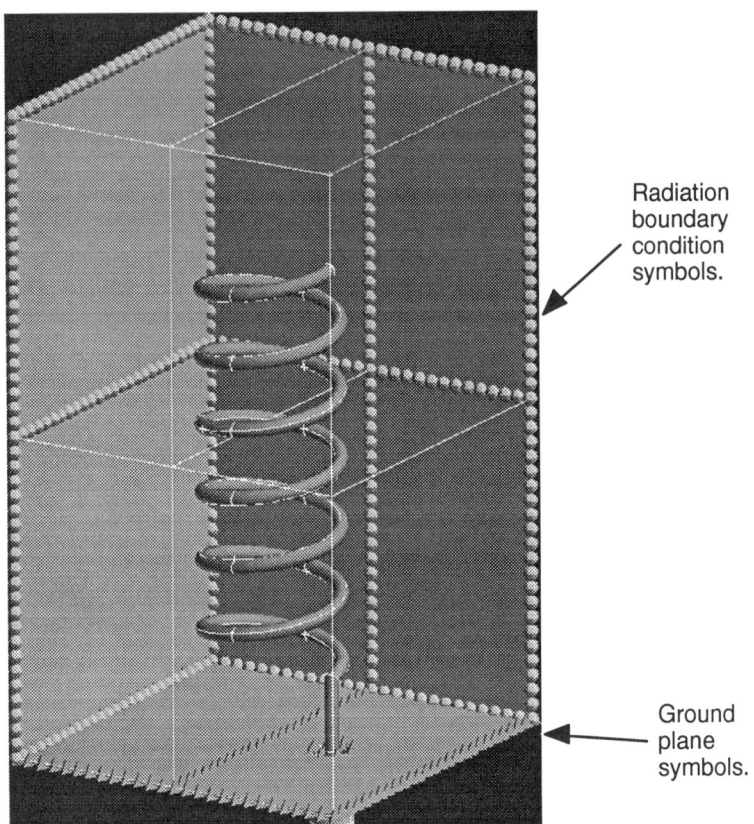

Figure 3: Example of geometry and boundary condition representation in MicroWaveLab. A helix may be generated by sweeping a circle along a helical spline. Boundary conditions are indicated using icons on the affected faces of the model.

The extrusion mesher also produces bricks and prisms, but it does so based on a 2-D mesh, not on solid geometry. Extrusions and revolutions are both supported.

Postprocessing support is provided for network parameters and antenna patterns. All standard network parameters are computed, including the scattering, impedance, and admittance matrices. Propagation constants and characteristic impedances of port modes are also generated. Degenerate modes are handled via a port calibration process, and the effect of lengthening or shortening port waveguides can be examined in postprocessing without re-running the analysis.

Antenna patterns and metrics are computed. Pattern data include orthogonal components of the electric far-field, far-field power density, and polarization ratios (axial and circular). Most standard metrics are computed, including gain, directivity, efficiency, sidelobe level, and beamwidth. Total radiated power is also computed.

APPLICATION EXAMPLE

This example was a challenge problem defined by the magazine Microwave Engineering Europe in 1995 [3]. It is a four-port waveguide power divider that divides power at the input port into approximately equal portions at the three output ports (see Fig. 4).

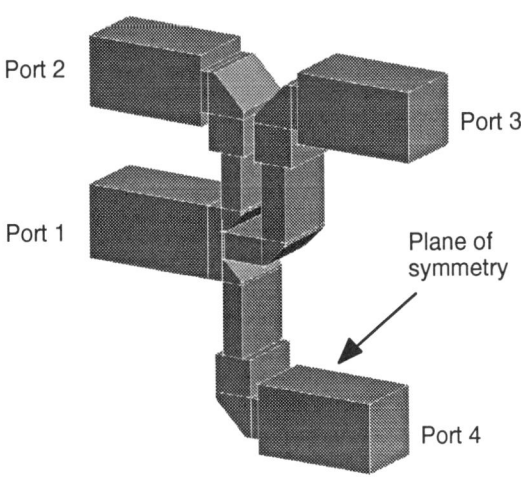

Figure 4: Four-port waveguide power divider. Power is input at Port 1, and split about equally between Ports 2, 3, and 4.

This device is operated in the fundamental mode, so a plane of symmetry (perfect magnetic conductor) may be used to reduce the model size. The material is air (or vacuum). The mesh is shown in Fig. 5. A mapped mesh was used because the field variation normal to the symmetry plane is the same throughout the device and can be adequately modeled with two elements, whereas rapid field variation necessitates a fine mesh in some areas parallel to the symmetry plane. Map meshing will usually yield the smallest element count that is adequate for a given problem when the mesh density requirements are anisotropic.

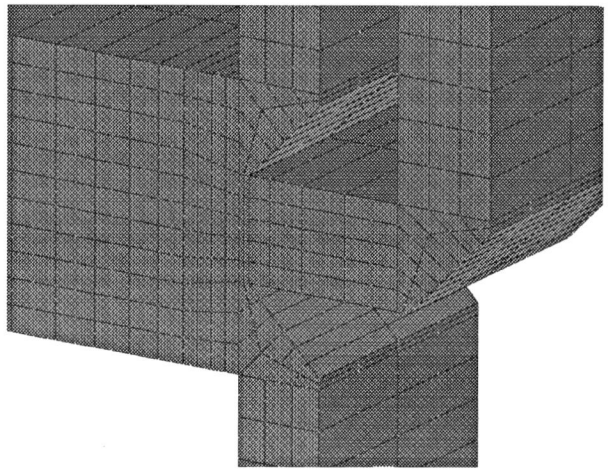

Figure 5: Portion of mesh of power divider. A mapped mesh was generated because the expected spatial field variation is rapid in the plane of the guides, and slow normal to the plane.

Scattering parameters for the operating frequency of 16.45 GHz for this model are shown in Table 1, along with measured data.

Table 1. Comparison of MicroWaveLab and published results[3].

Parameter	µWaveLab (dB)	Published(dB)
S_{11}	−32.09	−36.8
S_{12}	−5.09	−5.12
S_{13}	−4.06	−4.33
S_{41}	−5.27	−5.16

CONCLUSION

MicroWaveLab is an analysis product for use in computing electromagnetic fields in 3-D structures. Quantities derived from the fields can also be computed, such as network parameters and antenna patterns. The finite-element method is used to solve the Maxwell equations, and H-curl elements are utilized to eliminate the problem of spurious modes. Version 1 of MicroWaveLab shipped in September, 1995. Version 2 is presently in beta release.

REFERENCES

1. J.F. Lee, *Finite Element Methods for Modeling Passive Microwave Devices*, Ph. D. dissertation, Carnegie-Mellon University, 1989.
2. J. Webb, "Edge elements and what they can do for you," *IEEE Trans. on Magn.*, Vol. 29, No. 2, pp. 1460-1465, Mar., 1993.
3. "EM Simulation Benchmark," *Microwave Eng. Europe*, May 1995, pp. 23–25.

Status and perspectives of the PRIAM/ANTIGONE codes

G. Le Meur and F. Touze

Laboratoire de l'Accélérateur Linéaire, IN2P3-CNRS
et Université Paris-Sud. 91405 ORSAY CEDEX - FRANCE

Abstract. PRIAM/ANTIGONE are Maxwell finite element codes, developed at LAL. They make use of a coherent set of F.E. formulations suited to the Maxwell's equations, taking as unknowns the normal or tangential components according to the required continuities of the fields. A numerical investigation in frequency domain provides an example of the versatility and the complementarity of the different approaches. Some perspectives on applications of the domain decomposition methods in view of parallel computing are given.

PRESENTATION

In applying finite element methods to the Maxwell's equations, it is important to consider the continuity of the normal/tangential components, **D.n**, **B.n**, **E x n** or **H x n** (**n** being an unit vector normal to the considered interface or boundary) of the four vectors describing the electromagnetic field. There exits finite elements especially designed for achieving these continuities. In the 3D case the denoted "H(div)" tetrahedron element ensures the continuity of the normal components (with respect to the faces) which are actually the unknowns of the discretized problem. This finite element is suited, for example, in electrostatics in presence of different dielectric media and for getting the capacities which are directly related to the fluxes of **D** across the conductor surfaces. In "H(curl)", often called "edge element", the computed unknowns are the circulations of the calculated vector along the edges of the tetrahedra. This last element ensures the continuity of the tangential components.

It is out of the scope of this presentation to detail all the interesting properties of these elements in relation with the Maxwell's equations. For more theoretical information the reader is referred to [1] [2]. The codes PRIAM/ANTIGONE, developed at LAL, combine the use of these different finite elements H(div) (for **D**, **B**), H(curl) (for **E**, **H**) together with the classical P1-Lagrange element (for the potential V), in order to solve harmoniously the Maxwell's equations with respect to the particular assumption in concrete physical situations. For a review of the general topics of PRIAM/ANTIGONE one can refer to the Web (http://www.lal.in2p3.fr/SimNum/)

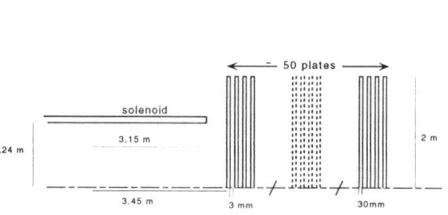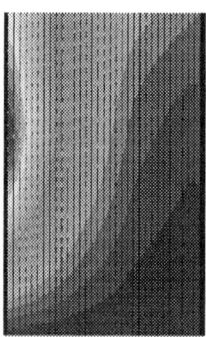

Figure 1: *This is an example of eddy current calculation in time domain (in 2D axisymmetric geometry) for the hadronic calorimeter of the ATLAS detector (for the future LHC machine). The problem was to evaluate the forces due to eddy current in the copper plates of the hadronic calorimeter resulting from a quench in the supraconducting solenoid. In the plates we have* **curl H** $= \sigma$**E** *(σ : conductivity); in the non-conducting domain we have* **curl H** $= 0$ *and **H** can be derived from a scalar potential. We are led to mix H(curl) elements (H-formulation) with P1-Lagrange (for representing the potential). On the left is given a scheme of the structure; on the right a contour plot of eddy current density in the first 30 plates after 9s (the current in the solenoid is considered to be decreasing exponentially with a relaxation time of 40s; the scale is different for the two schemes). The results were validated by a semi-analytical approach [6].*

or see ref [3]. Since this last reference, the code has been used in different applications either in detector design [4] or in accelerator physics [5], and also some developments have been provided (Figure 1). The frequency domain has been revisited and improved.

A NUMERICAL INVESTIGATION

In PRIAM/ANTIGONE, versatile modules are available in **frequency domain** either in 2D, 2D Fourier calculation ("transverses modes"), or in 3D ; with eventually periodic boundary conditions. Using algorithms combining the same integrals as in frequency domain, a more recent development allows the code to provide the **S-Matrix** of a structure having input and output ports, as well as the response of such a structure to a given **harmonic** input. For an explicitation of such algorithms see [7].

In frequency domain, PRIAM/ANTIGONE provides *two complementary formulations*, both based on the H(curl) element. The *E-formulation* starts from the eigenvalue equation: **curlcurlE** $= k^2$**E** ($k = \omega/c$; ω : pulsation, c : light velocity) with boundary conditions: **E** $\times$ **n** $= 0$ on conducting walls and **H** $\times$ **n** $= 0$ on magnetic walls. Using the H(curl) element we get the circulations of **E** along the edges of tetrahedra and therefore piecewise linear approximations for **E**. **B** is deduced from the relation : **B** $=$ (**curlE**)$/\omega$; **B** is piecewise constant. The *H-formulation* starts

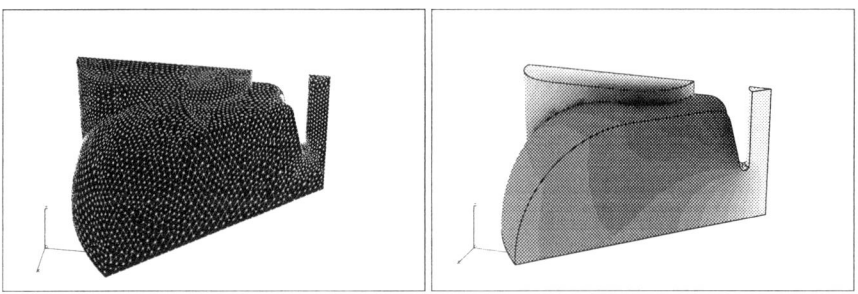

Figure 2: *TRISPAL cell: mesh d (see Fig. 3) and contour plot of moduli of the magnetic field. Two adjacent cells are magnetically coupled through four slots. Taking advantage of the symmetries of the cavity, the calculations have been carried out on 1/16 of a cell. One can see half of a coupling slot on the top of the figure. The axis is vertical, on the right.*

from the eigenvalue equation: **curlcurlH** = k^2**H**. A linear approximation of **H** is then got. **D** can be derived piecewise constant.

The method is free from any "spurious mode" pollution (see [8]).

The use of the two formulations provides complementary information. From the above considerations it follows that the quality factor (depending on H-moduli on the conducting boundary) should be better approximated with the H-formulation whereas the shunt impedance (depending on circulations or some similar integrals of **E**) should be better approximated in E-formulation. Nevertheless we will see that these arguments may have to be modulated.

We present here some results of calculations which were proposed to us by P. Balleyguier [9] concerning the optimized structure for the high intensity proton accelerator of the TRISPAL project (Figure 2). Results obtained with different refinements of meshes are plotted on Figure 3.

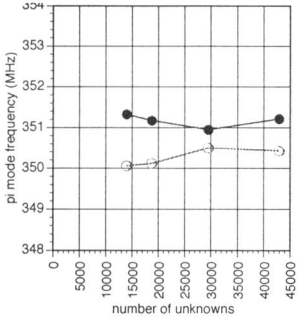 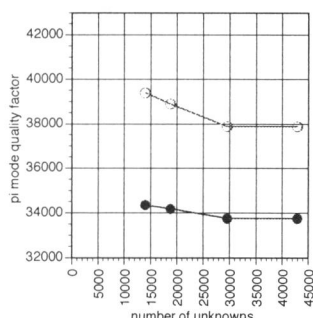

Figure 3: *TRISPAL cell: Frequency and quality factor as calculated for the π-mode using four meshes: a (13974 unknowns), b (18782 unknowns), c (29585 unknowns), d (42940 unknowns). Dark points are for H-formulation, light points for E-formulation.*

	π mode			zero mode		
	f(MHz)	Q	g(Ω)	f(MHz)	Q	g(Ω)
ANTIGONE (E)	350.43	37878	156.2	355.04	44478	152.8
ANTIGONE (H)	351.22	33753	151.3	356.33	42517	147.5
MAFIA	351.81	34114	154.5	356.72	40373	149.9
SOPRANO	352.01	38755	155.0	356.35	44754	150.5

	single cell (3D)				single cell (2D-axy.)		
	f(MHz)	Q	g(Ω)		f(MHz)	Q	g(Ω)
ANTIGONE (E)	360.17	40499	150.7	PRIAM (E)	360.49	40471	148.4
ANTIGONE (H)	361.69	40609	145.8				
MAFIA (3D)	361.72	36271	146.8	MAFIA (2D)	359.29	38578	150.6
SOPRANO	361.40	40058	147.	SUPERFISH	361.37	40888	148.7

Table 1: *TRISPAL cell: PRIAM/ANTIGONE and some other codes. Three cases have been systematically computed: π-mode, zero-mode and a "single cell" case (i.e. a single cell, with same dimensions but without coupling slots). The zero-mode and π-mode are obtained by considering the slot coupling plane respectively as electric or magnetic walls (from this fact, one may suppose, in PRIAM/ANTIGONE, that the zero-mode would be a little bit better approximated in E-formulation and π-mode in H-formulation). In the "single cell" case, 2D axisymmetric calculations have been made with 2D codes SUPERFISH, MAFIA-2D and PRIAM (MAFIA, SOPRANO and SUPERFISH results are from Balleyguier).*

Table 1 puts together our final results with those obtained with other programs by Balleyguier. The agreement between ANTIGONE H- and E-formulations is good for frequency and geometrical impedance. The results are more difficult to interpret for the quality factor. Results are close to each other for the "single cell" case, but, in other cases, there are quite important differences, especially in π-mode. The magnetic coupling seems to play a role in these differences, probably more important than the above continuity arguments on E- and H- approximation quality. It is then difficult to decide which is the best value for the quality factor in this situation (the given MAFIA and SOPRANO results do not help much in that goal!). A challenge is to complete information with further calculations on other types of structures in order to get a theoretical explanation of the observed phenomena.

PERSPECTIVES

The parallel architectures open new perspectives for the development of accelerator computing. The domain decomposition methods in view of parallel computing is a new goal for future PRIAM/ANTIGONE developments.

The general idea is to split the computational domain into "subdomains" and to assign one subdomain to each processor. In order to connect the subdomain results one has to reformulate the problem on the interface between the subdomains. This kind of method is widely known for elliptic problems (heat transfert, electrostatics etc.).

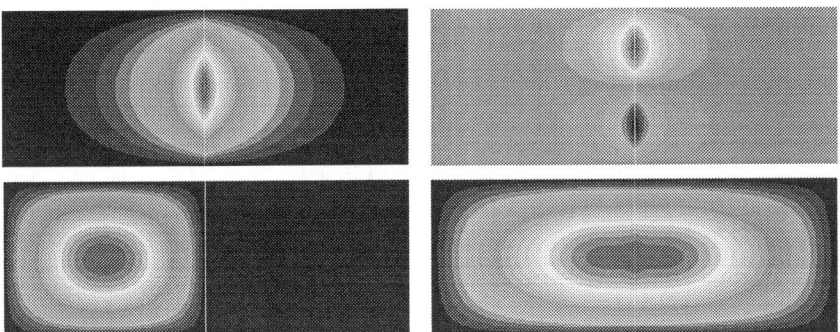

Figure 4: *Test for a modal synthesis method on a 2D eigenvalue problem ($\Delta u = k^2 u$). The rectangle is split into 2 subdomains. The 2 first coupling modes, the first fixed mode on the left subdomain and a combined solution (with 8 fixed modes on each subdomain and 4 coupling modes) are represented. The first two frequencies of the total structure is obtained with an accuracy better than one percent.*

The so-called "modal synthesis" method [10] adapts the domain decomposition technics to frequency calculations by using eigenmodes of subdomains (it has been successfully applied in mechanics). As a matter of fact, the general method for approximating an eigenmode of an operator is to search it as eigenmode of the matrix representing the considered operator in an approximation functional space with a reduced dimension. The idea of "modal synthesis", as proposed in the quoted reference, is to take as approximation space, a space spanned on one hand by (suited) extensions, on the whole domain, of subdomain eigenmodes (*fixed interface modes*; with fixed boundary conditions on the interface between subdomains) and on other hand by extensions of functions defined as eigenmodes of a certain interface operator (*coupling modes*; "connecting" the subdomains).

An example of the faisability of the method is shown Figure 4 on a very simple geometry. Although some questions remain to answer with this method applied to the electromagnetic eigemmode problems, it seems to be very promising.

These new developments in view of parallel computing imply the use of up-to-date programming methods, especially Object Oriented tools: they allow one to expect safety, versatility and quality of the code. In this goal we are helped by the programming rules we had already adopted in designing the current FORTRAN PRIAM/ANTIGONE which are close to O.O. prescriptions (encapsulation of data, with associated handling tools).

As programming language we have chosen the C++, associated with PVM for parallelization. An entirely O.O. C++ version of the topological description of a 2D mesh within PRIAM concepts has been achieved. A parallelized domain decomposition algorithm for Laplace's equation, both in P1 and H(div) formulation together with a preliminary modal synthesis module have been implemented.

THE PACKAGE

The algorithms of the PRIAM/ANTIGONE package are clearly separated from the mesh generating as well as from the post-treatment, through well identified interfaces which can be carried out very easy for any domestic configuration. The PRIAM/ANTIGONE package can be inserted in any existent finite element package. Because the used finite elements provide properties consistent with the nature of the physical fields (continuities) the algorithms can be easily extended to new problems without big programming investment. So that it is possible, with a unique set of meshing, graphical and numerical post-treatment tools, to use different finite element designs (mechanical, heat transfert etc.) together with PRIAM/ANTIGONE in the electromagnetic domain.

AKNOWLEDGMENT

We are debtful to P. Balleyguier from CEA (Bruyères-le-Châtel-France) for having submitted the TRISPAL cell calculation and for detailed discussions.

REFERENCES

[1] Bossavit A., *Electromagnétisme en vue de la modélisation*: Springer-verlag, 1991.

[2] Le Meur G., Touze F., "Implementation of a mixed finite element in a particle method," in *Proceedings of the Conference on mathematical and numerical aspects of wave propagation phenomena*, SIAM, 1991, pp. 752–754.

[3] Le Meur G., Touze F.,"PRIAM/ANTIGONE: A 2D/3D package for accelerator design," in *Proceedings of the 2nd European Particle Accelerator Conference*, London 1993, pp. 1321–1323.

[4] RD3 Collaboration, *Nuclear Instrumentation and Methods in Physics Research* **A330** 405–415 (1993).

[5] Thiery Y., Gao J., Le Duff J., *Nuclear Instruments and Methods in Physics Research* **A378** 21–26 (1996).

[6] Le Meur G., "Eddy currents in the plates of the ATLAS hadronic calorimeter," Atlas Internal Note LARG 3/1994.

[7] Wong M.F., Picon O., Fouad Hanna V., *J. Phys. III France,* **2** 2083–2099 (1992).

[8] Touze F., Le Meur G., "A non conforming finite element method for computing eigenmodes of resonant cavities," in *Proceedings of the 2nd European Particle Accelerator Conference*, Nice, 1990, pp. 1696–1698.

[9] Balleyguier P., "Coupling slots without shunt impedance drop" in *Proceedings of the LINAC96 Conference,"* Geneva, 1996.

[10] Bourquin F., d'Hennezel F., "Application of Domain Decomposition Techniques to Modal Synthesis for Eigenvalue Problems," *Proceedings of the fifth International Symposium on domain decomposition methods for PDE* , Norfolk 1991, Chapter 16.

PISCES II : 2.5-D RF Cavity Code

Yoshihisa Iwashita

Accelerator Laboratory, Nuclear Science Research Facility
Institute for Chemical Research, Kyoto University
Gokanosho, Uji, Kyoto 611, JAPAN

Abstract   The RF cavity code PISCES II can evaluate all the eigenfrequencies and fields for arbitrarily shaped axially symmetric RF cavities. The solutions include symmetric (m=0) and asymmetric modes (m>0) assuming the sin mθ and cos mθ dependencies. Using Vector Finite Element Method, the electric or magnetic components are calculated. The resulted eigenvalue system has many zero-eigenvalue solutions, which can be filtered out by zero-filter technique from the set of solutions. The eigensolutions of the specified number are obtained simultaneously from non-zero lowest frequency.

INTRODUCTION

The original PISCES code was written in early 1980's for studying DISK-AND-WASHER (DAW) structures which have dipole modes close to the operating frequency. Because URMEL uses rectangular mesh for the calculation, the approximated boundary is different from that of the SUPERFISH, and the comparison is not straightforward. Recently some difficulties in calculating vectorial fields by Finite Element Method (FEM) have been overcome(1,2,3) and PISCES II is rewritten for studying higher order modes including dipole modes. Second order mixed-interpolation-type (linear edge and quadratic nodal) element is implemented in the version 2.40.

FORMULATION

The differential equation for electric or magnetic field to be solved are (4,5),

$$\nabla\times\nabla\times\vec{E}+k^2\vec{E}=0, \ \nabla\cdot\vec{E}=0, \ \text{or} \ \nabla\times\nabla\times\vec{H}+k^2\vec{H}=0, \ \nabla\cdot\vec{H}=0 \ \ (\text{in } \Omega), \quad (1)$$

where $k^2=\omega^2\epsilon\mu$ and Ω is the entire volume. In vacuum space, $k^2=\omega^2/c^2$, where c is the speed of light. Boundary conditions are

$$\vec{E}\cdot\vec{n}=0 \ \text{ or } \ \vec{H}\cdot\vec{n}=0 \ \ \text{on magnetic boundaries (}\Gamma\text{m) for symmetry plane,} \quad (2)$$

$$\vec{E}\times\vec{n}=0 \ \text{ or } \ \vec{H}\times\vec{n}=0 \ \ \text{on electric boundaries (}\Gamma\text{e) for metal surfaces and} \quad (3)$$

$$\vec{E}_{left}=e^{i\varphi}\vec{E}_{right} \ \text{ or } \ \vec{H}_{left}=e^{i\varphi}\vec{H}_{right} \ \ \text{on periodic boundaries (}\Gamma\text{p),} \quad (4)$$

where $\vec{n}$ denotes the outward normal on the boundary, and φ is the phase advance in the problem(6,7). The periodic boundary is not a real boundary but only for a convenience of defining a problem. Because either $\vec{E}$ or $\vec{H}$ can be used as the field variable, only the electric field will be shown hereafter. Integrating Equ.(1) over Ω after multiplying by $\delta\vec{E}$ (virtual electric field), we get

$$\int_\Omega \delta\vec{E}\cdot\nabla\times\nabla\times\vec{E}\,dv = -k^2 \int_\Omega \delta\vec{E}\cdot\vec{E}\,dv, \qquad (5)$$

and applying Green's theorem, the following relations must hold for any δE:

$$\oint_\Gamma (\nabla\times\vec{E})\times\delta\vec{E}\,d\vec{s} - \int_\Omega (\nabla\times\vec{E})\cdot(\nabla\times\delta\vec{E})\,dv = -k^2 \int_\Omega \delta\vec{E}\cdot\vec{E}\,dv \qquad (6)$$

$$\vec{E}\times\vec{n}=0 \text{ and } \delta\vec{E}\times\vec{n}=0 \text{ on } (\Gamma e), \text{ or } \vec{E}\cdot\vec{n}=0 \text{ and } \delta\vec{E}\cdot\vec{n}=0 \text{ on } (\Gamma m) \qquad (7)$$

The term in the surface integration of Equ. (6) becomes zero on either (Γe) or (Γm) because of the boundary condition of Equ's (7).

FINITE ELEMENT MODEL

Because only the axisymmetric boundary problems are considered, we can assume sin mθ and cos mθ dependencies of E_r, E_z and E_θ components, and then the problem can be reduced to two-dimension problem:

$$\vec{E} = (E_\theta \sin m\theta,\ E_r \cos m\theta,\ E_z \cos m\theta). \qquad (8)$$

Then (E_θ, E_r, E_z) are functions of r and z only. The field variables are (rE_θ, E_r, E_z) for $m\geq 1$ and (E_θ, H_θ) for m=0. Only the case for $m\geq 1$ will be explained here.

The shape functions used are the mixed-interpolation-type triangular elements (3). Figure 1 shows the elements used in PISCES II. Only the tangential component in the rz-plane is assigned on the line. Either the lowest order (constant edge and linear nodal) or the second order (linear edge and quadratic nodal) element can be used. Both elements may have curved boundaries.

Then, $\vec{E}$ and $\nabla\times\vec{E}$ can be written as

$$\vec{E} = \begin{bmatrix} E_\theta \\ E_r \\ E_z \end{bmatrix} = \vec{N}\cdot\begin{bmatrix} rE_\theta \\ \vec{E}_{rz} \end{bmatrix}, \quad \nabla\times\vec{E} = \vec{N}'\cdot\begin{bmatrix} rE_\theta \\ \vec{E}_{rz} \end{bmatrix}, \qquad (9)$$

$$\vec{N} = \begin{bmatrix} \frac{1}{r}\vec{N}_\theta & 0 \\ 0 & \vec{N}_r \\ 0 & \vec{N}_z \end{bmatrix}, \quad \vec{N}' = \begin{bmatrix} 0 & \partial_z\vec{N}_r - \partial_r\vec{N}_z \\ \frac{-1}{r}\partial_z\vec{N}_\theta & \frac{-m}{r}\vec{N}_z \\ \frac{1}{r}\partial_r\vec{N}_\theta & \frac{m}{r}\vec{N}_r \end{bmatrix}, \qquad (10)$$

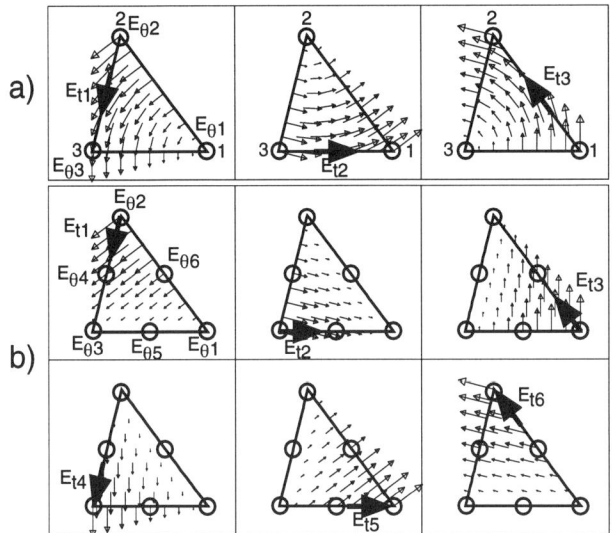

FIGURE 1. Mixed-interpolation-type triangular elements.
a) Constant edge and linear nodal elements.   b) Linear edge and quadratic elements.

where $\vec{N}$, $\vec{N}'$, $\vec{N}_\theta$, $\vec{N}_z$, and $\vec{N}_r$ are the shape functions, and $r\vec{E}_\theta$ and $\vec{E}_{rz}$ are the field variable. The element matrix equation is

$$\int_e \vec{N}'^T \cdot \vec{N}' \, rdrdz = -k^2 \int_e \vec{N}^T \cdot \vec{N} \, rdrdz, \tag{11}$$

where symbol e denotes the element volume. The integrations are performed numerically up to 11th order precision. The singularity in the integrand on the axis, is not serious because the real divergent terms are eliminated by the boundary condition on the axis. By assembling all element matrices and applying the boundary condition, finally we get the general eigenvalue equation.

$$\vec{M} \cdot \vec{x} = k^2 \vec{K} \cdot \vec{x}, \tag{14}$$

where $\vec{M}$ and $\vec{K}$ are large sparse symmetric matrices, and $\vec{x}$ is an eigenvector. Usually several eigensolutions starting from the smallest one but zero are of interest. Unfortunately, this eigenvalue problem has many zero-eigenvalue solutions, and then special care should be taken, which is stated in next section.

GENERAL EIGENVALUE SOLVER FOR LARGE SPARSE SYMMETRIC MATRICES

Because the matrices are sparse, only non-zero elements are stored by list vector technique. The eigenvalue solver is based on the subspace method[8] and the zero filter technique(9). Because $\vec{M}$ is not regular, all the eigenvalues are shifted and scaled by positive number α, and then the resulted eigenvalue equation is,

$$\vec{M}'\cdot\vec{x}=\lambda\vec{K}\cdot\vec{x},\,(\vec{M}'=\vec{M}/\alpha+\vec{K},\,\lambda=1+k^2/\alpha), \tag{15}$$

where α should be close to the lowest eigenvalue but zero. Because $\vec{M}'$ is positive definite and symmetric, the Preconditioned Conjugate Gradient Method (PCGM) can be safely used to solve the simultaneous linear equations. The zero filter operation is expressed as

$$\vec{x}_{n+1}=\vec{x}_n-\vec{M}'^{-1}\cdot\vec{K}\cdot\vec{x}_n. \tag{16}$$

By applying this operation in the subspace iterations with the appropriate frequency, the zero-eigenvalue solutions are well suppressed from the iterated vectors. This technique is also used to accelerate the convergence of the subspace method, by filtering the higher eigenvalue solutions.

COMPONENTS OF PISCES II

Three stages are required for PISCES II calculation, namely, preprocessor, field solver and post processor. The preprocessor prepares the mesh data for the field solver PISCES II. The mesh data can be read from TAPE35 file of POISSON / SUPERFISH Rel. 4 (PS4) (10). MESHNET reads TAPE35 from LATTICE and writes out an input file for PISCES II. NETREF can modify the input file to subdivide or modify the mesh at any place. Unlike the PS4, PISCES II can handle topologically non-uniform triangular mesh, because of the Finite Element Method. The mesh generator NET is planned to generate the mesh data directly from the input data for AUTOMESH. The post processor DISPLAY displays the graphical informations of the field interactively.

EXAMPLES

The relative frequency error of the second lowest mode in the spherical cavity with a 10 cm radius as a function of the number of unknowns is shown in Fig. 2. The mode corresponds to the first mode among the dipole solutions (m=1) in PISCES II calculation. The reference frequency is obtained analytically. The frequency of the spherical cavity converges within 8.6×10^{-6} at the mesh size of 0.125 cm, which corresponds to 11564 elements or 58120 unknown variables. While the curved boundary capability in the lowest order element is not remarkable on the accuracy, it improves the accuracy of the second order element case. Although the relative error for the second order element is smaller than that for the lowest order element, the convergence rate does not have much difference, which was not expected for such higher order elements. It is under investigation,.

Figure 3 shows the graphical output from DISPLAY at the mesh size of 1cm, which corresponds to 973 unknown variables. The contour plot of rH_θ, arrow plot in the r-z plane and the mesh plot are shown.

Figure 4 shows the field plots of the dipole modes in a Disk-and-Washer structure. The mesh is generated by AUTOMESH/LATTICE and converted by MESHNET. The mesh around the nose is modified by NETREF to be compatible with the curved boundaries.

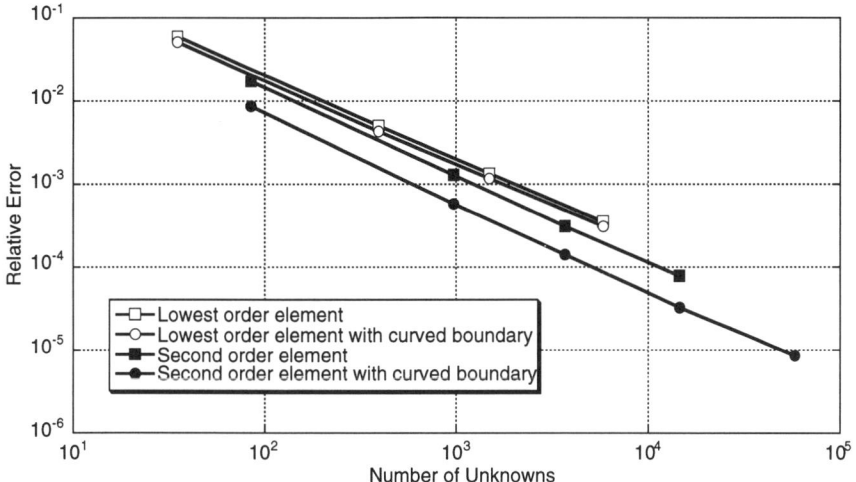

FIGURE 2 Relative frequency errors of the second lowest mode in a spherical cavity as a function of the number of unknowns.

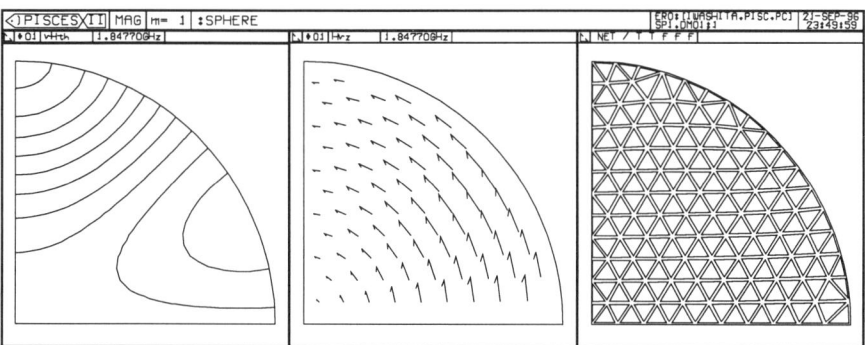

FIGURE 3. Right to left: contour plot of rH_θ, arrow plot in r-z plane and mesh plot.

CONCLUSION

Any mode in a cylindrically symmetric cavity can be calculated by Vector Finite Element Method with the mixed-interpolation-type element using 2.5D approach. The convergence problem on the high order elements is under investigation.

ACKNOWLEDGEMENTS

The author would like to thank Drs. R. K. Cooper, R. L. Gluckstern, R. A. Jameson, T. Weiland and T. Higo for their encouragement and discussions. He also is greatly indebted to Dr. Maruyama for his helpful information.

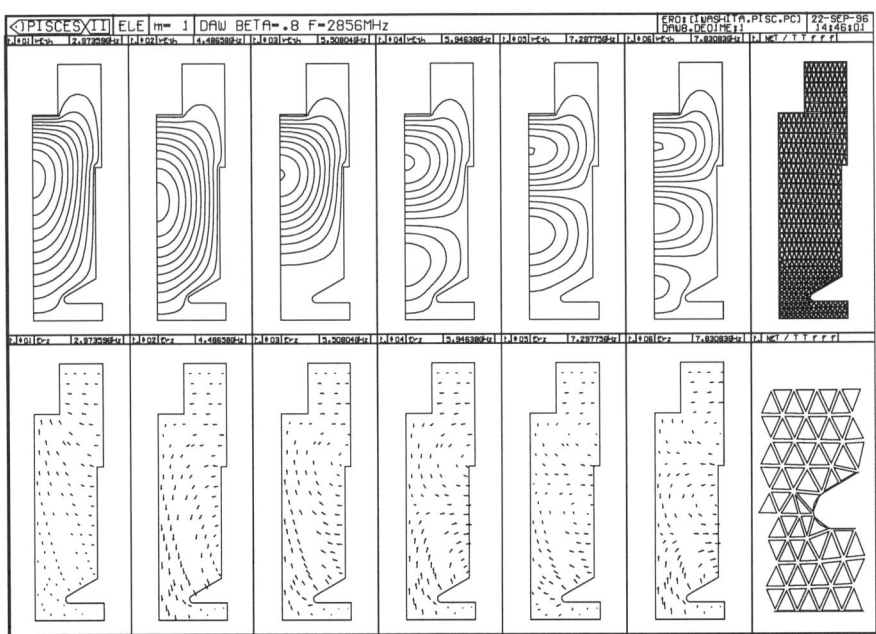

FIGURE 4. Field plots of the dipole modes in Disk-and-Washer structure. The mesh is generated by AUTOMESH/LATTICE and converted by MESHNET. The mesh around the nose is modified by NETREF to be compatible with the curved boundaries.

REFERENCES

1. F. Kikuchi, et.al., "A FINITE ELEMENT METHOD FOR 3-D ANALYSIS OF CAVITY RESONATORS", Distributed Parameter Systems: Modeling and Simulation, Elsevier Science Publishers B.V. (North-Holland) ©IMACS,1989
2. J. C. Nédélec, "A New family of Mixed Finite Elements in R3", Num. Math., Vol. 50 pp. 57-81, 1986
3. M. Koshiba, S. Maruyama and K. Hirayama, "A Vector Finite Element Method With the High-Order Mixed-Interpolation-Type Triangular Elements for Optical Waveguiding Problems", Journal of Lightwave Technology, Vol.12, No.3, March 1994, pp.495-502.
4. M. Hara, T. Wada, T.Fukasawa, and F. Kukuchi, "A three Dimensional Analysis of RF Electro-magnetic Fields by Finite Element Method", IEEE Trans., MAG-19 No. 6 Nov. 1983
5. K. H. Huebner and E. A. Thornton, " The Finite Element Method for Engineers", (J.Weiley, New York); and A.R.Mitchell and R.Wait, "The Finite Element Method in Partial Differential Equations" (J.Weiley, New York, 1977);
6. R. L. Gluckstern and E. N. Opp, "Calculation of dispersion curves in periodic structures", IEEE Trans. MAG-21 No. 6 Nov. 85 pp. 2344-2346
7. E. M. Nelson, "A Finite Element Field Solver for Dipole Modes", 1992 Linear Accelerator Conference Proceedings, AECL, p.814
8. K.J. Bathe, "Solution Methods for Large Generalized Eigenvalue Problem in Structural Engineering", Doctoral thesis, University of California, Berkeley, 1971
9. Y.Iwashita, "General Eigenvalue Solver for Large Sparse Symmetric Matrix with Zero Filtering", Bull. Inst. Chem. Res. Kyoto Univ. Vol. 67, No. (1989)
10. "User's Guide for the POISSON/SUPERFISH Group of Codes", LA-UR-87-115, Los Alamos National Lab.

"Thick-Slice" Simulation of Short Longitudinal-Scale Phenomena on a Space-Charge-Dominated Beam

I. Haber,* D. A. Callahan,** A. Friedman,** and D. P. Grote**

*Plasma Physics Division, Naval Research Laboratory, Washington, DC 20375
**Lawrence Livermore National Laboratory, Livermore, CA, 94550

Abstract. Simulations examining the propagation of a beam with a transverse temperature somewhat greater than the longitudinal temperature have shown evidence that the beam can evolve towards temperature isotropy via a collective electrostatic instability. Because the longitudinal scale of the unstable modes is comparable to the beam diameter, simulation of this instability with sufficient axial resolution can be costly, particularly in an alternating-gradient transport channel where the geometry is inherently three-dimensional. A "thick-slice" model has therefore been developed which extends the usual single-slice approximation to include a self-consistent solution for the longitudinal fields within the slice, while still assuming that the external forces act simultaneously on all particles in the slice. An example of the use of this method is presented to illustrate how the fundamental dynamics of the isotropization instability appear to be preserved even in the presence of the large excursions in the beam envelope which are present in a beam matching system.

INTRODUCTION

In the absence of an externally applied bunching force, acceleration of a beam of charged particles will generally result in a lengthening of the beam bunch. The adiabatic cooling which accompanies this increase in length results in a, usually large, temperature anisotropy in many accelerators, particularly in the source region. Furthermore, interparticle collision frequencies are generally low, so that significant thermodynamic equipartitioning of the resulting anisotropy is not expected to occur during typical beam lifetimes.

Three-dimensional simulations of a beam in an alternating-gradient transport system have, however, shown evidence of equipartitioning on a scale much faster than expected from beam collisions.[1,2] In order to understand the operative isotropization mechanism, simulations have been performed on a simpler, axisymmetric system, with uniform solenoidal focusing.[3-5] These axisymmetric simulations have identified an electrostatic instability which couples energy from the transverse to longitudinal directions, and which appears to be similar to the Harris mode[6] where the betatron motion replaces the cyclotron oscillations in that case and where the nature of the unstable mode is somewhat modified by the finite beam radius. The basic instability in the current case appears to occur when a longitudinal perturbation of the beam envelope radius, comparable in length to the beam radius, results in an electric field averaged over the betatron period which causes charge to move longitudinally so as to increase the magnitude of the perturbation.

Because of the generally kinetic nature of the dynamics of finite-emittance focused-beam equilibria, and the complexity of a generalized formulation of the relevant electrostatic modes on a warm bunch, especially for longitudinal wavelengths comparable to the beam radius, current understanding of the nature of the electrostatic temperature isotropization instability has relied on the results of simulation. In view of the difficulty of performing any analytic estimates of the behavior of such an unstable mode in the most interesting case of alternating-gradient transport, where the analytic characterization even of a realistic steady state distribution has proved elusive, simulations are likely to remain a primary tool for investigating the consequences of this instability. It is therefore desirable to obtain an experimental benchmark to compare against these simulations. However, the assumption of a uniform focusing system is not easy to achieve in an experimental system. Even if a solenoidal channel were available with sufficient length to examine the unstable behavior, such a channel almost always requires a matching section to transport the beam from the output of the gun into that channel, and any unstable behavior could be substantially modified by the traversal of this matching system.

"THICK-SLICE" METHOD

As discussed above, there is a need for the examination of the evolution of a long beam-bunch in a spatially varying transport system but with sufficient resolution to examine modes with longitudinal wavelengths comparable to the beam radius. Furthermore, since the most interesting case is the behavior of such a long-bunch in an alternating-gradient transport system, which is inherently three-dimensional, conventional simulations which follow the evolution of the entire bunch can be costly, particularly since it is necessary for the simulated bunch to be sufficiently long so that any waves launched at the bunch end do not propagate into the region of interest during the duration of the simulation.

A simple method for conducting simulations of phenomena which have characteristic longitudinal wavelengths comparable to the beam radius, and therefore much shorter than the characteristic lengths for the evolution of the beam envelope, is to extend the concept of a single-slice simulation to a "thick-slice" model. In this model, the same external forces are applied to every particle in the slice, just as in the single-slice model. However, the internal electrostatics in the slice are followed self-consistently, with periodic boundary conditions applied to the slice ends. In this way it is possible to include the influence of the long-length variations in the beam envelope, such as those due to variations in the external forces, or due to a beam mismatch, without conducting a full multi-dimensional simulation of a long beam with a highly resolved longitudinal mesh.

As will be discussed below, the particular case employed here for illustrating the characteristics of the thick-slice model, is the set of solenoidal matching lenses on the University of Maryland Transport Experiment, since it was desired to understand the role that the longitudinal anisotropy instability might play in causing the anomalously high longitudinal temperatures actually measured in this experiment.[7] This test case also has the advantage that the axisymmetry of the apparatus simplifies the simulations, which were conducted using the r,z WARP code.[8] This code is particularly suitable because it is written using the Basis system for run time control. The modifications necessary for implementing the thick-slice approximation were made entirely in the Basis interpreter.

MATCHING SECTION SIMULATIONS

Longitudinal beam temperatures measured after traversal of the matching lens section of the University of Maryland Transport Experiment have been reported[7] which are somewhat higher than what would be expected from known source characteristics. In order to investigate the possibility that the electrostatic isotropization instability discussed above could be responsible for the high longitudinal temperature observed, a series of simulations was performed to examine the behavior of the beam as it traverses the two solenoidal matching lenses. Figure 1 is a plot of the envelope-equation solution for the beam radius as the beam traverses the matching section.

Figure 2 is a plot of the evolution of v_{zrms} as the beam traverses the matching lens section. The deviation from the smooth exponential growth which was seen in previous simulations, employing a uniformly focused beam, reflects the influence of the beam envelope excursions. The 5 KeV, 7.3 mA, electron beam in this simulation was given an initially semi-Gaussian transverse distribution (uniform in space and Gaussian in velocity), with a 17.3 mm·mrad unnormalized emittance. This run employed 200 K particles and 64 by 256 grid cells to resolve the 19 mm pipe and 10 mm slice thickness respectively. Using 200 timesteps to traverse the 0.4 m matching section, the variation in the rms radius during traversal of the transport system accurately tracks the envelope code solution shown in Fig. 1. The behavior of the simulation was essentially unchanged with variations in slice length, timestep, and grid size, provided the number of particles per longitudinal cell was kept unchanged. It should be noted that a reference simulation with a uniform applied external focusing field sufficient to keep the beam at a constant 2 mm radius showed approximately the same growth when integrated for 0.4 m, although the unstable wavelength was slightly larger than in the present case.

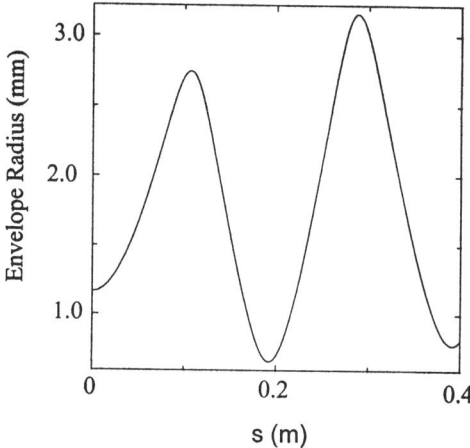

Fig. 1. Envelope equation solution for the variation in beam radius during traversal of the 0.4 m matching section.

A series of tests did, however, reveal a very clear $n^{-1/2}$ dependence of the growth in the v_{zrms}. Changing the number of particles per longitudinal cell was found to affect the magnitude of the value of v_{zrms} but not the shape of the growth curve, so that if the curves were rescaled they could be overlaid. This suggests that the dependence on particle number was due to a change in the seed amplitude of the initial excitation rather than any collisional effect on the mechanism for growth itself, i.e. the system was acting as a linear amplifier of the initial random seed.

To test this assertion, the system was seeded with an initial sinusoidal modulation of the envelope radius within the slice. As expected, the values of v_{zms} along the matching section varied linearly with the amplitude of the initial sinusoid. Figure 3 is a plot of the evolution of v_{zrms} for an initial sinusoidal seed with amplitude corresponding to 8% of the envelope radius. As additional evidence that the numerical dependence of the growth on the number of particles resulted exclusively from the seed amplitude dependence on the initial statistics, the growth in v_{zrms} for the non-randomly seeded runs showed virtually no dependence on particle number.

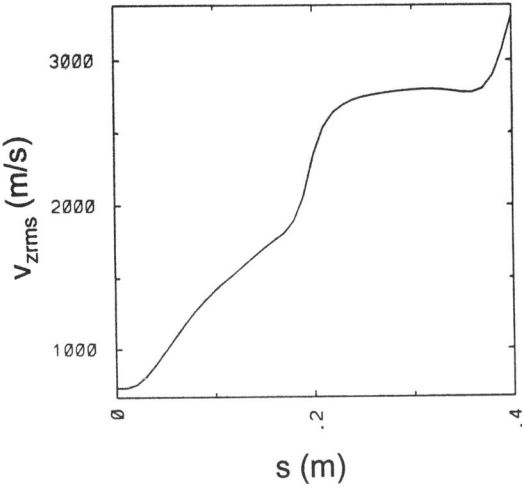

Fig. 2. Evolution of the v_{zrms} as the beam with an initial transverse temperature much greater than the longitudinal temperature traverses the matching lens section. This evolution differs from the smooth exponential growth seen for a similar beam propagating in a transport system with uniform focusing.

From the behavior of the seeded runs an estimate of the approximate seed amplitude which would be required to explain the experimental data can be obtained. For an 8% initial seed, the final v_{zrms} corresponds to a temperature of approximately 2.4 eV. Therefore, even such a large seed amplitude results in somewhat less of a final temperature than the 16 eV reported experimentally. It is therefore unlikely that the current mechanism is the sole source of the observed temperature anomaly.

SUMMARY

Simulations in the "thick-slice" approximation have been presented to demonstrate evidence that the transverse-longitudinal electrostatic isotropization instability under investigation appears to be robust in the sense that it is not strongly influenced by large envelope excursions during beam propagation. Further investigation using this technique should allow examination of more complex three-dimensional alternating-gradient focusing systems including a benchmarking of this method against previously undertaken fully three-dimensional simulations.[1,2]

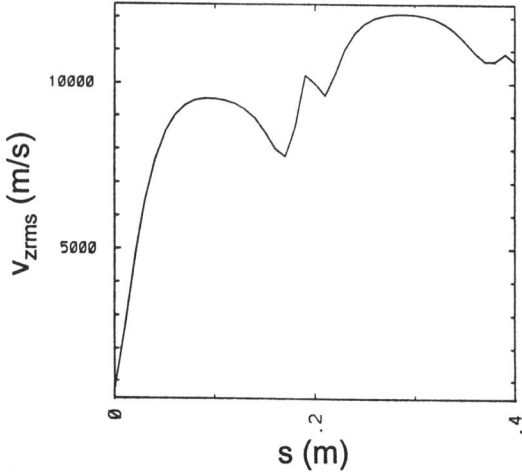

Fig. 3. Evolution of v_{zrms} during traversal of the matching lens section when the beam envelope is given an initial 8% sinusoidal modulation.

ACKNOWLEDGMENT

The authors acknowledge support for the research reported here by the United States Department of Energy at the Naval Research Laboratory under contracts DE-AI02-93ER40799, and DE-AI02-94ER54232, and at the Lawrence Livermore National Laboratory under contract W-7405-ENG-48.

REFERENCES

1. Alex Friedman, David P. Grote, and Irving Haber, "Three-Dimensional Particle Simulation of Heavy-Ion Fusion Beams," *Phys. Fluids B* **4** (1992) 2203.
2. A. Friedman, D. A. Callahan, D. P. Grote, A. B. Langdon, and I. Haber, "Studies of Equilibration Processes in Heavy Ion Beams," *Bull. Am. Phys. Soc.,* **35**, 2121 (1990).
3. I. Haber, D. A. Callahan, A. Friedman, D. P. Grote, and A. B. Langdon, "Transverse-Longitudinal Energy Equilibration in a Long Uniform Beam," *Proc. of the 1995 Particle Accelerator Conf.*, 3283, (IEEE, 1995).
4. I. Haber, D. A. Callahan, A. Friedman, D. P. Grote, and A. B. Langdon, "Transverse-Longitudinal Temperature Equilibration in a Long Uniform Beam," to be published in *Proc. of the International Symposium on Heavy Ion Fusion*, Princeton, NJ, Sept 6-8, 1995, *Journal of Fusion Engineering Design*, 1996.
5. I. Haber, D. A. Callahan, C. M. Celata, W. M. Fawley, A. Friedman, D. P. Grote, and A. B. Langdon, "PIC Simulation of Short Scale-Length Phenomena," *Space Charge Dominated Beams and Applications to High Brightness Beams*, S. Y. Lee, Ed., AIP Con. Proc. # 377,(AIP, New York, 1996) 244.
6. E. G. Harris, Unstable Plasma Oscillations in a Magnetic Field, *Phys. Rev. Letters*, **2** (1959) 34.
7. N. Brown, M. Reiser, D. Kehne, D. X. Wang, J. G. Wang, "Longitudinal Kinetic Energy Spread from Focusing in Charged Particle Beam," *Proc. of the 1993 Particle Accel. Conf.*, 62, (IEEE, 1993).
8. Debra A. Callahan, A. Bruce Langdon, Alex Friedman and Irving Haber, "Longitudinal Beam Dynamics for Heavy Ion Fusion," *Proc. 1993 Particle Accel. Conf.*, 730 (IEEE, 1993).

BPERM Version 3.0
— A 2D Wakepotential/Impedance Code

Therese Barts* and Weiren Chou[†]

*SSC Laboratory, 2275 Highway 77 N., Waxahachie, TX 75165
†Fermilab, P.O. Box 500, Batavia, IL 60510

Abstract. BPERM 3.0 is an improved version of a previous release (1). The main purpose of this version is to make it more user friendly. Following a simple 1-2-3 procedure, one obtains both text and graphical output of the wakepotential and impedance for a given geometry. The calculation is based on a boundary perturbation method (2-5), which is significantly faster than numerical simulations. It is accurate when the discontinuities are small. In particular, it works well for tapered structures.

INTRODUCTION

The program **bperm** is a 2-dimensional code and can be employed for periodic structures with rotational symmetry. The input is one complete period of the structure described in the **r** (radial) and **z** (longitudinal) plane as an array of points. The output is the wakepotential and impedance associated with a traversing Gaussian bunch. This code and its user's guide was released in June, 1994 (1). Since then, we have received a number of comments from the users. The current version 3.0 is aimed at accomodating these comments and being more user friendly.

In this version, the input file format remains unchanged. The intermediate output can be ignored by the user. The final output contains two files: One is an ASCII file for text output, another is a postscript file for plots. The code is written in Fortran 77. It runs on Unix systems as well as on VAX/VMS. The source code, executable and examples are free to the public. The graphics interface is GNUPLOT, which is also a free software.

This paper briefly discusses some basic features of **bperm** and the execution procedure, using Unix systems as an example. For more detailed information about this code, the reader is referred to Ref. 1.

INPUT FILES

The input file must be named **bperm.in**. The following is an example:

```
dataset=test
title=bperm Test Problem
sigma=0.5
```

© 1997 American Institute of Physics

```
pmax=128
smax=100
shape
1.6 0.0
2.0 0.4
2.0 4.0
1.6 4.4
1.6 8.0
end
```

The meanings of the keywords are as follows: **dataset** and **title** define the name of the output files and the title on the plots, respectively; **sigma** is the rms length of a Gaussian bunch in centimeters (cm); **pmax** is the number of interpolated coordinates used for the structure; **smax** is the region of the wakepotential calculation in units of **sigma**; and the pair of numbers between **shape** and **end** are the (**r,z**) coordinates in cm describing the structure (similar to the way in TBCI, ABCI and URMEL). Only **shape**, **end** and the coordinates in between are required; all other keywords are optional.

COMPILE, EXECUTION AND OUTPUT

The source code **bperm.f** can be compiled using any f77 compiler. The following is an example:

```
f77 -o bperm -dn -fast -O3 bperm.f
```

This will generate the executable **bperm**. To use the code, follow the 1-2-3 steps:

1. Write a **bperm.in** and put it in the same folder where **bperm** is;

2. Click on the **bperm** icon to execute;
   (This will generate a number of intermediate files, which can be ignored by the user.)

3. Type gnuplot `filename.gp` to get the plots and a postscript file.

The intermediate files will disappear and the final output contains just two files: `filename.out` (ASCII) and `filename.ps` (postscript).

EXAMPLES

1. A tapered periodic structure:
   Figure 1 is the structure described by the above input file, and the longitudinal and transverse wakepotentials and impedance.

2. A small bump:
The difference from the above example is the following section:

```
shape
1.6 0.0
2.0 0.4
2.0 0.8
1.6 1.2
1.6 8.0
end
```

Figures 2 is the plots of the structure, wakes and impedance.

3. A small iris:
The shape/end section is as follows:

```
shape
1.6 0.0
2.0 0.4
2.0 7.2
1.6 7.6
1.6 8.0
end
```

Figures 3 is the plots of the structure, wakes and impedance.

A useful feature of **bperm** is that the three input files can be concatenated into one (simply by putting the three shape/end sections in sequence) and all the results can be obtained from one run.

Interesting applications of **bperm** include: (i) The study of the different behaviors of impedance in high frequency regions for a periodic structure and for a single discontinuity. (ii) The study of the similar behaviors of impedance for a bump and for an iris. Results of these studies will be published elsewhere.

REFERENCES

1. T. Barts and W. Chou, *A User's Guide for the Computer Code BPERM — A Boundary Perturbation Method for Wakepotential and Impedance Calculations*, SSCL-MAN-0035, SSC Laboratory (1994).
2. Z. H. Zhang, Acta Physica Sinica, V 28, p. 563 (1979).
3. M. Chatard-Moulin and A. Papiernik, Proc. Particle Accelerator Conference, San Francisco, 1979, IEEE Trans Nucl. Sci. V 26, p. 3523 (1979).
4. R. K. Cooper, S. Krinsky and P. L. Morton, Particle Accelerators, V 12, p. 1 (1982).
5. W. Chou, Argonne National Laboratory, Light Source Note LS-149 (1990).

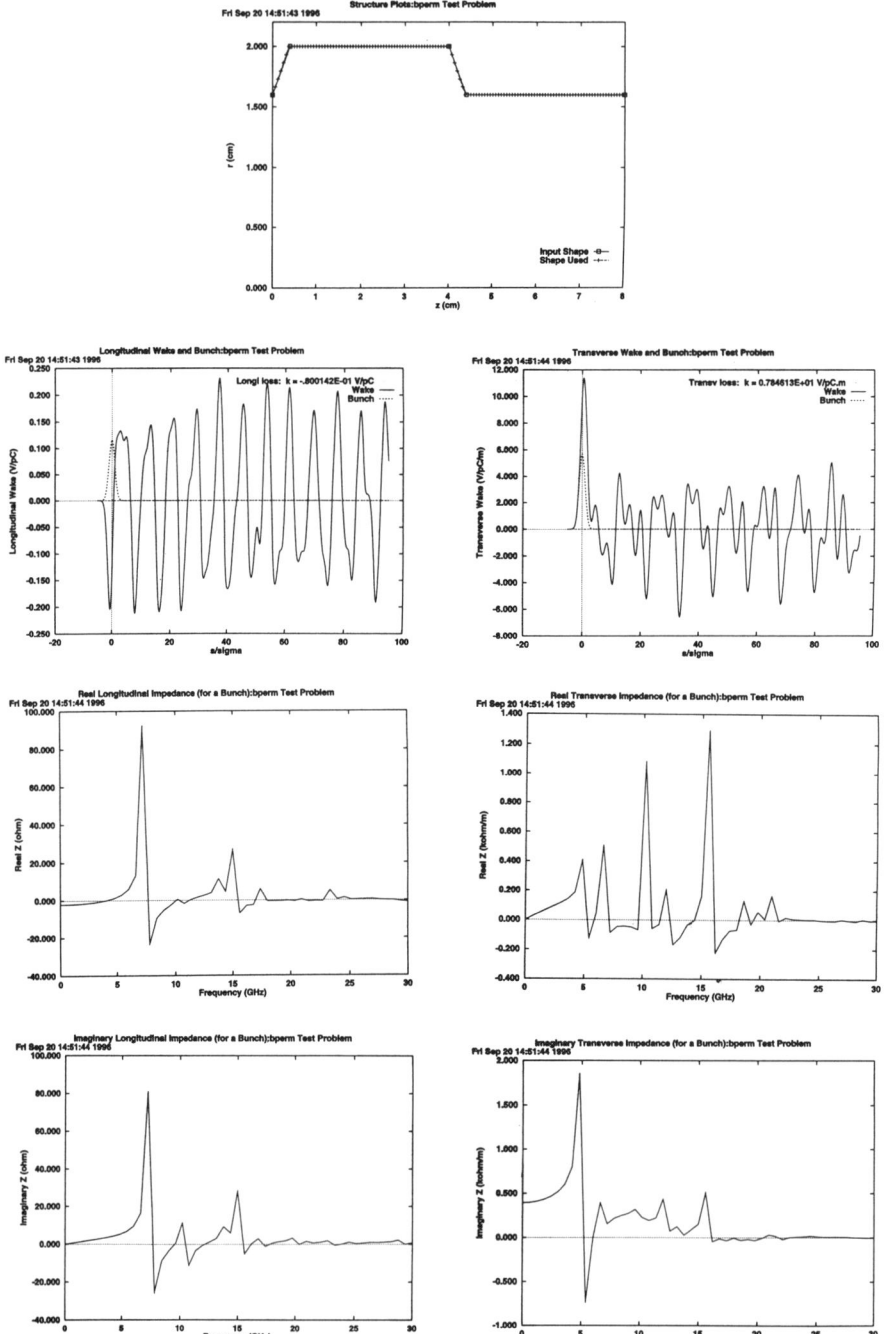

Figure 1. BPERM plots for Example 1: structure, longitudinal and transverse wakepotentials, and impedance.

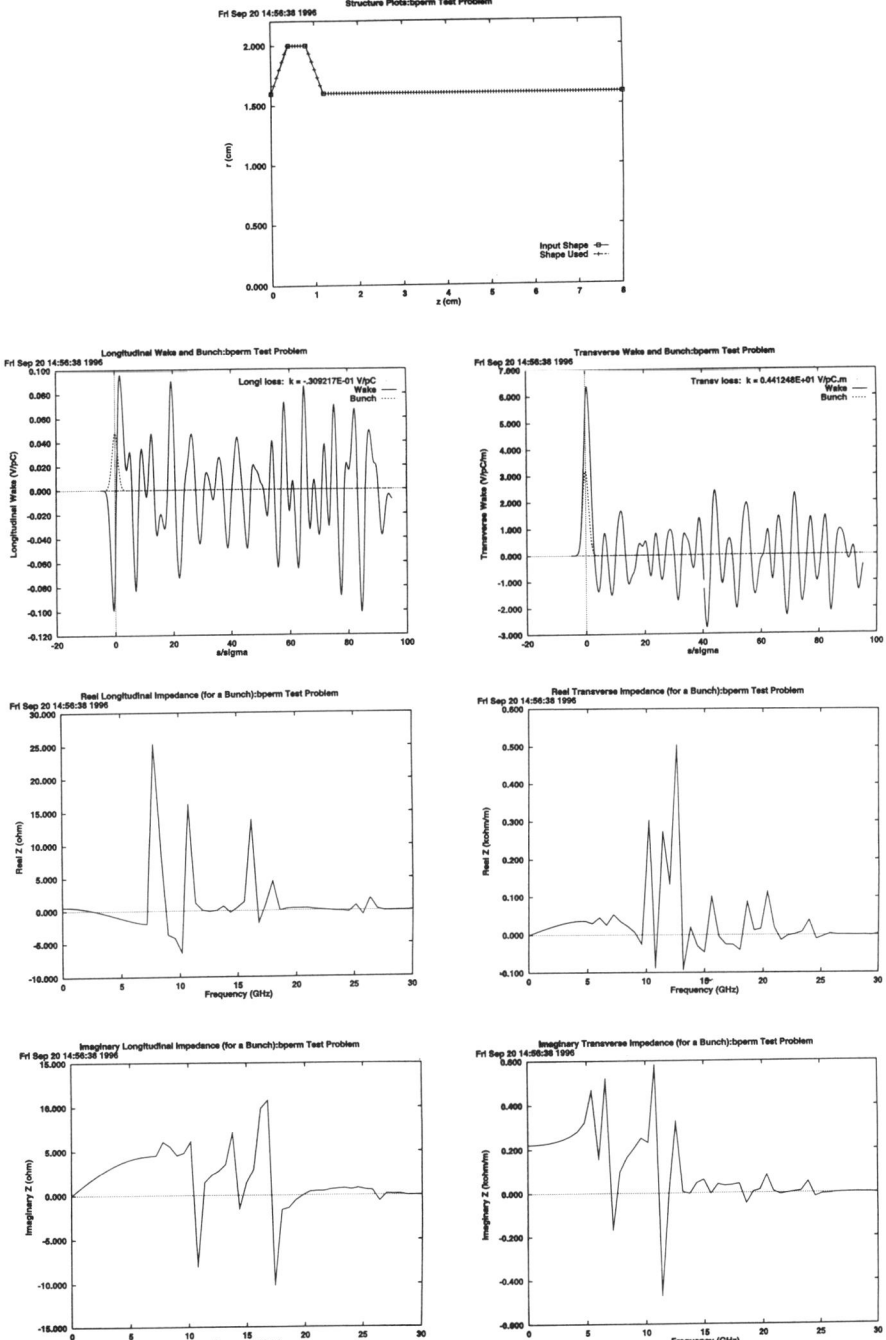

Figure 2. BPERM plots for Example 2: structure, longitudinal and transverse wakepotentials, and impedance.

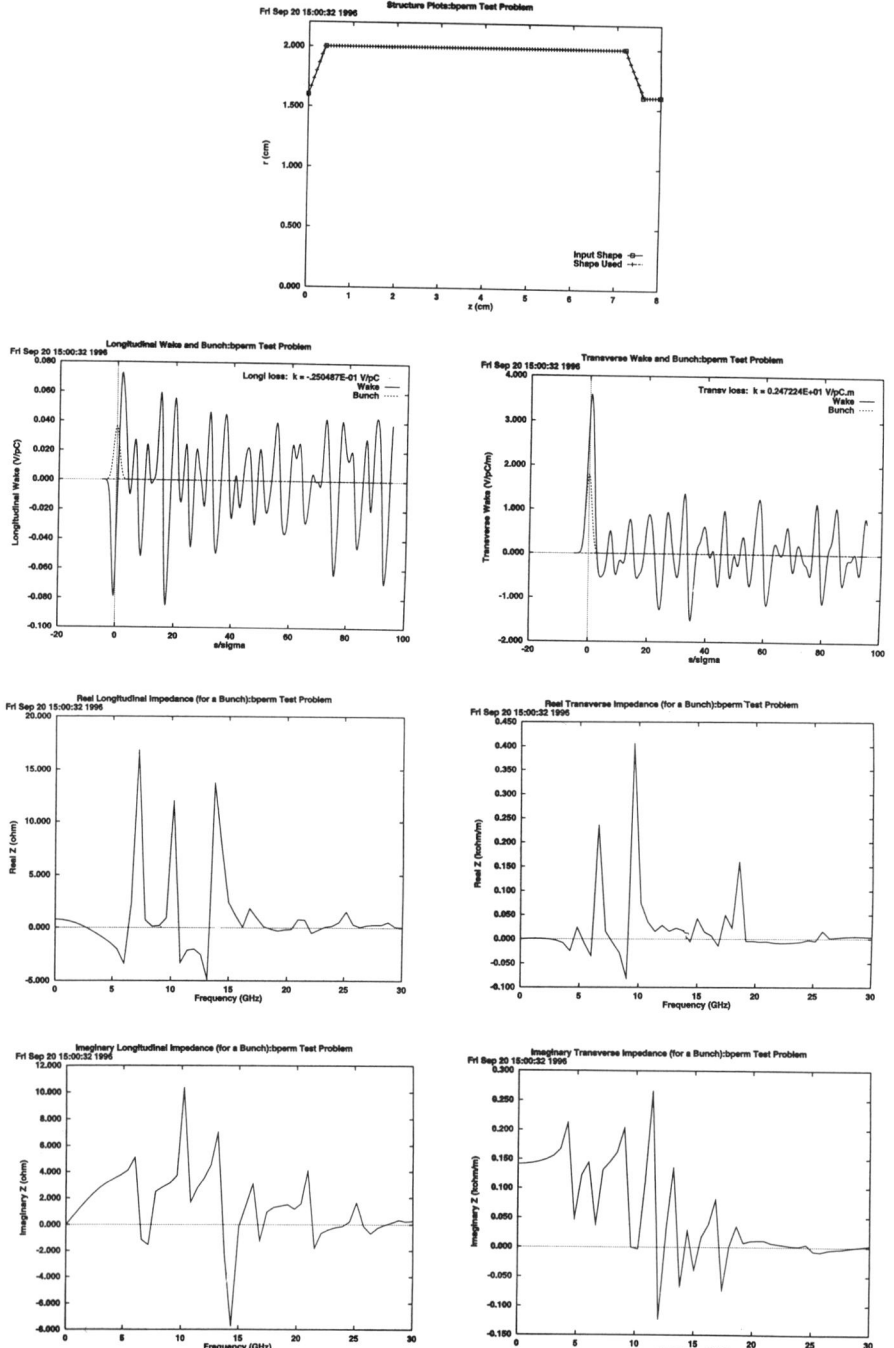

Figure 3. BPERM plots for Example 3: structure, longitudinal and transverse wakepotentials, and impedance.

Renoir, a Numerical Simulation Code for the Study of Halo in Intense Charged Particle Beams

Alain Piquemal

Département de Physique Théorique et Appliquée,
CEA/DRIF/DPTA, BP12, 91680 Bruyères-le-Châtel, France

Abstract. Renoir was developed for the study of halo in the transport of intense charged particle beams. The numerical schemes, used for the field calculation and the particle transportation, was designed and optimized specialy to study the physics of these beams : tiny details of the phase space like resonances or K.A.M. surfaces, space charge waves generation on the density profile, thermalization mechanisms. The geometry can be built with elementary active structures, in which are taken into account boundary conditions for electromagnetic field and particles. Diagnostics were implemented which give statements like particle, motion, energy conservation, temperature, heat flux...and boundary losses. The model used in this simulation is justified and some results are presented.

INTRODUCTION

Simulation codes are intensively developed for the study of beam dynamics in high intensity accelerators. CW operation of such future machines requires a better control of the beam losses to minimize the radio-activation of the structures. This is why it is important to study the halo physics which is responsible for these losses, and to acquire a better understanding of the mechanisms which drive its formation. A microscopic model of such systems is hardly tractable and computations are costly; it is thus necessary to make approximations, trying to restitute as best as possible the physics of the problem. We show, in this paper, that the Vlasov-Maxwell Particle-In-Cell codes like Renoir are well suited for this particular study, justifying approximations from a parametric study.

FINE-GRAINED MODEL

A complete description of a system composed of N_0 identical charged particles, would be one in which the position $x_k(t)$ and the velocity $v_k(t)$ of each particle are given as a function of time; since we are dealing with quasi-point particles, we can completely describe the system in the $6N_0$-dimensional phase space, with the Klimontovich distribution function :

$$N(X,t;X_1,X_2,...,X_{N_0}) = \sum_{1 \le k \le N_0} \delta(X - X_k(t)) \qquad (1)$$

where $X=(x,v)$ is the 6-dimensional phase space variable and $\int N(X,t)dX = N_0$.

If we consider that the system is collisionless, the distribution function $N(X,t)$ satisfies the conservation of particles in phase space and its space-time evolution is given by the Klimontovich-Dupree equation :

$$\frac{dN}{dt} = \frac{\partial N}{\partial t} + v \bullet \frac{\partial N}{\partial x} + \frac{q}{\gamma m}\{E + v \times B\} \bullet \frac{\partial N}{\partial v} = 0 \qquad (2)$$

where q and m are respectively the charge and mass of the particles, γ is the Lorentz factor, $E=E(x,t)$ and $B=B(x,t)$ are respectively the local electric and magnetic fields and consist of two separate contributions : those applied from the external sources (x), and those produced by the microscopic fine-grained distribution of the charged particles (s)

$$E = E_x + e_s, \quad B = B_x + b_s \qquad (3)$$

The microscopic fields $e_s = e_s(x,t)$ and $b_s = b_s(x,t)$ are to be determined from the Maxwell equations, or in a "smooth" approximation from the Poisson and Ampere equations; we have :

$$\nabla \bullet e_s = \frac{q}{\varepsilon_0}(\int N(x,v,t)dv - 1) \qquad (4)$$

$$\nabla \times b_s = q\mu_0 \int v N(x,v,t)dv \qquad (5)$$

where we omitted in (4) the electrical field due to the particle located at point (x,v) of the phase space.

Each particle has its own trajectory, which can be described by the Hamilton equations :

$$\frac{dx_k}{dt} = v_k = \frac{\partial H}{\partial p_k} \qquad (6)$$

$$\frac{dp_k}{dt} = q(E_k + v_k \times B_k) = -\frac{\partial H}{\partial x_k} \qquad (7)$$

where $p_k = m\gamma_k v_k$ is the particle motion, and E_k, B_k are respectively the electric and magnetic fields applied to the particle k and have the same definition as in (3). With equations (1) through (7) the description of the system in the 6-N dimensional phase space is exact; but, in practice, this is not very useful for a large number of particles, because we have too much informations to analyze.

We have thus to reduce the fine-grained quantities of the system, and try to obtain some coarse-grained quantities which could be observed more easily.

COARSE-GRAINED MODEL

To establish the connection between the microscopic and macroscopic levels, we need to introduce an averaging process over the 6N-dimensional phase space; once we know the detailed location $\{X_k\}$ of each particle, we may define a distribution function $D(\{X_k\},t)$ which verifies $\int D(\{X_k\},t)d\{X_k\}=1$ and gives the probability that at time t the position and velocity of the particles have the values $\{X_k\}=(X_1,...,X_{No})$ in the range $dX_1...dX_{No}$ [see 1].

The function D evolves in time from a given starting distribution $D(\{X_k\},0)$, and satisfies the Liouville equation at any time t.

Now, we can average any quantity G depending of the Klimontovich function, with respect to this Liouville distribution function D, we have :

$$<G(N_\alpha,N_\beta,...)> \equiv \int D(\{X_k\},t)G(N_\alpha,N_\beta,...)d\{X_k\} \qquad (8)$$

The problem is that the reduced quantity $<G>$ provides a much less complete description of the system than does the full quantity G; in particular, we have lost informations about initial conditions and microscopic evolution of the system.

We will see that the understanding of the relations between the microscopic (G) and the macroscopic ($<G>$) levels governs the model which will be used.

Use of the averaging process (8) on the phase space density N_α gives the one-particle reduced distribution function $f_\alpha(X,t)$:

$$<N_\alpha(X,t\ ;\{X_i\})> = f_\alpha(X,t) \qquad (9)$$

Similarly, the product $N_\alpha N_\beta$ gives the two-particle reduced distribution function $f_{\alpha\beta}(X,X^*,t)$; it can be shown that to a good approximation [2] :

$$<N_\alpha(X,t\ ;\{X_i\})N_\beta(X^*,t\ ;\{X_i\})> = f_{\alpha\beta}(X,X^*,t)+\delta_{\alpha\beta}\delta(X-X^*)f_\alpha(X,t) \quad (10)$$

Finally, the direct averaging of equation (2) gives the equation for the time development of f_α :

$$\frac{df_\alpha}{dt} = \frac{\partial f_\alpha}{\partial t} + v\bullet\frac{\partial f_\alpha}{\partial x} + \frac{q_\alpha}{\gamma m}<\left\{(E_x+v\times B_x)+(e_s+v\times b_s)\right\}\bullet\frac{\partial N_\alpha}{\partial v}> = 0 \quad (11)$$

which can be rearranged under the following form [2] :

$$\frac{\partial f_\alpha}{\partial t} + v \cdot \frac{\partial f_\alpha}{\partial x} + \frac{q}{\gamma m}\{E + v \times B\} \cdot \frac{\partial f_\alpha}{\partial v} = \qquad (12)$$

$$-\frac{q_\alpha}{\gamma m}\left\{<\{(E_x + v \times B_x) + (e_s + v \times b_s)\} \cdot \frac{\partial N_\alpha}{\partial v}> - \{(E_x + v \times B_x) + (E_s + v \times B_s)\} \cdot \frac{\partial f_\alpha}{\partial v}\right\}$$

where the macroscopic self-fields E_s and B_s are now the average of all microscopic fields, considering indistinctly all the particles:

$$\nabla \cdot E_s = \frac{\rho_\alpha}{\varepsilon_0} = \frac{q_\alpha}{\varepsilon_0} \int f_\alpha(x,v,t) dv \quad , \quad \nabla \times B_s = \mu_0 J_\alpha = q_\alpha \mu_0 \int v f_\alpha(x,v,t) dv \qquad (13)$$

where $\rho_\alpha = \rho_\alpha(x,t)$ and $J_\alpha = J_\alpha(x,t)$ are respectively the macroscopic charge and current densities. We can now oversee what is contained in (12):
 • the left hand side has the Vlasov form where only the action of the averaged fields is considered; we suppose that particles are interacting smoothly via long range and collective coulombian forces ($|x_\alpha - x_\beta| \geq \lambda_D$),
 • the right hand side looks like a collision term which represents the exact contribution of the binary+ternary+...short range coulombian diffusion ($|x_\alpha - x_\beta| < \lambda_D$).

The evaluation of the right hand side is important, because it will indicate the validity of a coarse-grained model, in which is retained only the averaged field in the Vlasov operator, and are neglected all the correlations with a range shorter than the Debye length λ_D.

In this aim, we can integrate (4) by Green technique assuming the magnetic field is negligible, and use the Mayer cluster expansion to develop $f_{\alpha\beta}(X,X^*,t)$

$$f_{\alpha\beta}(X,X^*,t) = f_\alpha(X,t) f_\beta(X^*,t) + g_{\alpha\beta}(X,X^*,t) \qquad (14)$$

where $g_{\alpha\beta}(X,X^*,t) \approx O(g)$ represents the correlation between two particles α and β, and we suppose a priori that the correlations between three particles or more are negligible $g_{\alpha\beta\gamma} \approx O(g^2)$... We then obtain finally:

$$\frac{\partial f_\alpha}{\partial t} + v \cdot \frac{\partial f_\alpha}{\partial x} + \frac{q_\alpha}{\gamma m}\{E + v \times B\} \cdot \frac{\partial f_\alpha}{\partial v} =$$

$$\sum_\beta \frac{q_\alpha q_\beta}{\varepsilon_0 \gamma m} \int \frac{\partial}{\partial x} \frac{1}{|x-x^*|} \cdot \frac{\partial}{\partial v} g_{\alpha\beta}(x,v,x^*,v^*,t) \, dx^* dv^* \qquad (15)$$

where all the correlation effects were concentrated in the right hand side.

But, in classical plasmas, we know that the correlations $g_{\alpha\beta}$ are small compared with $f_\alpha f_\beta$ as long as the plasma parameter g verifies the condition $g \ll 1$.

We made an estimation of $g(r)$ for an intense ion beam with : a kinetic energy T_0=500 keV, a total current I_0=50 mA, a r.m.s. radius R_0=0.001 m assuming a gaussian distribution for the density profile $n(r)$, a normalized emittance ε_{nx}=0.6 π.mm.mrad; the radial temperature $T(r)$ and the thermal velocity $v_{th}(r)$ were calculated from emittance, using the relation $T(r)=C\,n(r)$ where C is a constant [4].

TABLE 1. Plasma parameters in an intense ion beam

r/r_0	0	1	2	3	4
n/n_0	1.0	3.6x10^{-1}	1.8x10^{-2}	1.2x10^{-4}	1.1x10^{-7}
ω_p (rad.s^{-1})	1.3x10^{+8}	7.9x10^{+7}	1.7x10^{+7}	1.4x10^{+6}	4.3x10^{+4}
$\tau_p=2\pi/\omega_p$ (s)	7.6x10^{-9}	1.2x10^{-8}	5.7x10^{-8}	6.8x10^{-7}	2.3x10^{-5}
$\delta z=\beta c \tau_p$ (m)	7.6x10^{-2}	1.2x10^{-1}	5.7x10^{-1}	6.8	2.3x10^{+2}
v_{th} (m.s^{-1})	2.4x10^{+5}	1.6x10^{+5}	3.4x10^{+4}	2.9x10^{+3}	8.6x10^{+1}
$\delta x=v_{th}\Delta t$ (m)	4.9x10^{-5}	3.1x10^{-5}	6.9x10^{-6}	5.9x10^{-7}	1.7x10^{-8}
$\lambda_D=v_{th}/\omega_p$ (m)	1.9x10^{-3}	1.9x10^{-3}	1.9x10^{-3}	1.9x10^{-3}	1.9x10^{-3}
$g=1/(n\lambda_D^3)$	1.5x10^{-8}	4.0x10^{-8}	8.1x10^{-7}	1.2x10^{-4}	1.2x10^{-1}
$\Lambda=4\pi/g$	8.3x10^{+8}	3.1x10^{+8}	1.5x10^{+7}	1.0x10^{+5}	1.0x10^{+2}
$\vartheta_c/\omega_p=\ln\Lambda/\Lambda$	2.5x10^{-8}	6.3x10^{-8}	1.1x10^{-6}	1.1x10^{-4}	5.0x10^{-2}

The results reported in Table 1 show that the condition $g \ll 1$ is always verified; as this agreement increases still with the beam current, **we can conclude that the short range correlations are negligible in intense ion beams.**

WHAT DOES A COARSE-GRAINED MODEL

Now, we have to go back and try to analyze the loss of information in the averaging process between the microscopic and the macroscopic levels; it is clear that this loss of information is due to high frequencies which are generated at the microscopic level and are restituted only in part by the coarse-grained quantities.

We saw that **the evolution of the averaged quantities could be well restituted by the Vlasov-Maxwell system of equations**, but we know also that our coarse-grained model is limited to linear or weakly non-linear effects; it is thus important to verify in what type of regime is the beam :

• when short range collisions are rare ($v_c \ll \omega_p$, they act just like local perturbations of the system, which will be damped by collective space charge waves; in this case the beam tends rapidly to a local thermodynamic equilibrium,

• when short range collisions are the dominant process ($v_c \gg \omega_p$), the system cannot relax to a some equilibrium, because it has no time to absorb perturbations; the beam is in strong turbulence.

We calculated the ratio ϑ_c/ω_p for an intense ion beam; we can verify in Table 1 that **we are in the regime of local thermodynamic equilibrium**.

In this regime, **well described by the Vlasov-Maxwell equations**, the beam can still be excited by an external frequency; this is in fact what happens with the external fields of confinement, where the beam is regularly excited by a frequency $\omega_0 \approx \omega_p \gg \nu_c$, and we have to deal with a quiescent equilibrium or a transition to a weak turbulent equilibrium, where all excited modes are in the vicinity of ω_p.

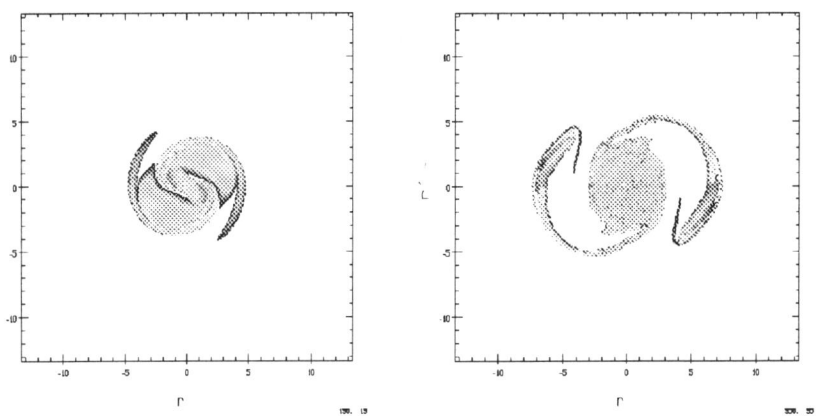

FIGURE 1. Phase Space (r-r') for a proton beam (I=30mA, η=0.8, $T(r,0)=C$)

CONCLUSION

As a conclusion, we will comment the two plots of the Figure 1, which were obtained with Renoir, and represent two stages in the transport of the same beam in a continuous magnetic channel. We can easily observe in the left plot, the particles which are trapped in the 1/2 resonances. In the right plot, the central bag of energy is near a weak turbulent equilibrium, and resonances have disappeared; the peripheric particles are creating a new bag of energy, where still exist resonances.

ACKNOWLEDGMENTS

The author wish to express his thanks to S.Joly, G.Laval, J.L.Lemaire, P.A.Raviart and J.Segre for their advices and constant interest in this work.

REFERENCES

1. J.L.Delcroix, "Physique des Plasmas",Tome1, Dunod-Paris(1963)
2. N.A.Krall, A.W.Trivelpiece, "principles of Plasma Physics", Mc Graw-Hill Book Comp. (1973)
3. S.Ichimaru, "Basic Principles of Plasma Physics", Frontiers in Physics, W.A. Benjamin (1973)
4. A.Piquemal, proceedings EPAC'96 Sitges-Spain (1996)

GATOR: A HYBRID SPECTRAL/PIC FORMULATION OF THE INTERACTION OF ELECTRON BEAMS WITH SLOW-WAVE STRUCTURES

H.P. Freund* and E.G. Zaidman[†]

*Science Applications International Corp., McLean, VA 22102
[†]Naval Research Laboratory, Washington, D.C. 20375

Abstract. A time-dependent nonlinear analysis of a helix traveling wave tube is presented for a beam propagating through a sheath helix surrounded by a conducting wall. The effects of dielectric- and vane-loading are included as is the tapering of the helix pitch. Dielectric-loading is described when the gap between the helix and the wall is uniformly filled. Vane-loading describes the insertion of vanes running the length of the helix. The field is represented as a superposition of azimuthally symmetric waves *in vacuo*. An overall explicit sinusoidal variation is assumed, and the polarization and radial variation of each wave is determined by the boundary conditions in a vacuum circuit. The propagation of each wave as well as the interaction with the beam is included by allowing the amplitudes of the waves to vary in z and t. A dynamical equation is derived analogously to Poynting's equation, and solved in conjunction with the 3-D Lorentz force equations. The model is compared with linear theory.

INTRODUCTION

The development of the traveling wave tube (TWT) extends over several decades since the pioneering work of Pierce and coworkers (1). More recently, complete field theories of beam-loaded helix TWTs have been developed for both sheath (2) and tape (3) helix models, and dielectric-loading has also been incorporated into the sheath helix analyses (4). Nonlinear theories can be grouped into two classes dealing with steady-state and time-dependent models. A review of the steady-state models has been given by Rowe (5). Time-dependent models rely on particle-in-cell (PIC) simulations. The most general PIC formulation of a TWT to date is in 2-D (6).

The approach we adopt differs from a PIC formulation. As for the 2-D PIC formulation, we assume azimuthal symmetry and a sheath helix. However, we treat the fields in terms of a spectral decomposition of the normal modes of the vacuum helix. An overall sinusoidal variation of the form $\exp(ikz - i\omega t)$ is assumed for each wave, where ω denotes the angular frequency determined from the vacuum sheath helix dispersion equation corresponding to wavenumber k. The polarization and radial variation of each wave is given by the normal mode solutions of the vacuum sheath helix boundary conditions. The evolution of each wave is described by allowing the amplitudes to vary in both axial position and time. The dynamical

© 1997 American Institute of Physics

equation is analogous to Poynting's equation which includes the coupling to the electron beam and, hence, the intermodulation between the waves themselves. In conjunction with the equations for the fields, the trajectories of an ensemble of electrons are integrated using the 3-D Lorentz force equations.

This spectral approach provides a good model for the dispersion and radial variation of the field, but requires an explicit choice of the waves of interest. It has two advantages over the 2-D PIC models, however, in that the technique (1) can be readily generalized to deal with more realistic tape helix models which include higher harmonic components, and (2) is much less computationally demanding.

THE GENERAL FORMULATION

The general formulation treats multiple waves in a dielectric- and vane-loaded sheath helix in the presence of an electron beam. This is a fully time-dependent problem The physical configuration is of an electron beam propagating through a dielectric and vane-loaded helix. Azimuthal symmetry is assumed, and the vanes are positioned radially. We use R_h and R_g to denote the radii of the helix and the outer cylinder, R_v denotes the inner radius of the vanes which extend to the outer wall, and ε_0 is the dielectric constant for the material which uniformly fills the space between the helix and the outer wall. The unit vector describing the pitch of the helix is $\hat{e}_\phi = \hat{e}_\theta \cos\phi + \hat{e}_z \sin\phi$, where $\tan\phi = 1/k_h R_h$ for a helix wavenumber k_h [$\equiv 2\pi/\lambda_h$, and λ_h denotes the helix period].

The normal modes for this circuit structure can be determined analytically (7). The dispersion equation fort the azimuthally symmetric modes is

$$\frac{\omega_n^2}{c^2} = \frac{p_n q_n}{k_h^2 R_h^2} \frac{J_0(\rho_n) W_{0,0}(\chi_n,\xi_n)}{J_1(\rho_n) W_{1,1}(\chi_n,\xi_n)} \frac{\left[p_n J_0(\rho_n) W_{1,1}(\zeta_n,\xi_n) - q_n J_1(\rho_n) W_{1,0}(\zeta_n,\xi_n)\right]}{\left[q_n J_1(\rho_n) W_{0,0}(\chi_n,\xi_n) + \varepsilon_0 p_n J_0(\rho_n) W_{1,0}(\xi_n,\chi_n)\right]}, \quad (1)$$

where k_h [$\equiv 2\pi/\lambda_h$] is the helix wavenumber, $\rho_n \equiv p_n R_h$, $\zeta_n \equiv q_n R_g$, $\xi_n \equiv q_n R_h$, $\chi_n \equiv q_n R_v$, for $p_n^2 \equiv \omega_n^2/c^2 - k_n^2$, $q_n^2 \equiv \varepsilon_0 \omega_n^2/c^2 - k_n^2$, and frequency and wavenumber (ω_n, k_n). In addition, $W_{m,n}(x,y) \equiv Y_m(x) J_n(y) - J_m(x) Y_n(y)$, and J_n and Y_n denote the Bessel and Neumann functions. The corresponding fields are

$$\begin{pmatrix} \delta\mathbf{E}(\mathbf{x},t) \\ \delta\mathbf{B}(\mathbf{x},t) \end{pmatrix} = \sum_n \left[\delta\hat{E}_n^{(1)} \begin{pmatrix} \mathbf{e}_n(\mathbf{x},t) \\ \mathbf{b}_n(\mathbf{x},t) \end{pmatrix} + \delta\hat{E}_n^{(2)} \begin{pmatrix} \mathbf{e}_n^*(\mathbf{x},t) \\ \mathbf{b}_n^*(\mathbf{x},t) \end{pmatrix} \right], \quad (2)$$

where the summation is over the modes to be included, and

$$\begin{pmatrix} \mathbf{e}_n(\mathbf{x},t) \\ \mathbf{e}_n^*(\mathbf{x},t) \end{pmatrix} \equiv \left\{ R_n(r)\hat{e}_r \begin{pmatrix} \sin\varphi_n \\ \cos\varphi_n \end{pmatrix} + \left[\beta_{ph}(k_n)\Theta_n(r)\hat{e}_\theta - Z_n^{(e)}(r)\hat{e}_z\right] \begin{pmatrix} -\cos\varphi_n \\ \sin\varphi_n \end{pmatrix} \right\},$$
(3)

$$\begin{pmatrix} \mathbf{b}_n(\mathbf{x},t) \\ \mathbf{b}_n^*(\mathbf{x},t) \end{pmatrix} \equiv \left\{ \Theta_n(r)\hat{e}_r \begin{pmatrix} \cos\varphi_n \\ -\sin\varphi_n \end{pmatrix} + \left[\beta_{ph}(k_n)R_n(r)\hat{e}_\theta - Z_n^{(b)}(r)\hat{e}_z\right] \begin{pmatrix} \sin\varphi_n \\ \cos\varphi_n \end{pmatrix} \right\},$$

where $\varphi_n \equiv k_n z - \omega_n t$ for wavenumber $k_n \equiv n\Delta k$ and Δk is the separation in wavenumber. We assume that both $\delta \hat{E}_n^{(1,2)}$ vary in z and t. In addition, $\beta_{ph}(k_n) \equiv v_{ph}(k_n)/c = \omega_n/ck_n$ denotes the normalized phase velocity of each wave. The polarization functions $R_n(r)$, $\Theta_n(r)$, $Z_n^{(e,b)}(r)$ are given in ref. (7) for the regions within the helix, between the helix and the inner vane radius, and from the inner vane radius to the outer wall.

The dynamical equations for a tapered helix can be expressed as (7)

$$\left(\frac{\partial}{\partial t} + v_{gr}^{(n)} \frac{\partial}{\partial z}\right) \binom{P_n^{1/2} \delta \hat{E}_n^{(1)}}{P_n^{1/2} \delta \hat{E}_n^{(2)}} = \frac{2\Delta k P_n^{1/2}}{cU_n} \int_0^{2\pi/\Delta k} dz' \iint_{A_h} d\mathbf{x}_\perp \mathbf{J}(\mathbf{x}_\perp, z', t) \cdot \binom{\mathbf{e}_n(\mathbf{x}_\perp, z', t)}{\mathbf{e}_n^*(\mathbf{x}_\perp, z', t)}, \quad (4)$$

where $\mathbf{J}$ denotes the source current. The time-averaged power flux for each mode over the entire cross-section of the cylinder and helix is given by $S_n = P_n \delta \hat{E}_n^2$, and the time-averaged energy density per unit axial length over the entire cross-section of the helix and cylinder is given by $W_n = U_n \delta \hat{E}_n^2$, where $\delta \hat{E}_n^2 = \delta \hat{E}_n^{(1)2} + \delta \hat{E}_n^{(2)2}$. Finally, $v_{gr}^{(n)} = \partial \omega_n / \partial k_n = P_n/U_n$ denotes the group velocity of each wave. The detailed expressions for P_n, U_n, and $v_{gr}^{(n)}$ are given in ref. (7).

The electron dynamics are treated using the 3-D Lorentz force equations. Azimuthal symmetry is imposed in the sense that (1) the beam distribution upon entry to the helix is azimuthally symmetric, and (2) each of the components of the electromagnetic fields varies only in (r,z,t). Hence,

$$\frac{d}{dt}\mathbf{p} = -e\delta\mathbf{E}(\mathbf{x},t) - \frac{e}{c}\mathbf{v} \times \left[B_0 \hat{e}_z + \delta\mathbf{B}(\mathbf{x},t)\right]. \quad (5)$$

Electrons are injected at uniform time intervals to simulate a continuous beam.

NUMERICAL ANALYSIS

The dynamical equations are solved for a system of length L on a grid with spacing Δz and time step Δt using the MacCormack method (8). The current is

$$\mathbf{J}(\mathbf{x},t) = -q \sum_{i=1}^{N} \mathbf{v}_i(t) \, \delta[x - x_i(t)] \, \delta[y - y_i(t)] \, S[z - z_i(t)], \quad (6)$$

where N is the number of electrons, $(\mathbf{x}_i, \mathbf{v}_i)$ represents the location and velocity of the ith electron q is the charge per electron, and S is the triangular shape function

$$S(z - z_i) = \frac{1}{\Delta z^2} \begin{cases} z - z_i + \Delta z & ; z_i - \Delta z \leq z \leq z_i \\ -z - z_i - \Delta z & ; z_i < z \leq z_i + \Delta z \end{cases}. \quad (7)$$

The charge per electron is $q = I_b \Delta t / N_{\Delta t}$, where I_b is the beam current and $N_{\Delta t}$ is the number of electrons injected per time step. Open boundary conditions are imposed. The orbits are integrated via a 4th order Runge-Kutta algorithm. In order to be consistent, we linearly interpolate the field from the grid to the particles.

In our algorithm, we first calculate the sources by accumulating charge to the grid and average over the appropriate scale length. We then step the fields using the

MacCormack method. Once the updated fields are calculated, we advance the electron trajectories. This procedure is repeated over the time scale of interest.

TWTs are typically operated as amplifiers in which a signal is amplified over the length of the helix. Hence, we specify an algorithm for the injection of a signal(s), and we assume that a pulse is injected at $z = -\Delta z$ and allowed to propagate into the interaction region. The pulse has a smooth temporal shape given by

$$\delta \hat{E}_n^{(1)}(z=-\Delta z, t) = \delta \hat{E}_0^{(1)} \begin{cases} \sin^2\left[\dfrac{\pi}{2} \dfrac{(t-\tau_{start})}{\Delta \tau_{rise}}\right] & ; \tau_{start} \leq t < \tau_{rise} \\ 1 & ; \tau_{rise} \leq t < \tau_{flat} \\ \cos^2\left[\dfrac{\pi}{2} \dfrac{(t-\tau_{fall})}{\Delta \tau_{fall}}\right] & ; \tau_{flat} \leq t \leq \tau_{fall} \end{cases} \quad (8)$$

and $\delta \hat{E}_n^{(2)}(z=-\Delta z, t) = 0$ for all t, where $\delta \hat{E}_0^{(1)}$ is chosen to describe the peak power, τ_{start} denotes the start time of the pulse, $\Delta \tau_{rise}$ is the rise time of the pulse, $\tau_{rise} = \tau_{start} + \Delta \tau_{rise}$ is the time at which the pulse has risen to its peak value, $\tau_{flat} - \tau_{rise}$ is the time interval over which the pulse retains a constant magnitude, $\Delta \tau_{fall}$ is the time interval over which the pulse falls to zero, and $\tau_{fall} = \tau_{flat} + \Delta \tau_{fall}$ is the time after which the injected pulse vanishes. Note that injection of power at $z = -\Delta z$ requires the inclusion of a guard cell in the grid outside of the interaction region.

Electrons are injected at the start of each time step, and we allow for an arbitrary current rise time by choosing a current of the form $I_b(t) = I_b \sin^2(\pi t/2\tau_r)$ for $t \leq \tau_r$, and I_b for $t > \tau_r$, where τ_r denotes the rise time of the beam.

The sources are determined by first mapping the charge from each electron to the two nearest-neighbor grid points and then averaging over a length $\Delta \lambda$. The specific procedure we employ for this is a "moving-window" average in which the sources at the ith grid cell are determined by averaging over those grid cells within a length $\pm \Delta \lambda/2$ on either side. For example, if $\Delta z = \Delta \lambda/N$ for N even, then the sources at the ith grid cell are determined by averaging the charge over all grid cells within $i \pm N/2$. This necessitates the inclusion of $N/2$ guard cells corresponding to $z < 0$ and $z > L$. Thus, electrons are injected at $z = -\Delta \lambda/2$ and allowed to propagate ballistically until they reach $z = 0$, at which point the interaction with the radiation is "turned on". Similarly, the electrons also propagate ballistically [i.e., the interaction with the radiation is "turned off"] when the electrons exit the interaction region at $z = L$. This is illustrated schematically in Fig. 1. Electrons are ejected form the simulation whenever they pass beyond $z = L + \Delta \lambda/2$ or intersect the helix.

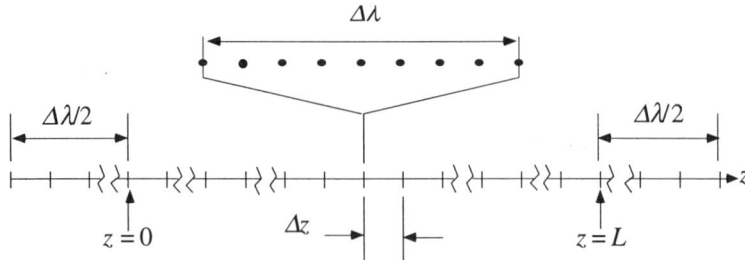

Fig. 1 Schematic illustration of the mapping of charge to the grid and the current average.

Since charge is mapped onto the two nearest neighbor grid cells, the end cells of the average corresponding to the ith cell [i.e., the grid cells at $i \pm N/2$] will contain contributions from charges outside the length $\Delta\lambda$. This introduces an additional oscillation with a period of $\Delta\lambda + 2\Delta z$ into the sources which must be filtered out.

The circuit parameters for the specific case we choose to study correspond to a helix TWT built at Northrop-Grumman Corp. (9). The helix and wall radii were 0.12446 cm and 0.2794 cm respectively, and the helix period was 0.010137 cm. The helix was supported by three dielectric rods with rectangular cross-sections running the length of the helix. The dielectric constant of the rods was 6.5 and the rod dimensions were 0.0508 cm × 0.14732 cm. No vanes were used in this structure. The cold dispersion solutions have been compared with the measured dispersion of this TWT in ref. (4) in which it was found that close agreement with the measured dispersion properties over a broad band were obtained for an effective dielectric constant $\varepsilon_0 = 1.75$. As described in ref. (4), this effective dielectric constant can be estimated by a simple volume-weighted average.

The nonlinear simulation is compared with both a linear theory of the interaction derived in ref. (4) and with a phase trapping estimate of the saturation efficiency first described by J.C. Slater (10). The linear theory includes the effect of space-charge forces on the interaction, which are not included in the present nonlinear model; hence, we must apply the linear theory in the absence of space-charge effects. The phase-trapping estimate of the saturation efficiency is

$$\eta \approx \frac{2\gamma^3}{\gamma-1} \frac{v_b}{c} \left(\frac{v_b}{c} - \frac{v_{ph}}{c} \right). \qquad (10)$$

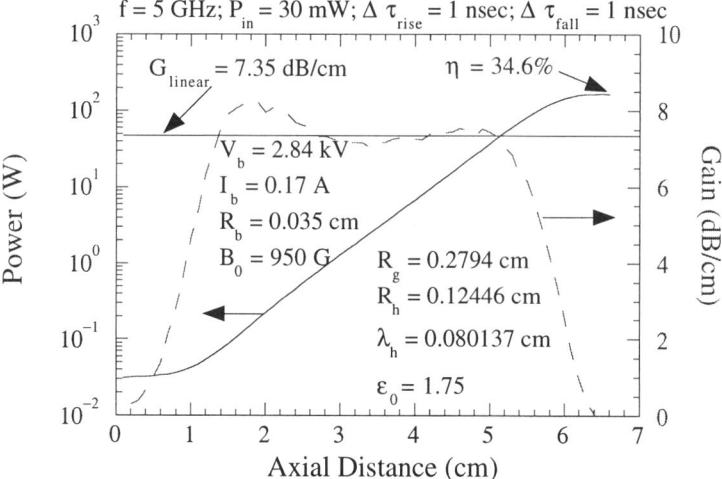

Fig. 2 Plots of the gain and power versus axial distance from the nonlinear simulation.

We now turn to the nonlinear simulation, and choose the same circuit parameters as the aforementioned tube at Northrop-Grumman. The beam has a voltage and current of 2.84 kV/0.17 A, and we inject an annular beam with a radius of 0.035 cm and a rise time of 1 nsec. A magnetic field of 950 G is used. We inject a single pulse at 5 GHz with a 30 mW peak power starting after 1 nsec with a 1 nsec rise and fall time. The gain and power are shown as functions of axial position in Fig. 2. Saturation occurs after ≈ 6.5 cm at a power ≈ 167 W for an efficiency of 34.6 %. This is in good agreement with the efficiency estimate of about 36% at 5 GHz. As shown in the figure, the linear gain is also in substantial agreement with the prediction of the linear theory of a gain of 7.35 dB/cm at 5 GHz.

SUMMARY AND DISCUSSION

We have described a nonlinear formulation of the interaction of an electron beam and multiple waves in a dielectric- and vane-loaded sheath helix TWT in the time domain that can treat multiple radiation pulses. Comparison with a linear theory of the interaction as well as with an estimate of the nonlinear saturation efficiency show good agreement. The essential characteristics of the interaction are well-described by the simulation once the effective dielectric constant and vane radius have been determined.

The fundamental dynamical equation for the fields is quite general in form, and relies largely on a knowledge of the dispersion, polarization, energy density, and Poynting flux for the waves under consideration. Thus, the technique is readily generalized to other configurations and structures. In view of the computational efficiency and generality of the present formulation, the technique has advantages over PIC simulations of a variety of structures.

ACKNOWLEDGMENTS

This work was supported by the Office of Naval Research, and by a grant of High Performance Computing time from the DoD HPC Shared Resource Center of the Corps of Engineers Waterways Experiment Station (CEWES).

REFERENCES

1. J.R. Pierce, *Traveling-Wave Tubes* (Van Nostrand, New York, 1950).
2. H.P. Freund, M.A. Kodis, and N.R. Vanderplaats, *IEEE Trans. Plasma Sci.* **20**, 543 (1992).
3. H.P. Freund, N.R. Vanderplaats, and M.A. Kodis, *IEEE Trans. Plasma Sci.* **21**, 654 (1993).
4. H.P. Freund, E.G. Zaidman, M.A. Kodis, and N.R. Vanderplaats, *IEEE Trans. Plasma Sci.* **24**, 895 (1996).
5. J.E. Rowe, *Nonlinear Electron-Wave Interaction Phenomena* (Academic, New York, 1965).
6. B. Goplen, D. Smithe, K. Nguyen, M.A. Kodis, and N.R. Vanderplaats, "MAGIC simulations and experimental measurements from the emission gated amplifier I & II experiments," International Electron Devices Meeting Tech. Digest, pp. 759-762, 1992.
7. H.P. Freund, E.G. Zaidman, and T.M. Antonsen, Jr., Phys. Plasmas, 2, 3871 (1996).
8. K.A. Hoffman, *Computational Fluid Dynamics for Engineers* (Engineering Education System, Austin, TX, 1989), p. 171.
9. G. Groshart, Northrop-Grumman Corp. (personal communication 1996).
10. J.C. Slater, *Microwave Electronics* (Van Nostrand, New York, 1950).

Large-Timestep Techniques for Particle-In-Cell Simulation of Systems with Applied Fields that Vary Rapidly in Space*

Alex Friedman and David P. Grote

Lawrence Livermore National Laboratory
L-440, P.O. Box 5508, Livermore CA 94550

Abstract. Under conditions which arise commonly in space-charge-dominated beam applications, the applied focusing, bending, and accelerating fields vary rapidly with axial position, while the self-fields (which are, on average, comparable in strength to the applied fields) vary smoothly. In such cases it is desirable to employ timesteps which advance the particles over distances greater than the characteristic scales over which the applied fields vary. Several related concepts are potentially applicable: sub-cycling of the particle advance relative to the field solution, a higher-order time-advance algorithm, force-averaging by integration along approximate orbits, and orbit-averaging. We report on our investigations into the utility of such techniques for systems typical of those encountered in accelerator studies for heavy-ion beam-driven inertial fusion.

I. INTRODUCTION

We are exploring techniques for enhanced efficiency in PIC simulations of beams and plasmas. A simple such technique, used in the WARP code[1,2] from its inception, is the "residence correction," whereby the impulse imparted by an (idealized) sharp-edged element is corrected to reflect the fraction of the velocity-advance step during which the particle resides within the element. It is desirable to extend this concept to cases where the effects of extended fringe fields and other smooth but rapid variations must be accurately captured. Techniques which may be applicable include: a) a subcycled leapfrog advance—taking N substeps between each major step on which the self-consistent field is computed and applied; b) a family of high-order symplectic advances—for smooth forces, these converge quickly as the step size is reduced, but it will be nontrivial to take advantage of that property in a practical PIC code; c) a newly-invented family of force-averaged velocity advances—effectively, these amount to subcycling on the velocity advance only; using integration along an approximate orbit to do the averaging; however, "special" weightings may afford higher order, and again the challenge is to make the methods practical; and d) orbit-averaging—this is basically a noise-reduction strategy which may allow use of a smaller number of particles, and probably must be non-symplectic; it is discussed briefly here but has not been studied in detail. In Section II immediately below these methods are described. A model problem was devised to evaluate the methods; the problem and some results are described in Section III. We are beginning to test the most promising of these methods via full 3-d PIC simulations using the WARP3d code; some initial results, and a concluding discussion, are presented in Section IV.

II. METHODS CONSIDERED
A. Subcycled Leapfrog

In this algorithm, the external force is applied every substep, while a space charge "kick" is applied every N substeps, after the source term is accumulated and the field equation solved. It is necessary to apply the self-force only at those times at which the source is calculated, to avoid spurious self-forces, unless special steps are taken as in orbit-averaging. In the absence of space charge the subcycled advance is identical to ordinary leapfrog with a step size 1/N as large.

B. High-Order Symplectic Integrators

These methods preserve Hamiltonian structure and avoid spurious damping or excitation, but are not energy conserving.[3] Nonrelativistically, defining the position "x", velocity "v", and acceleration "a", the general explicit algorithm is:

$$x' = x + c_i \Delta t\, v; \quad v' = v + d_i \Delta t\, a; \quad \text{repeated for } i = 1, ..., k. \tag{1}$$

The c_i and d_i are chosen so that the final composed mapping satisfies the Taylor series expansion of the solution up to order Δt^n, for an n-th order scheme. The Candy-Rozmus (C-R) scheme has the advantage of offering 4th-order accuracy with 3 force evaluations per step. We suspect that rotations of v in an external magnetic field applied at each substep won't break the invariants, since they are 1:1 maps, but have yet to prove this. Coefficients for some schemes are:

Leapfrog: $\quad c_1 = 0,\ c_2 = 1,\ d_1 = d_2 = 1/2$ (2)

Ruth 3rd-order: $\quad c_1 = 7/24,\ c_2 = 3/4,\ c_3 = -1/24$ (3)
$\quad\quad\quad\quad\quad\quad d_1 = 2/3,\ d_2 = -2/3,\ d_3 = 1$

Candy & Rozmus: $\quad c_1 = c_4 = 1/[2(2-2^{1/3})],\ c_2 = c_3 = [1-2^{1/3}]/[2(2-2^{1/3})]$ (4)
$\quad\quad\quad\quad\quad\quad d_1 = d_3 = 1/(2-2^{1/3}),\ d_2 = -2^{1/3}/(2-2^{1/3}),\ d_4 = 0$

C. Force-Averaged Velocity Advance

The original concept was to integrate the external force over the velocity-advance step along an approximate orbit, for an improved impulse. This generalizes the "residence correction" used in WARP3d on entry to or exit from a sharp-edged element. The averaged applied force is: $\langle F_{ext} \rangle = \Sigma\, W_j F_{ext}(x_j)$, where the x_j are computed at temporal offsets δ_j relative to the middle of the velocity-advance step, and the W_j are "weights." Letting "sc" denote space charge, * a temporary quantity, "h" the half-level, and acceleration a = F/m, a timestep is:

$$v_h := v + \tfrac{1}{2}\Delta t\, \langle a \rangle$$
$$x := x + \Delta t\, v_h$$
$$a^* := a_{ext}(x) + a_{sc}(x)$$
$$v^* := v_h + \tfrac{1}{2}\Delta t\, a^*$$

$$\langle a_{ext}\rangle := 0 \qquad (5)$$
$$\text{for } 1 \leq j \leq k: \qquad x_j := x + v^* \delta_j + {}^1\!/_2 a^* \delta_j^2 ;$$
$$\langle a_{ext}\rangle := \langle a_{ext}\rangle + W_j\, a_{ext}(x_j)$$
$$\langle a\rangle := \langle a_{ext}\rangle + a_{sc}(x)$$
$$v := v_h + {}^1\!/_2 \Delta t\, \langle a\rangle$$

During the evaluation of force-averaged methods using the model problem, it was discovered that certain simple quadratures work "better" than a detailed integration using many points (very large k, all W's equal). One "good" choice, found via experimentation for k = 3, is: W = $\{^1\!/_3,^1\!/_3,^1\!/_3\}$, $\delta = \Delta t\{-^1\!/_2,0,^1\!/_2\}$. The simplest "good" scheme, found via the analysis outlined below for k = 2, is: W = $\{^1\!/_2,^1\!/_2\}$, $\delta = \Delta t\{-^1\!/\sqrt{6},^1\!/\sqrt{6}\}$. However, it is difficult to start the calculation so as to preserve accuracy, and we have not yet learned how to do so reliably. The difficulty arises when the trajectory calculation is initiated at an axial position where the applied force has a nonzero axial gradient.

To analyze the force-averaged velocity advance, we assumed harmonic motion at frequency ω in a well of natural frequency ω_0, i.e., $x^{n+1} = e^{-i\omega\Delta} x^n$ and $a^n = -\omega_0^2 x^n$ (superscripts denote time levels). We also insist that $\Sigma W_j = 1$, $\Sigma \delta_j = 0$, and $\Sigma W_j \delta_j = 0$ (W's symmetric about the middle of the set, δ's antisymmetric). Computing ω in terms of ω_0, Δt, W_j, and δ_j, one obtains the condition (satisfied by the "good" cases mentioned above, and others) for fourth-order agreement between ω and ω_0, i.e., $\omega / \omega_0 = 1 + O(\omega\Delta)^4$:

$$\Sigma W_j \delta_j^2 = (\Delta t)^2 / 6 . \qquad (6)$$

D. Orbit-averaging

In orbit averaging, the field equation is solved every N substeps using a net charge that comes from a sum over depositions done each substep; the field is applied every substep using interpolation in time, in a predictor-corrector loop. In general, the substep will be set by the need to resolve external field gradients. In comparison with leapfrog, there can be a gain in efficiency due to less-frequent field-solving. In comparison with subcycling, there can be reduced self-field noise, at the expense of extra deposition time and the need for predictor-corrector iteration. A disadvantage is that such methods are unlikely to be symplectic, but this is probably not a problem for our applications, and in any event results can be validated by varying the step size. Recent research in France exploring this technique has been described as encouraging [4]

III. MODEL PROBLEM

This problem models transport through a lattice period (a pair of quadrupole lenses of alternating polarity), tracking a particle on the principal axis y = 0. Linear fringe fields ($\propto$ x) are included, but higher multipoles and pseudo-multipoles are not included. "Exact" solutions (needed for the computation of errors) were obtained by use of very small steps in a leapfrog advance. The error for each run was defined as the absolute value of the difference between the "exact" and computed x values at the end of the problem. Three elements were varied, making a series of 12 tests. In each test, 5 algorithms were run at 7 step sizes ranging from 16 to 1024 steps per lattice period (256 is a typical value used in WARP3d leapfrog runs). The elements varied were:

(1) Three models of the force profile (shown in Fig. 1):

"sine^5": $F(x,z) = - x\, F_0 \sin^5[(z + z_{off})\,\pi/L]$
"model": $F(x,z) = \pm 1/2\, x\, F_0\{\cos[(z + z_{off})\,\pi/\chi] + 1\}$, or 0, pieced together
"tabulated": Linear interpolation into a table of values derived analytically

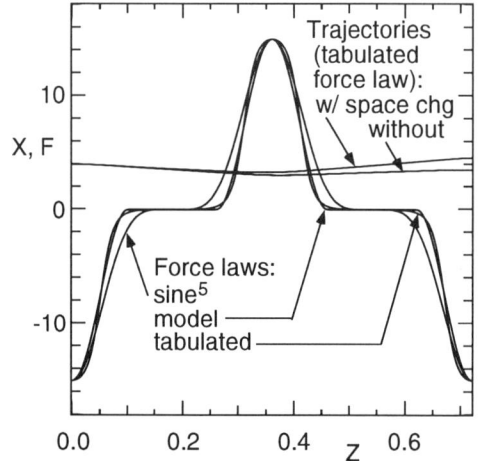

FIGURE 1. Force profiles and trajectories, $z_{off}/hlp=0.5$

Here, L is the half-lattice period length, and χ the physical magnet length. Of these, "sine^5" is perfectly smooth, while "model" has a discontinuous second derivative, and "tabulated" has a discontinuous first derivative due to abrupt termination of the list of 300 tabulated values describing each "peak," which were offset and rescaled slightly so that the last tabulated force value was zero.

(2) Two options for modeling space charge: spacechg=0, no space charge force; and spacechg=1, a linear kick applied at end of each full cycle of the advance.

(3) Two offsets of the initial position relative to the center of a "drift space": $z_{off}/hlp=0.5$, with the test particle launched from the center of a focusing quad; and $z_{off}/hlp=0.4$, with the particle launched from the "side of the hill."

Some results are shown in Fig. 2. In (a), the "canonical" smooth sine^5 case with the particle launched from the center of a quad and no space charge, the subcycled leapfrog is identical to ordinary leapfrog with a 4x smaller step, while Ruth's scheme converges cubically, and both the Candy-Rozmus and the force-averaged scheme converge quartically. When (b) the tabulated force is used, all of the improved schemes do better than leapfrog, but none clearly converges more rapidly than quadratic; we conclude that a tabulated force with the granularity

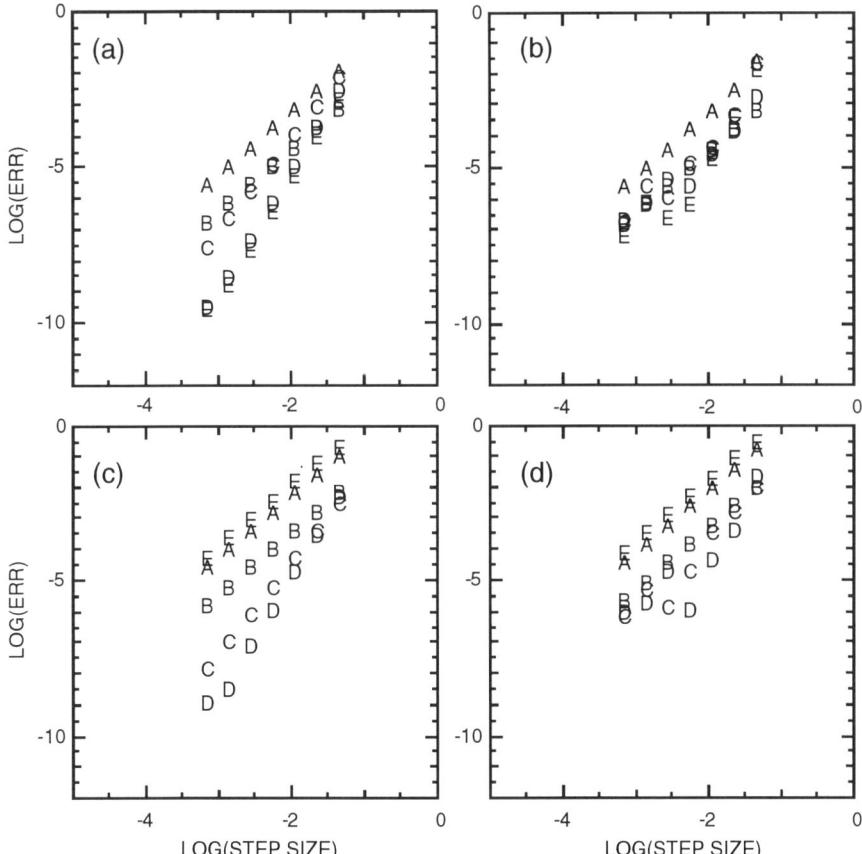

FIGURE 2. Error vs. step size: (a) $z_{off}/hlp=0.5$, $sine^5$ force; (b) $z_{off}/hlp=0.5$, tabulated force; (c) $z_{off}/hlp=0.4$, $sine^5$ force; (d) $z_{off}/hlp=0.4$, tabulated force. Points marked A were computed with leapfrog, B with 4:1 subcycled leapfrog, C with the Ruth scheme, D with the Candy-Rozmus scheme, and E with the optimized two-point force-averaged scheme.

typically used is insufficiently smooth with linear interpolation, and expect that a cubic spline interpolation is needed for full benefit from the high-order schemes. When (c) the $sine^5$ force is used with the particle launched from the side of the hill, the leapfrog error is considerably worsened, and the utter failure of the optimized force-averaged scheme is evident. The high-order schemes do not degrade the way leapfrog does. In this case the high-order advance performed so well that the "exact" answer was not quite "exact enough"; the quartic convergence of C-R actually persists down to the 1024-steps case. When (d) the tabulated force is used with a side-of-the-hill start, none of the high-order schemes does significantly better than subcycled leapfrog.

When the $sine^5$ force is used with (non-subcycled) space charge kicks turned on (not shown), subcycled leapfrog is not as good as leapfrog with a 4x smaller step. The high-order schemes do somewhat better than subcycled leapfrog, but are ultimately just second order because the space-charge error eventually dominates. When the "model" force is used with $z_{off}/hlp=0.4$, the Ruth scheme does better than the Candy-Rozmus one, and is anomalously fourth-order. Leapfrog does poorly on that problem as well. A number of other interesting features are evident in the runs, but space does not permit their explication here.

IV. SIMULATIONS AND DISCUSSION

Experiments at LLNL studying the bending of space-charge-dominated beams are simulated using WARP3d, a 3-d particle-in-cell code developed for heavy-ion fusion accelerator studies. The beamline includes seven electrostatic quadrupoles, eight permanent magnet quads, and a lattice of five electric dipoles interleaved with five permanent magnet quadrupoles. This serves as a realistic test of new methods. To date, 5:1 subcycling has been tested and seems to work well, with little distinguishable difference in the output. The speed-up in the case tested was relatively minor because a fast field-solver and a large particle number were used.

The utility of high-order advances is application-dependent. Such methods are often used in the absence of strong space-charge, and work well for the applied fields in typical applications. However, in a PIC code with multilinear interpolation (such as WARP3d), the requisite force smoothness may be absent— the pairwise interparticle force is continuous as a function of interparticle separation, but its derivative is "almost always" discontinuous, [5] and we don't yet know if multilinear interpolation is "smooth enough" in practice. If it is not, a smooth, higher-order spatial interpolant could be used. The cost of interpolation would be increased, but this is likely to be usually unimportant. Since a finer grid would be undesirable in 3-D, one should avoid enlarging the effective particle size. "Sharpening" operators can be used to offset any spreading. [6] For the 4th-order advance, a three times larger step size would be needed for net gain if the self-consistent field were computed at each substep. However, in many problems that field might be computed and applied only once per step, as in subcycling.

We conclude that there are clear advantages to using one or more of these methods to model space-charge-dominated beams. Subcycling is simple and seems to work well, while the payoff from a sophisticated scheme may ultimately be larger. The "startup" question needs to be addressed. We plan to explore these options systematically, while taking advantage of what we have already learned.

REFERENCES

*Work performed under the auspices of the U.S. Department of Energy by Lawrence Livermore National Laboratory under contract W-7405-ENG-48.
1. A. Friedman, D. P. Grote, and I. Haber, *Phys. Fluids B* 4, 2203 (1992).
2. D. P. Grote, A. Friedman, I. Haber, and S. S. Yu, in *Proc. Int. Sympos. on Heavy Ion Inertial Fusion*, Princeton, Sept. 6-9, 1995; to be publ. in *Journal of Fusion Eng. Design*, 1996.
3. H. Yoshida, *Celestial Mech. and Dynamical Astron.* **56**, 27-43 (1993).
4. A. Piquemal, private communication, June 1996.
5. C. K. Birdsall and A. B. Langdon, *Plasma Physics via Computer Simulation*, New York, McGraw-Hill, 1981, 70-72.
6. R. W. Hockney and J. W. Eastwood, *Computer Simulation Using Particles*, Bristol, Adam Hilger, 1988, 137-140.

3-D Electromagnetic Modeling of Wakefields in Accelerator Components [*]

Brian R. Poole, George J. Caporaso, Wang C. Ng,
Clifford C. Shang, and David Steich

Lawrence Livermore National Laboratory, Livermore, California 94550

Abstract. We discuss the use of 3-D finite-difference time-domain (FDTD) electromagnetic codes for the modeling of accelerator components. Computational modeling of cylindrically symmetric structures such as induction accelerator cells has been extremely successful in predicting the wake potential and wake impedances of these structures, but fully 3-D modeling of complex structures has been limited due to the substantial computer resources required for a fully 3-D model. New massively parallel 3-D time domain electromagnetic codes now under development using conforming unstructured meshes allow a substantial increase in the geometric fidelity of the structures being modeled. Development of these new codes will be discussed in the context of their applicability to accelerator problems. A variety of 3-D structures are tested with an existing cubical cell FDTD code and the wake impedances are compared with simple analytic models for the structures. These results will provide a set of benchmarks for testing the new time domain codes. Structures under consideration include a stripline beam position monitor as well as circular and elliptical apertures in circular waveguides. Excellent agreement for the monopole and dipole impedances with the models are found for these structures below the cutoff frequency of the beam line.

INTRODUCTION

As a charge bunch passes a perturbation in an accelerator structure a wakefield is generated which can interact with the charged particle beam. This process can lead to degradation of the beam, or worse, can lead to beam breakup instabilities. A substantial amount of work in the analysis of cylindrically symmetric accelerator structures has been done using the 2-D finite-difference time-domain (FDTD) code, AMOS (azimuthal mode simulator) (1), for the calculation of the wake potential and wake impedances of these structures. More complicated

---

[*] This work performed under the auspices of the U. S. Department of Energy by Lawrence Livermore National Laboratory under contract No. W-7405-Eng-48

accelerator structures do not have the cylindrical symmetry required by AMOS for solution of the wake properties of the structure. We use a 3-D FDTD code, TSAR (temporal scattering and response) to model more general structures that do not necessarily possess cylindrical symmetry.

TSAR uses a 3-D grid composed of cubical cells for the FDTD analysis, with each cell being assigned a given material property. The cell size is typically chosen to resolve the shortest wavelength expected in the problem and is chosen small enough to accurately represent the smallest dimension in the model. In addition, to satisfy stability requirements for the FDTD field solver, the Courant condition for the time step must be satisfied. This condition is defined in Equation (1)

$$dt < \frac{1}{2c} dx \qquad (1)$$

where dx is the length of a cell edge. It is also important to resolve the lowest frequency that the electron beam may excite which determines the total number of time steps to be run for a specific problem.

The concept of the wake potential and wake impedance is traditionally used as a measure of the interaction of the beam with the structure it is traversing and has been discussed elsewhere in the literature (2). TSAR is used to perform 3-D modeling of the wakefields associated with a stripline beam position monitor and beam line aperture to determine the longitudinal monopole ($m = 0$) and transverse dipole ($m = 1$) impedances.

Beam Position Monitor

The stripline beam position monitor is discussed in some detail by Ng (3). The beam position monitor consists of 2 pairs of strip transmission lines shown schematically in Figure (1). The dimensions of the structure are shown in the figure. The 25 Ω terminating resistors were chosen based on a frequency domain calculation of the TEM impedance of a single stripline in the enclosure that resulted in a 25 Ω characteristic impedance. Since the four striplines are not strongly coupled due to the proximity of the striplines to the enclosure and the large angular sector between them, 25 Ω is a good estimate of the impedance for each stripline. For modeling purposes, the terminating resistors represent external coaxial transmission line feeds to the strip lines. Even to represent the gross geometric characteristics of the structure using cubical cells requires a cell size of 0.5 mm making the computational size of this problem 8×10^6 cells and requires a time step of .83 ps as dictated by the Courant condition. To accurately represent the details of the coaxial feeds would have required a much smaller cell size to have adequate geometric fidelity in the feed region and therefore it was decided to use simple wire resistors to model this region to minimize the computer memory requirements for the model. The length of the striplines was 10 cm which indicates a quarter-wavelength resonance of 750 MHz allowing us to sufficiently resolve the low frequency behavior using 16000 time steps. Approximately 25 cm of beam transport line was used on each side of the beam position monitor to

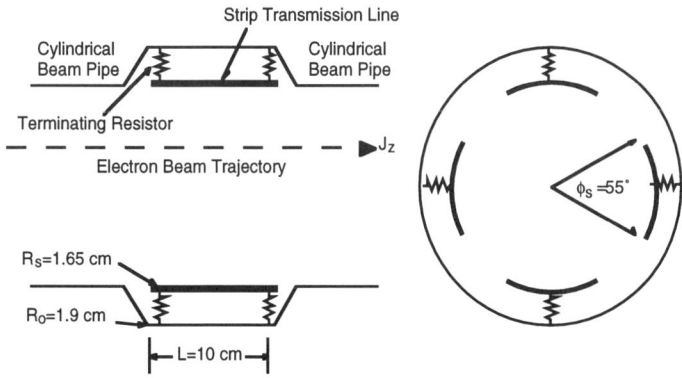

FIGURE 1. Model of beam position monitor for determining wake impedances

allow evanescent waves to decay and minimize the effects of the MUR radiation boundary conditions applied to the ends of the problem space. The structure is excited with a current distribution of the form

$$J_z = I_0 \delta(r - r_0) \delta(\phi - \phi_0) \delta(ct - z) \qquad (2)$$

where I_0 represents the strength of the current source, (r_0, ϕ_0) represent the transverse location of the relativistic electron beam, and $\delta(ct - z)$ represents the motion of a relativistic point charge. Analytic expressions for the longitudinal monopole ($m = 0$) and transverse dipole ($m = 1$) impedances were found for a matched transmission line model of the beam position monitor by Ng (3) as

$$Z_\parallel = Z_s \left(\frac{\phi_s}{2\pi}\right)^2 \left[\sin^2\left(\frac{\omega L}{c}\right) + j \sin\left(\frac{\omega L}{c}\right) \cos\left(\frac{\omega L}{c}\right)\right] \qquad (3)$$

$$Z_\perp = \frac{8cZ_s}{\pi^2 R_s^2}\left(\frac{1}{\omega}\right) \sin^2\left(\frac{\phi_s}{2}\right) \left[\sin^2\left(\frac{\omega L}{c}\right) + j \sin\left(\frac{\omega L}{c}\right) \cos\left(\frac{\omega L}{c}\right)\right] \qquad (4)$$

Figures (2) and (3) show comparisons of the analytic models with results obtained numerically through the TSAR modeling code for the longitudinal and transverse impedances respectively. As is seen there is excellent low frequency agreement. The deviation at higher frequencies may possibly be attributed to the high frequency impedance of the wire resistors.

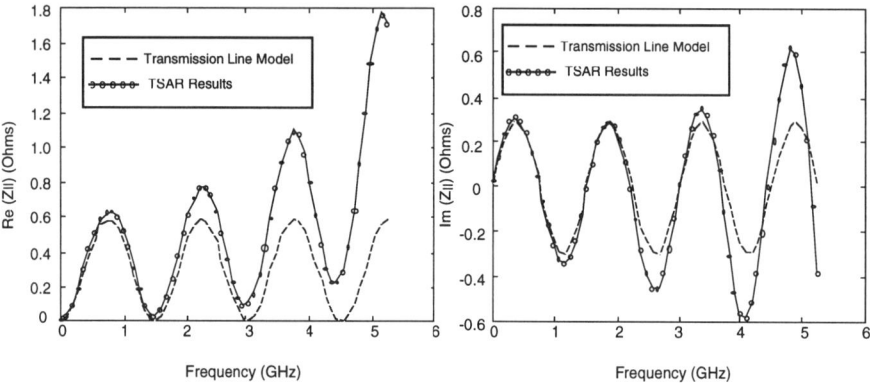

FIGURE 2. Real and Imaginary Longitudinal Impedance of the Beam Position Monitor

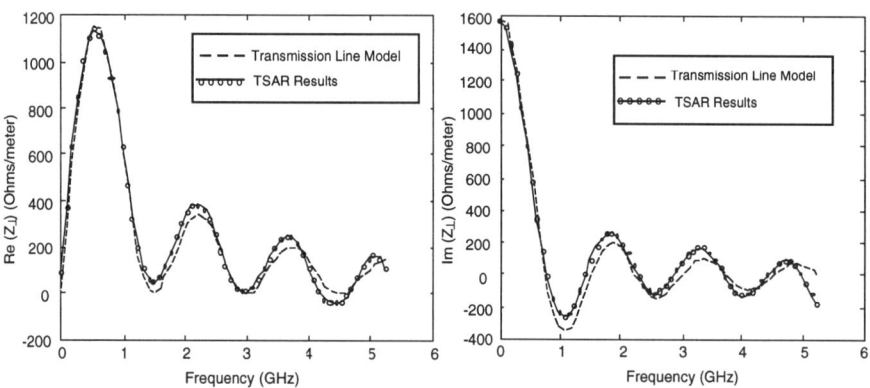

FIGURE 3. Real and Imaginary Transverse Impedance of the Beam Position Monitor

Circular Aperture in Circular Beam Line

The impedance of coupling apertures in beam transport lines has been treated in the literature by Gluckstern (4) and Kurennoy (5). In this section we consider a 3 cm diameter cylindrical beam line with either an 8 mm or a 3 cm circular waveguide connected at 90° to the beam line as shown in Figure (4). The 3 ports of the problem space are terminated with MUR radiation boundary conditions. A cell size of 0.5 mm is used to model both aperture problems. The length of the beamline is about 30 cm and the coupling waveguide is 3 cm long resulting in a mesh of about 1.2×10^7 cells.

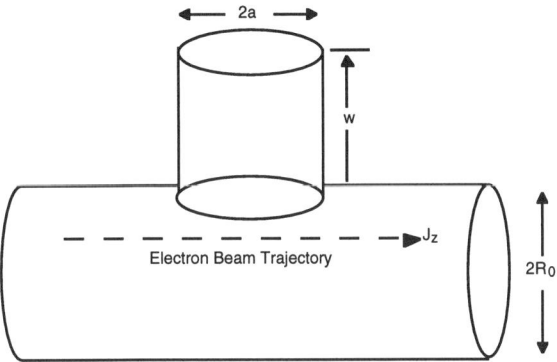

FIGURE 4. Geometry of the Beam Line Aperture Problem

An analytic expression for the low frequency transverse dipole impedance is given by Equation (5).

$$Z_\perp = j \frac{2a^3 Z_0}{3\pi^2 R_0^4} f(w) \qquad (5)$$

In Equation (5) Z_0 is the impedance of free space and the factor of $f(w)$ is a correction factor based on the length of the coupling guide and has a value of 0.562 for $w/a > 2$ as defined in Gluckstern (4). In our model, $w = 3$ cm. For the 8 mm coupling guide Equation (5) gives an inductive transverse impedance of 18.1 Ω and the TSAR model predicts an inductive transverse impedance of 18.3 Ω for frequencies up to 6 GHz. Figure (5) shows the TSAR results for the 3 cm large aperture case for the transverse impedance along with the low frequency asymptote for the reactive component of the impedance. The real part arises from the fact that the secondary coupling waveguide is only 3 cm long in the model and does not let low frequency fields evanesce sufficiently before reaching the MUR boundary condition on the coupling guide. This accounts for the resonance at the TE_{11} cutoff at 5.9 GHz.

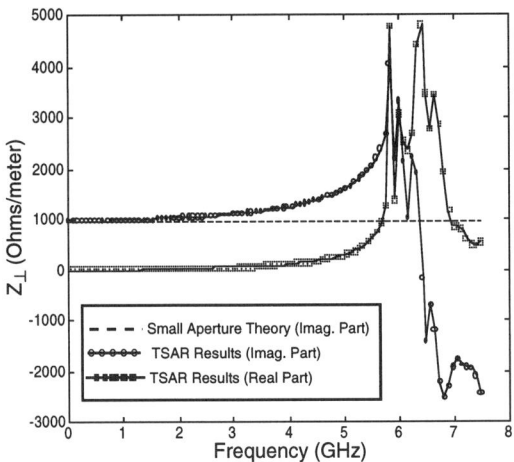

FIGURE 5. Transverse Impedance of the 3 cm Diameter Coupling Waveguide

SUMMARY AND CODE IMPROVEMENTS

While excellent results were obtained using TSAR for our geometries, 3-D cubical cell FDTD codes are extremely limited in the problems that can be modeled. New 3-D codes are being developed that will extend the complexity of structures that can be modeled. These new codes include such features as conforming grids, hybrid structured/unstructured grids, massively parallel capability, and new radiation boundary conditions. Conforming grids eliminate stair-casing errors associated with cubical cell codes. However, conforming unstructured nonorthogonal grids require typically 30 times the memory overhead compared to structured orthogonal grids. Hybrid grids can overcome this limitation by only using conforming grids near appropriate surfaces and structured grids elsewhere. Advanced radiation boundary conditions can also reduce the computational effort by allowing a closer truncation of the problem space.

REFERENCES

1. DeFord, J. F., Craig G. D., and McLeod, R. R. (1990), *The AMOS Wakefield Code*, Lawrence Livermore National Laboratory, Livermore, California, UCRL-102731.
2. DeFord, J. F., Craig G. D. *Particle Accelerators,* **37–38**, 111–121 (1992).
3. Ng, K-Y, *Particle Accelerators,* **23**, 93–102 (1988).
4. Gluckstern, R. L., *Phys. Rev. A*, **46**, 1106–1109 (1992).
5. Kurennoy, S. S., *Particle Accelerators*, **39**, 1–13 (1992).

Beamtracking in Cylindrical and Cartesian Coordinates

B. Schillinger, T. Weiland

*Technische Hochschule Darmstadt, Fachbereich 18, FG TEMF,
Schloßgartenstr. 8, 64289 Darmstadt, Germany*

Abstract. For the design of devices with circular optical axes, e. g. bending magnets or spectrometers, the use of cylindrical coordinates for field calculations could be favourable. Additionally, in case of applications like bending systems with nonorthogonal entry and exit faces, the coupling of cylindrical and cartesian coordinates improves the simulation of fringe fields.

In this context we have implemented a consistent coupling between the two coordinate systems and have extended the tracking code of the electromagnetic simulator MAFIA to cylindrical coordinates.

This extensions could be of interest for the calculation of transfer maps of ionoptical devices using the tracked particle orbit as reference trajectory and including fringe field effects in a more general manner.

We will give a short introduction to the extensions and show some examples for bending systems with nonorthogonal entries.

INTRODUCTION

For the design or analysis of charged particle optical systems the use of simulated electromagnetic field data becomes necessary, if the corresponding accelerating or bending fields are not given analytically. In this context the tracking mode of the MAFIA TS3-module has the facility of performing particle orbit calculations in arbitrary static or timeharmonic electromagnetic fields being calculated by the static module S or the eigenmode solver E (1).

In order to get a better treatment of bending systems or circular accelerator structures with azimuthal dependence we have now extended the tracking code to cylindrical coordinate systems. In addition, especially for bending magnets with nonorthogonal entry and exit faces, we introduce a coupling between cartesian and cylindrical coordinate systems.

In the following we first give a small introduction into the coupling between the coordinate systems, explain the main features of the tracking algorithm and give an application and some examples for orbit calculations.

© 1997 American Institute of Physics

COUPLED COORDINATE SYSTEMS

The Finite Integration Method on which MAFIA is based is well documented in the literature (2,3). To demonstrate the possible consistent coupling between a cartesian and a cylindrical coordinate system we only look at the Maxwell Grid Equations in the magnetostatic case:

$$Ch = j$$
$$\tilde{S}b = 0$$
$$b = \tilde{D}_\mu h \qquad (1)$$

The transformation of Maxwell's equations to a set of matrix equations based on a dual orthogonal grid system maintains all analytical properties of the electromagnetic fields. The so called curl matrix C and source matrix $\tilde{S}$ describe the topology of the grid G and the dual grid $\tilde{G}$. h represents the magnetic voltage and is defined as integral of a magnetic field strength along a mesh edge and b is the magnetic flux, which is the integral of a magnetic flux density over a mesh cell surface. The diagonal matrix $\tilde{D}_\mu$ contains the material properties.

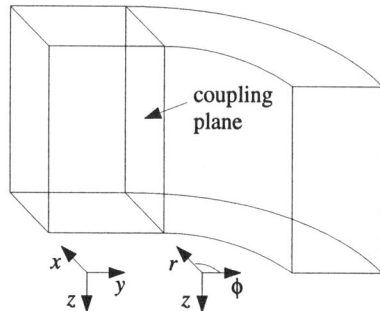

FIGURE 1: Possible combination of cartesian and cylindrical calculation domains.

Figure 1 introduces a possible combination of two calculation domains with different coordinate systems. The left side responds to a cartesian the right side to a cylindrical grid.

For both coordinate systems the curl and source matrices have up to the origin the same structure, so that no changes have to be made. Figure 2 shows a cut through the combined calculation domain with renamed directions u, v, w and w =const. Obviously the only change is the area of the shaded dual mesh cell, which means a modification of the flux b values in u and w direction located at the coupling plane.

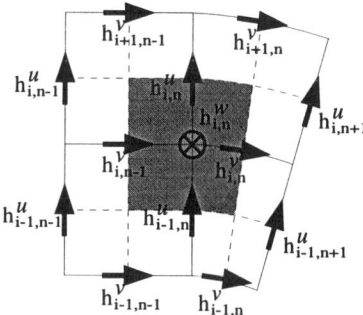

FIGURE 2: Location of the magnetic voltages at the coupling plane.

TRACKING METHOD

The integration of the motion equation

$$\frac{\partial}{\partial t}\vec{p} = q\left(\vec{E} + \vec{v} \times \vec{B}\right)$$

implemented in MAFIA for cartesian coordinate systems is documented in (4) and is based on the relativistic method introduced by Birdsall and Langdon (5). The algorithm is energy-preserving.

The extension to cylindrical coordinates only requires a modification of the position advance in the $r\phi$-plane. The velocity advance for static or time-harmonic electromagnetic fields is handled for both coordinate systems in the same way.

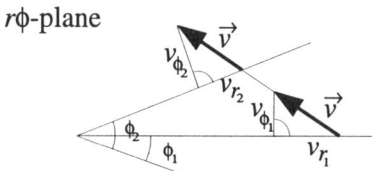

FIGURE 3: Position advance in the $r\phi$-plane.

Figure 3 shows the projection of the velocity vector in the $r\phi$-plane. The velocity vector is defined relative to a local cartesian coordinate system located at the particle position. After the position advance the local coordinate system has to be rotated according to the new position (5).

CALCULATION OF THE TRANSFER MAP

As an application of particle tracking we can use a single particle orbit, calculated in the midplane of a bending system, as reference trajectory and determine the ion optical transfer map of the system numerically. This is of great interest if the contributing fields, e.g. fringe fields, cannot be described analytically, but are available by numerical simulation. The method and examples are described in (6).

EXAMPLES

In the following we give two examples for particle orbit tracking: a bending magnet with nonorthogonal exit faces and a quadrupole lens calculated in a cylindrical coordinate system including the origin.

Bending Magnet with Nonorthogonal Exit

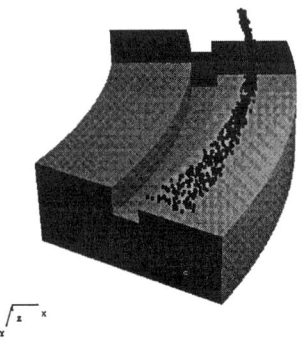

FIGURE 4: Deflected particle beam with a Gaussian gamma distribution.

Figure 4 shows a the yoke of a bending magnet with a nonorthogonal entry face (behind). The entry is modelled in cartesian and the middle part in cylindrical coordinates. Because of symmetry (we assume the entry and exit faces have equal angles) only a quarter of the structure has to be modelled. The magnetic field was calculated, solving the equation system (1), where the flux b includes the coupled geometry information.

To show a deflection, e.g. depending on the particle velocity, we calculated a beam with a Gaussian gamma distribution.

Quadrupole Lens

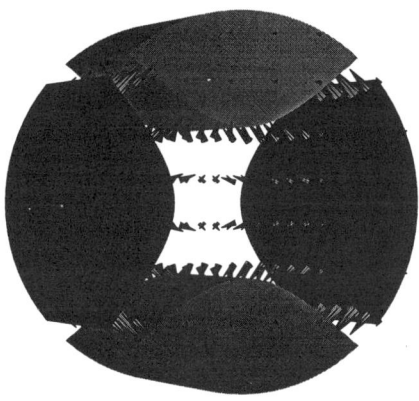

FIGURE 5: Quadrupole with electrostatic field.

Figure 5 shows the simple structure of a quadrupole with hyperbolic shaped pole faces generating an electrostatic field. The distance of a pole face to the center is 2 cm. We have calculated the focusing of an electron beam with diameter $d = 1$cm and $\gamma = 1.05$. Figure 6 shows the shape at the center. The phase space plots (figure 7) of the beam in front of and in the quadrupole illustrate the vertical focusing and horizontal defocusing effects.

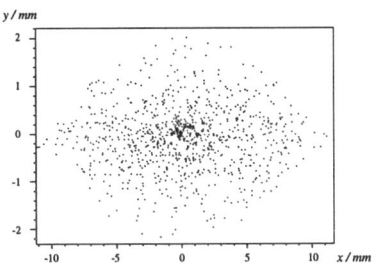

FIGURE 6: Beamshape at the center of the quadrupole.

CONCLUSION

We have introduced an extension of the MAFIA tracking code to cylindrical coordinates and combined calculation grids. Particle tracking can be done in static or timeharmonic electromagnetic fields. Additionally, the tracking code can be used for the calculation of the ion optical transfer map of a bending magnet.

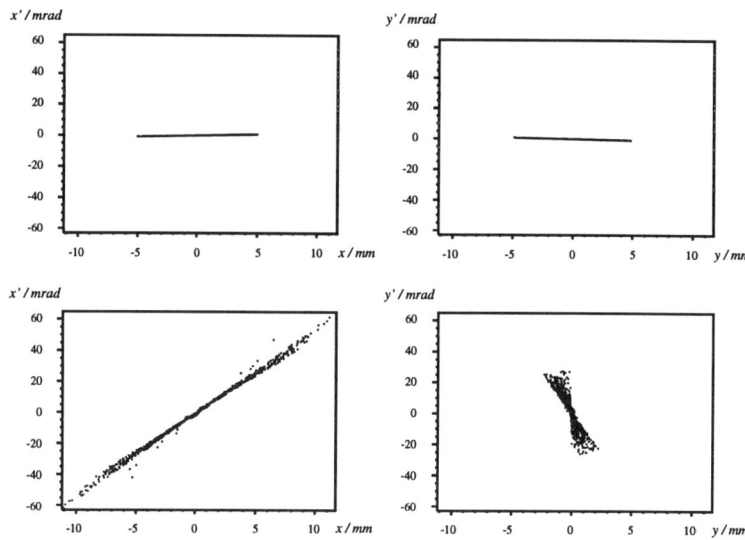

FIGURE 7: Phase space plots in front of (above) and at the center (below) of the quadrupole.

REFERENCES

1. The Mafia collaboration, *User's Guide MAFIA Version 3.20*, CST GmbH, Lauteschlägerstr.38, D-64289 Darmstadt.

2. Weiland, T., "A Discretization Method for the Solution of Maxwells Equations for Six-Component Fields", Electronics and Communication (AEÜ), Vol. 31, 1977, pp. 116.

3. Weiland, T., "On the Unique Solution of Maxwellian Eigenvalue Problems in three Dimensions", Particle Accelerators, Vol. 17, 1983, pp. 227–242.

4. Schütt, P., *Zur Dynamik eines Elektronen-Hohlstrahls*, Dissertation Hamburg, DESY M-88-03, 1988.

5. Birdsall, C. K., Langdon, A. B. Plasma Physics via Computer Simulation, Adam Hilger, IOP Publishing Ltd, 1991.

6. Schillinger, B., Weiland, T., Langenbeck, B., "Ion Injection through fringe fields of dipole magnets", presented at EPAC '96, Sitges, Spain, 1996.

Space Charge Effects With Periodic Focusing

Nathan Brown

G. H. Gillespie Associates, Inc.
PO Box 2961, Del Mar, CA 92014

Abstract. The dielectric response of a charged particle beam to a periodic focusing field enhances the effective focusing strength of the channel, reducing the matched beam radius and affecting the motion of halo particles. The dielectric response depends on the shape of the beam, the type of focusing and the ratio of the plasma frequency of the beam to the frequency of the focusing. TRACE 3-D is used to find the change in the effective focusing strength for a uniform-density beam in a quadrupole channel, showing agreement with theory for enhancements in the effective focusing strength of up to about 10%.

INTRODUCTION

The effective focusing strength of a periodic channel is an important factor for accelerator applications requiring high beam intensities, such as heavy ion inertial fusion, radioactive waste transmutation, spallation neutron sources, tritium production and muon production. Limiting currents have been found in the past using the smooth approximation [1] to find the effective focusing strength of a periodic channel which, along with the aperture, determines the current that can be transported through a given channel [2]. Accurate knowledge of the effective focusing strength is also important for matching. Transverse mismatch has been shown to be an important cause of halo production and the resulting particle losses [3-5]. Fractional current losses as small as 10^{-8} /m can result in radioactivation, inhibiting routine maintenance [6]; this can also be the limiting factor in the transport of intense, high-energy beams [7].

The dielectric response of a plasma to the periodic field of a Paul trap was recently shown to enhance the effective focusing strength of the trap [8]. The dielectric response results from the correlation between the oscillations in the space charge field and the periodic focusing field [9]. The dielectric response is shown here to increase the effective focusing strength of the channel, by an amount that depends on the shape of the beam, the type of focusing, and the ratio of the plasma frequency of the beam, ω_p, to the frequency of the focusing, ω. Results of the linear theory are compared to simulation results using TRACE 3-D.

THEORY

A beam in a periodic focusing channel experiences a fluctuating electric field $E(r, s_0)$, which consists of the fluctuating component of the focusing field, $E_{cf}(r,$

s_0), and small fluctuations in the space charge field, $E_{sf}(r, s_0)$. The position relative to the center of the beam is r, and the focusing is periodic in s, the longitudinal distance along the channel. Although particles with different longitudinal positions within the beam are at different phases in the periodic field, it is assumed that the effects of this are negligible so that the fluctuating fields can be written as periodic functions of the longitudinal position of the beam center along the channel, s_0. The focusing field and the space charge field are each divided into two parts so that the fluctuating components have an average value of zero and the steady-state components vary slowly or not at all with s_0.

The frequency of the focusing is $\omega = 2\pi v_B/S$, where S is the period of the focusing along the longitudinal direction and v_B is the beam velocity. In general there are three periods (S_x, S_y and S_z) and three frequencies (ω_x, ω_y and ω_z), one for each of three directions in Cartesian coordinates (x and y are transverse and z is parallel to the beam axis; for most practical applications $S_x = S_y$). The focusing field can be the result of electrostatic or magnetic quadrupole lenses, induction-acceleration gaps, and magnetic solenoids (if the beam is considered in the Larmor frame). It can also be the result of focusing by electromagnetic fields which are periodic in time and space, as in the case of radio-frequency quadrupole (RFQ) focusing. The focusing field is written as an electric field with the approximation that particle motion in the beam frame is nonrelativistic, so that magnetic focusing can be represented by equivalent electrostatic fields. The force resulting from the magnetic field of the beam is included in the self electric field (the space charge field) with the same approximation. Unless otherwise stated, all quantities are considered in the lab frame. With RFQ focusing and induction-acceleration gaps, it is assumed that acceleration along the longitudinal direction is slow enough that it can be treated as adiabatic, and that the beam is in phase with the time-varying field so that the focusing field can be treated as periodic only in longitudinal distance along the channel.

The effective focusing field can be found from the average field of a particle due to its motion in the periodic field [10]. The motion of a particle in the periodic field is first found with the fluctuating field as a function of position fixed at $E_f(r, s_0) = E_f(r_0, s_0)$, where r_0 is the position of the particle averaged over a period. The resulting particle position is $r_0 + \delta r$; the first-order variation in the position of the particle resulting from the fluctuating field is δr. The effective field that results from the fluctuating field is then found to first order from

$$E_{\textit{eff}} = \langle E_f(r, s_0) \rangle = \langle (\delta r \cdot \textbf{\textit{DEL}}_0) E_f(r_0, s_0) \rangle, \qquad (1)$$

where $\textbf{\textit{DEL}}_0$ is the gradient with respect to r_0 and the brackets represent averages over a focusing period. The effective field of Equation (1) has previously been derived for a Paul trap without space charge [10] and for a periodic focusing channel without space charge fluctuations [11]. Solving for δr from the fluctuating field and substituting into Equation (1) gives an effective field of

$$E_{\textit{eff}} \approx \frac{q}{\gamma m v_B^2} \left\langle \left(\int_0^{s_0} \int_0^{s_0'} E_f(r_0, s_0'') ds_0'' \, ds_0' \cdot \textbf{\textit{DEL}}_0 \right) E_f(r_0, s_0) \right\rangle, \qquad (2)$$

where q and m are respectively the particle charge and mass, and $\gamma = (1-v_B^2/c^2)^{-1/2}$ is the relativistic factor. In a quadrupole channel the steady-state component of the transverse focusing field is zero, so the field of Equation (2) is the total effective focusing field. For transverse focusing by solenoids or longitudinal focusing by induction-acceleration gaps, the steady-state component of the focusing field is typically much larger than the effective field of Equation (2), so that the dielectric response, which affects only the fluctuating component of the field, has much less effect than in a quadrupole channel with the same frequency and focusing strength.

The dielectric response occurs through the effect of space charge fluctuations on $E_f(r_0, s_0)$. This will be found first for the core and halo of a uniform-density continuous beam with a diffuse halo in a quadrupole channel with average axial symmetry. The same theory has previously been applied to the core of a uniform-density bunched beam in a quadrupole channel [9].

The electric field in a transverse direction (x) of a continuous, uniform elliptic beam with current I and velocity v_B is $E_{sx} = Ix/(\pi\varepsilon_0\gamma^2 v_B x_m(x_m+y_m))$, where x_m and y_m are respectively the beam envelopes in the x and y directions, and ε_0 is the permittivity of free space [12]. In a quadrupole channel which has average axial symmetry, the fluctuations in the two transverse directions have the same magnitude and functional form, and are out of phase by π. The beam envelopes can then be written as $x_m = x_{m0} + \delta x_m$ and $y_m = x_{m0} - \delta x_m$, where x_{m0} is nearly independent of s_0 and δx_m has an average value of zero. The electric field can then be split into a steady-state component and a fluctuating component with a linear expansion in δx_m. The resulting fluctuating field component is

$$E_{sfx} = \frac{-I\delta x_m x}{2\pi\varepsilon_0\gamma^2 v_B x_{m0}^3}. \tag{3}$$

Using Equation (3), setting $E_{fx} = E_{sfx} + E_{cfx}$, where E_{fx}, E_{sfx} and E_{cfx} are respectively the x components of the fluctuating parts of the effective focusing field, the space charge field and the focusing field, and solving for δx_m, gives

$$E_{fx} = E_{cfx}/\varepsilon, \tag{4}$$

in which ε (by definition) is the dielectric constant. The dielectric constant for this case is

$$\varepsilon = 1 - \Gamma\frac{\omega_p^2}{\omega^2}, \tag{5}$$

in which $\Gamma = 1/2$, $\omega_p = (q^2 n_s/\varepsilon_0\gamma m)^{1/2}$ is the plasma frequency, and n_s is the particle number density. Equation (5) can be used for other types of beams and for halos with different values for Γ, depending on the geometry.

In deriving Equations (4) and (5) it was assumed that $\omega_p^2/\omega^2 \ll 1$, and that fluctuations in the focusing fields and space charge fields occur sinusoidally with the same frequency. For most focusing channels, the fluctuating component of the focusing field is not a sinusoidal function of longitudinal distance along the channel. In order to define the dielectric constant, the fluctuations are approximated as sinusoidal functions of s_0. Small deviations in the functional form are assumed not to have a significant effect on the dielectric response of the beam.

Since ε<1, Equation (4) represents an enhancement of the periodic focusing field. This effect results from the fact that the beam has maxima in its extent along any axis, and minima in the magnitude of its space charge field, at longitudinal positions along the channel where the focusing field along that axis is at a maximum. Likewise, the beam has maxima in the magnitude of its space charge field where the focusing field is at a minimum. Fluctuations in the space charge field are therefore correlated with the focusing so that they enhance the effective focusing field.

Substituting Equation (4) into Equation (2) leads to the conclusion that the effect of the dielectric response of the beam is to increase the effective transverse focusing field of a quadrupole channel by the factor $1/\varepsilon^2$. For example, a continuous beam in a quadrupole channel with $\omega_p/\omega = 0.2$ has a dielectric constant of 0.98. The dielectric response increases the effective focusing field of this channel by about 4%.

The same technique can be used to find the effect of the dielectric response on halo particles surrounding the uniform-density core of a continuous beam. The beam has average axial symmetry, and variations in x_m and y_m are out of phase by π. The dielectric response of halo particles arises from the periodic motion of the particles relative to the beam axis, and also from the periodic variations in the shape of the core. The position of a halo particle is written as $(x, y) = (x_0 + \delta x, y_0 + \delta y)$, and the envelopes are again $x_m = x_{m0} + \delta x_m$ and $y_m = x_{m0} - \delta x_m$. Using the electric field along a transverse direction outside of a continuous, uniform-density elliptic beam core [12], the self electric field can be written in terms of a steady-state component and a fluctuating component with linear expansions in the fluctuating quantities. Solving for the resulting particle motion by the same method as in the previous case, the fluctuating field is again described by Equations (4) and (5). In this case

$$\Gamma = \frac{x_{m0}^2\left(x_0^2 - y_0^2\right)}{\left(x_0^2 - y_0^2\right)^2} + \frac{x_{m0}^4\left(3y_0^2 - x_0^2\right)}{\left(x_0^2 + y_0^2\right)^3}. \tag{6}$$

TRACE 3-D RESULTS

Any simulation that uses periodic focusing should exhibit this effect if the beam has a sufficient space charge field and the frequency of the focusing is low enough. TRACE 3-D is used here to find changes in the effective focusing strength with changes in the frequency of the focusing with fixed beam current and energy. The approximation of linear space charge forces is also used in TRACE 3-D with a uniform-density ellipsoidal beam. Variations in the space charge field, however, are assumed to be sinusoidal in the theory and are treated self-consistently with variations in the beam size and shape in TRACE 3-D. The periodic focusing was also assumed to be sinusoidal in the theory, and consists of step function (square wave) lenses in TRACE 3-D. Comparison of the theory and simulation therefore provides tests of both of these approximations.

A continuous beam is simulated with TRACE 3-D with a current of $4I/3\beta$, where I is the continuous beam current and $\beta = v_B/c$, and with a bunch that is much

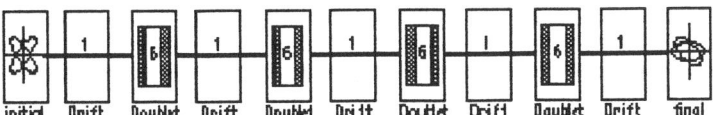

Figure 1: A periodic quadrupole channel from MacTrace™

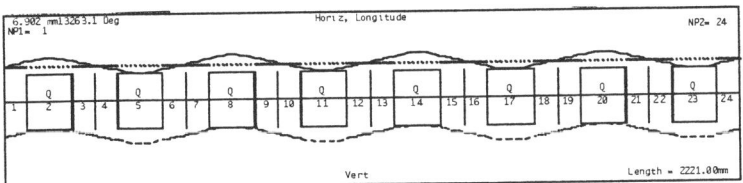

Figure 2: Horizontal (solid line) and vertical (dashed line) envelopes from TRACE 3-D in the channel from Figure 1

Table 1: TRACE 3-D results compared with the linear theory

ω_p/ω	Period Length	TRACE-3D k_0^2	% change, theory	% change, TRACE-3D
0	-	2.76 /m²	-	-
0.083	0.22 m	2.79 /m²	+1%	+1%
0.21	0.56 m	2.87 /m²	+5%	+4%
0.29	0.74 m	2.98 /m²	+9%	+8%

longer than the channel. Figure 1 shows a channel created with MacTrace™ [13], consisting of four quadrupole doublets with drifts. Figure 2 shows the horizontal and vertical envelopes for a matched beam in the same channel. In order to see the effect of the dielectric response, the channel period was varied and the lens fields were adjusted in each case to maintain the same effective focusing in the absence of space charge. This effective focusing field can be found from the first-order particle motion in the periodic field using Equation (2). For a channel consisting of step function quadrupole lenses, the resulting focusing strength is

$$k_0^2 = \frac{q^2 B_0'^2 S^2 f^2}{8 m^2 v_B^2 \gamma^2} \left(\frac{1}{2} - \frac{f}{3} \right), \qquad (7)$$

where B_0' is the magnetic field gradient, S is the length of a channel period, and f is the fraction of the channel length that is occupied by lens fields, which are assumed to be evenly spaced.

In the presence of space charge, the focusing strength is found from the matched beam envelope equation using the average beam radius. The fractional change in the focusing strength is shown in Table 1 as a function of the ratio of the plasma frequency of the beam to the frequency of the focusing. The beam in each case consists of 2 Mev protons, and has a current of 100 mA, an emittance of 2 mm-mrad, and an average radius of approximately 2.8 mm. Good agreement with the linear theory is found for enhancements of the effective focusing strength of up to about 10%.

CONCLUSION

Two uniform-density beams with the same energy, current, space charge tune depression, and aspect ratio will have different matched beam properties if one is in a periodic quadrupole channel and one is in a channel with continuous focusing, if both channels have the same effective focusing strength in the absence of space charge. With space charge, the effective focusing strength of the periodic channel is increased over that of the continuous channel, resulting in a smaller phase advance per period, a higher average beam density and a smaller average beam radius.

Simulations using TRACE 3-D have shown that the linear theory, in which particle oscillations, space charge field oscillations and lens fields are taken as sinusoidal, is accurate with predicted enhancements in the effective focusing strength of up to about 10%.

ACKNOWLEDGMENT

This work is supported by G. H. Gillespie Associates, Inc., internal research and development funds.

REFERENCES

1. M. Reiser, Part. Accel. **8**, 167 (1978), and references therein.
2. M. Reiser, J. Appl. Phys. **52**, 556 (1981), and references therein.
3. C. L. Bohn, Phys. Rev. Lett. **70**, 932 (1993).
4. T. P. Wangler, *Proceedings of the 1993 Computational Accelerator Physics Conference*, edited by R. Ryne (American Institute of Physics, New York, 1993), p9.
5. R. L. Gluckstern, Phys. Rev. Lett. **73**, 1247 (1994).
6. J. M. Lagniel, Nucl. Inst. Methods Phys. Res., Sect. A **345**, 46 (1994).
7. R. A. Jameson, *Advanced Accelerator Concepts*, edited by J. Wurtele (American Institute\ of Physics, New York, 1993), p969.
8. K. Avinash, A. K. Agarwal, M. R. Jana, A. Sen and P. K. Kaw, Phys. Plasmas **2**, 3569 (1995).
9. N. Brown, "Dielectric Response of Particle Beams to Periodic Focusing," Physical Review E, in press.
10. G. Schmidt, *Physics of High Temperature Plasmas* (Academic, New York, 1979), p47.
11. M. Reiser, *Theory and Design of Charged Particle Beams* (Wiley, New York, 1994), Section 4.4.
12. O. D. Kellogg, *Foundations of Potential Theory* (Dover, New York, 1953), p194.
13. G. H. Gillespie and B. W. Hill, "A Graphical User Interface for TRACE 3-D Incoporating Some Extert System Type Features," *1992 Linear Accelerator Conference Proceedings*, AECL-10728, 787 (1992).

SCD-Beam Main Regularities In Beginning Part of High-Current Proton Linac

Boris I. Bondarev, Alexander P. Durkin

Moscow Radiotechnical Institute
Warshawskoe shosse, 132, 113519, Moscow, Russia

Abstract. Space charge-dominated (SCD) beam main regularities in the beginning part of high-current proton linac are presented. Charge redistribution during beam transport and bunching are considered. Code tools ZHALO and KERN+HALO are used for physical process better understanding and scientific visualization of different factor influence on SCD-beam transport and bunching.

The main problem for high-current, high-energy CW ion linac is beam acceleration without particle losses. Even small losses of high power beam leads to intolerable radioactive pollution of linac making impossible service activities of the linac. It is why intense beam physics investigations are a major focus of interest for all accelerator centers. In present report the authors state in detail their own understanding of space charge-dominated (**SCD**) beam problem which working out during recent MRTI investigations with LANL attention and support [1-2].

SCD-BEAM TRANSPORT

We have a clear knowledge of matched (or ideal) beam only in case when space charge does not change a transverse oscillations frequency. The matched (or ideal) beam is defined as a beam which transverse behavior is repeated (or has a smooth change) from one period to the next.

The question *"What a matched beam imply?"* is arisen when the above definition is extended on the case when transverse oscillation frequency changes essential due to space charge. We can denote that ideal beam has well-known K-V distribution and this distribution must be used for focusing field calculation and for a choice of bore radius. But real beam distributions differ from K-V one and its redistribution during beam transport leads to beam size and emittance growths. It means that such ideal beam definition results in an serious error of bore radius.

The **SCD**-beam investigations in channel with different transverse distributions of input beam were made by authors in order to answer on the above question. A simple model must be chosen for better understanding of process of **SCD**-beam redistribution such as a continuous cylindrical beam transporting in longitudinal magnetic field. In this case a Coulomb force calculation is very much simplified. Within the context of this model there are only a few characteristics for understanding a cause and an effect of halo formation. Only distributions with beam density growth from the origin to outlying area were considered as initial ones [1].

The equations of motion have a unitless form [1] after transition to unitless variables $z = L\tau$, $x = A\widetilde{x}$, $y = A\widetilde{y}$, where L is focusing period length, $A = (\varepsilon L/(\beta\gamma))^{1/2}$, ε is beam emittance

$$\begin{cases} \widetilde{x}'' - 2\Lambda\widetilde{y}' - \alpha \dfrac{Q(\widetilde{r})}{\widetilde{r}^2}\widetilde{x} = 0 \\ \widetilde{y}'' + 2\Lambda\widetilde{x}' - \alpha \dfrac{Q(\widetilde{r})}{\widetilde{r}^2}\widetilde{y} = 0 \end{cases}$$

$$\Lambda = \frac{eB(z)L}{2m_0 c\beta\gamma}, \qquad \alpha = \frac{2IL}{I_0 \varepsilon (\beta\gamma)^2}$$

This set of equations will be considered as base for further considerations. The magnitude of $B(z)$ is constant inside solenoidal lens and equal zero outside of them.

It is evident from general form of equations that there are only two parameters Λ and α which define the **SCD**-beam transverse form in the case when a structure of focusing period is preset. Uniform and Gauss distributions as well as its different combinations are used as initial ones.

The main regularities of **SCD**-beam transporting can be formulated as ten following statements [1]:

Statement 1. High-density core and low-density halo with particle active interchange are established in every **SCD**-case. The value of a beam radius in point where Coulomb force takes its maximum is defined as "core radius".

Statement 2. Most of core particles are "ex-halo" or "coming-halo" ones which income from halo in previous instant of time or will emerge from core in next instant of time. Core residents are approximately one third of all particles.

Statement 3. Uniform charge distribution inside core is established in every SCD-case.

Statement 4. Final steady states with uniform distribution of core charge are states with minimal potential energy of Coulomb field.

The transition from the SCD-beam initial state into a final steady state is accompanied by particle kinetic energy increasing and emittance growth.

Statement 5. A steady state with small core size oscillations can be established in a **SCD**-case. Such beam we will be nominated as matched one.

Generally a procedure of matched beam redistribution has been going on the following manner. At the first stage a redistribution from initial to the steady state with an uniform core distribution take place. The potential energy difference between initial and final states (always positive for considered class of distributions) transforms into kinetic energy and is accompanied by an emittance growth. In the steady state only small fraction of particles (about 30%) never escapes the core, other particles (about 70%) can turn up in core or in halo.

The core radius is kept constant during beam redistribution. It means that the core radius can be determined from initial redistribution. The matched focusing field value can be determined from a core radius and a core emittance. In the point of a crossover (an average radial velocity is equal to zero) the core emittance is equal to an straight ellipse area with r and $k\beta_r$ as semi-axes, where β_r is rms value of radial velocity and $k^2 = 3$ is a square of boundary value to rms one ratio for uniform distribution. As is evident from the foregoing the beam kinetic energy is increasing in the steady state on the difference of potential energies between the initial and final space charge distributions. This difference will be called as "*a beam heating up*".

The unmatched beam investigations offer a clearer view of how the core-halo is formatted as well as further regularities of the process:

Statement 6. During the process of charge redistribution in mismatched SCD beam core oscillations are drastically damping.

Statement 7. If potential Coulomb energy of input beam is much larger that the same energy of beam steady state, core is automatically matched with channel. There is a damping of core oscillation in continuos magnetic field. In periodical field only one clearly defined harmonic (in close agreement with external force harmonic) stays in core oscillation spectrum.

Statement 8. A factor of 4 is a sufficient estimate for halo-core radii ratio.

It is evident that a halo is formatted from particles which increase their kinetic energies during transverse oscillations. A separate particle motion inside the field of oscillated uniform core was considered in order to study of mechanism of particle energy increasing. A comparison of particle potential energy with its kinetic one before and after oscillated core passing gives a possibility to state that:

Statement 9. Energy mechanism for halo formation is kinetic energy growth in the case when a halo particle have passed through core concurrently with core ultimately decreasing. A potential energy decreasing is much smaller then kinetic energy increasing and total energy is increased with core size decreasing. The duration of particle being outside of the core is increased with increasing of the particle total energy. That is the phase of the core envelope oscillation will be changed in the next core traversal. It means there is a mechanism which desynchronized the oscillations of core envelope and halo particles. It limits a kinetic energy of halo particle and as a result limits a transverse size of the halo.

Statement 10. The choice of focusing channel parameters must be made from a desired size of a core and the choice of a bore radius from a value four times over (halo size maximum).

SCD-BEAM BUNCHING

An adiabatic bunching section must be presented at a high-current ion linac for beam loss prevention. The problem is to choose an optimal length of the section as well as optimal values for synchronous phase and accelerating field amplitude. In order to clear up the main regularities of the bunching process the simplest model was used by analogy with the previous case. In the context of this model the beam without transverse motion and with fixed transverse size is moving inside a channel with bore radius R. Synchronous phase is equal -90 degrees and the beam is bunching without acceleration. During bunching process the uniform beam gets a longitudinal charge density modulation. The distribution of potential can be obtained in the form of an expansion into a series by solving a Poisson equation for the beam with periodical charge density modulation and uniform density in any transverse crossection [3,4].

Using unitless variables $\psi = \dfrac{2\pi(z-z_s)}{L}$, $L = \beta\lambda$, τ - phase of longitudinal oscillations, equation for bunching process analysis can be obtained [2]

$$\frac{d^2\psi}{d\tau^2} = -H(\tau)\cdot\sin\psi + E_{coul}$$

where function $H(\tau)$ describes the general rule of the first harmonic of accelerating field growth up to maximum value E_m.

Detail analysis developed in [2] shows that

$$E_{coul} = \frac{2ID_0}{\varepsilon_0 c\beta^2 \lambda E_m}\sum_{k=1}^{\infty} q_k F_k(R_b/L, R_b/R_a)\cdot\sin(k\psi)$$

where R_a - aperture, R_b - beam radius, q_k - harmonic coefficient in the expansion of unitless density $\dfrac{\partial}{\partial z}\rho(0,z)$, $F_k \sim \left(\dfrac{L}{R_b}\right)^2$ and $0 \leq F_k \leq 1$, $D_0 = \dfrac{1}{4}\left(1 - 2\ln(R_b/R_a)\right)$

The general view of the above equation shows that in the context of given R_b/L and R_b/R_a the bunching process depends on the single parameter

$$\alpha = 2ID_0 / (\varepsilon_0 c\beta^2 E_m \lambda)$$

By this means the task is reduced to the optimal form of function $H(\tau)$ which allows maximal α for the bunching process without beam losses.

The introduced Coulomb parameter α differs from commonly accepted one. It is evident to factor out the term L/R_b from the equation for bunching process analysis and join it to α. The new approach underlines that the Coulomb field determines by parameter α but relations R_b/L and R_b/R_a shows only geometrical similarity. The Coulomb parameter decreased with an energy growth whereas harmonic factors F_k are increasing. But their growth is limited because the values are between 0 and 1.

The image-based simulation code package **ZHALO** was generated for investigating space charge-dominated beam bunching. In the context of the bunching model considered here, the Poisson equation is solved with initial conditions that correspond to a mono-energy beam with a uniform phase distribution across the full RF period. The function $H(\tau)$ was taken in the form

$$H(\tau) = E_0/E_m + (1 - E_0/E_m)(\tau/\tau_m)^2$$

where the coordinate τ_m and beginning level of field E_0/E_m are given.

To determinate R_d/L and R_b/R_a a beam with parameters typical for a high-current CW linac was chosen: $f = 350$ MHz, $R_a = 3$ mm, beam energy $W = 0.1$ MeV, $R_b/R_a = 0.5$. (In the context of the chosen parameters, the value ($\alpha = 1$ corresponds to current 0.4 A if $E_m = 1$ MV/m). When the investigation was made with a growing RF field, $E_0/E_m = 0.2$ was taken.

A visual inspection of a **SCD**-beam bunching with increasing of α parameter step by step gives us a general view of the process depending on α.

It is evident that particles with initial positions near the bunching center achieve this center at an earlier instant of time and form a Coulomb potential barrier for other particles. During the first quarter of the phase oscillation period this barrier is increasing and each next particle meets a bigger opposition then the previous one. It means that during the first quarter of the phase oscillation period a high-density core is formed inside the bunch.

Starting from some value of α the energy needed for a particle to overcome the Coulomb barrier may be comparable with the kinetic energy that the RF field has given to this particle. The typical points of inflection appear in the phase portrait. For this and for all bigger α-values the beam can be called a **SCD**-beam. On further α growth particle energy loss becomes larger as the center is approached. In its turn, the phase oscillation amplitude can be increased owing to added momentum which a particle has assumed in escaping from the core. As it will be shown below at this stage the fraction of particles which escape the phase period rises sharply. We call such particles "lost" ones.

The final stage of α growth leads to the situation where particles are completely decelerated by the core Coulomb field and the particle cannot pass through the core.

Nevertheless, as also indicated below beam bunching can take place in this case as well. To do this, it is necessary to "match" the rate of Coulomb barrier growth with the growth of particle kinetic energy. The "matching" can be made by the choice of E_o/E_m and τ_m values.

A comparison between beam bunching in a channel where RF amplitude is constant (case A) and in a channel where RF amplitude is build-up (case B) shows that in the first case:

- deformations of phase portrait start from $\alpha = 0.15$;
- the kinetic energy becomes comparable with the potential energy for α near to 0.25;
- for $\alpha > 0.5$ lost particle fraction rises sharply.

In the case B the essential influence of the core starts from a smaller kinetic energy of particles than in case A. In case B the total core charge is smaller. For the same α the added momentum from the core is small and it can not raise the amplitude of particle oscillations by a large margin. Although the Coulomb potential energy exceeds the particle kinetic energies, the bunching process takes place practically without losses up to $\alpha = 1.2$ (for $E_o/E_m = 0.2$ and $\tau_m = 1$). It means that the limit current is doubled as compared to case A. The increasing of τ_m up to 1.5 increases the α limit up to 1.4 but no more. Particles can not pass the core but the attraction momentum does not let them escape from the stability zone of bunching.

A second trend has emerged from observation of the Coulomb field harmonic spectrum for a "frozen" beam. With increase in α the first harmonic of the Coulomb field far exceeds the other harmonics. Its amplitude approximates that of the external field. It may be concluded that the **SCD**-beam automatically makes Coulomb field distribution similar to external field distribution. Such a high-current effect can be considered as a beam "auto-matching" with a channel. Along with this for small α the higher harmonics are comparable with the first one and local resonance effects can be caused in the beam transverse motion.

REFERENCES

1. B.I.Bondarev, A.P.Durkin, B.P.Murin, "Halo Production in Charge-Dominated Beams. Single-Particle Interactions". Contract 9-XG3-5167H-1 between LANL and MRTI, Phase 3, Moscow 1993.

2. B.I.Bondarev, A.P.Durkin and R.A.Jameson, "Space Charge-Dominated Beam Bunching", presented at Second International Conference on Accelerator-Driven Technologies and Applications, Kalmar, Sweden, June 3-7, 1996.

3. B.I.Bondarev, A.P.Durkin and B.P.Murin, "Intense Ion Beam Transport and Space Charge Redistribution" presented at Linac96 Conference, Geneva, August 26-30, 1996.

4. B.I.Bondarev, A.P.Durkin and B.P.Murin, "Study of Space Charge-Dominated Beam Bunching" presented at Linac96 Conference, Geneva, August 26-30, 1996.

High-Order Space Charge Effects Using Automatic Differentiation

Michael F. Reusch and David L. Bruhwiler
Northrop Grumman Corporation, Advanced Technology and
Development Center, 4 Independence Way, Princeton NJ 08540
Computer Accelerator Physics Conference Williamsburg, Va. 1996

Abstract

The Northrop Grumman Topkark code has been upgraded to Fortran 90, making use of operator overloading, so the same code can be used to either track an array of particles or construct a Taylor map representation of the accelerator lattice. We review beam optics and beam dynamics simulations conducted with TOPKARK in the past and we present a new method for modeling space charge forces to high-order with automatic differentiation. This method generates an accurate, high-order, 6-D Taylor map of the phase space variable trajectories for a bunched, high-current beam. The spatial distribution is modeled as the product of a Taylor Series times a Gaussian. The variables in the argument of the Gaussian are normalized to the respective second moments of the distribution. This form allows for accurate representation of a wide range of realistic distributions, including any asymmetries, and allows for rapid calculation of the space charge fields with free space boundary conditions. An example problem is presented to illustrate our approach.

* Work supported by Northrop Grumman Corporation

Introduction

Since the seminal work of Berz [1] and Forest [2], automatic differentiation has become a staple of accelerator design but the incorporation of space charge effects in such codes has never been easy. Amplifying on ideas in Ryne's Charlie code [3], we are proposing a strategy for computing high order space charge effects for complex particle distributions in a new Fortran 90 Taylor-map version of the Northrop Grumman Topkark code. We are implementing this method and will benchmark it against the ray tracing version of the code under known situations.

Ultimately, the method may prove useful in the study of beam halo formation in the 3D bunched beams relevant to APT. It is very difficult to simulate this problem with a ray tracing code since the fraction of particles in the halo is only about one in 10^8. In contrast, the collective space charge effects of a 3D bunched beam may be well represented by a properly generated Taylor map. This map and the associated particle distribution function then allow a quasi-analytical study of the halo orbits and halo formation.

Two versions of Northrop Grumman's Topkark code [4] exist, a ray tracing version that pushes individual particle orbits and a Taylor map version that uses automatic differentiation to do arbitrary order perturbation analysis. Both the ray-tracing and map versions of Topkark code have recently been rewritten in Fortran 90 to take advantage of the operator overloading capabilities of this language.

We have consistently attempted to keep both versions of the code similar in capability and on par with each other in terms of implemented accelerator elements since each code has unique and complementary areas of application in the optical design of a given accelerator. This process has been not always been easy, as rarely was there time to implement a novel diagnostic or accelerator element in both codes. However, using the operator overloading capabilities of Fortran 90 this task is much simplified as the identical routine can often be used in both codes with only minor changes. A second advantage in moving to Fortran 90 is the enormous simplification that it allows in Topkark's automatic differentiation library. The application of automatic differentiation to the treatment of three dimensional space charge forces, using what we believe to be a novel technique, is the main subject of this paper.

Explanation of the technique

Any function of compact support can be written as the product of a multivariate Taylor series or, what is equivalent a series of Hermite polynomials, times a Gaussian or Maxwell-Boltzmann distribution, since the Hermite polynomials form a complete set of functions in this space. Thus, we assume that the particle distribution can be written as

$$f(\varsigma) = T(\varsigma) \; \exp\left(-\frac{1}{2}\sum_{i,j} \sigma^{-1}_{i,j} \varsigma_i \varsigma_j\right). \qquad (1)$$

Here, $\varsigma_i = x, P_x, y, P_y, z,$ or P_z as $i = 1,...,6$, are the 6 independent position and momentum phase space variables and. $T(\varsigma)$ is a truncated Taylor series, σ^{-1} is the inverse of the positive definite and symmetric "sigma" matrix whose entries may be set to the second moments of the distribution. The charge density $\rho(x,y,z)$ is the elementary charge q times the integral of $f(\varsigma)$ over the momentum variables. Evaluating integrals of this type when the distribution is Gaussian or a Gaussian times a Taylor series is made simple by a differentiation-under-the-integral-sign trick. For even n,

$$I_n = \frac{1}{\sqrt{2\pi}a} \int_{-\infty}^{\infty} dx \; x^n e^{-\frac{1}{2}\left(\frac{x}{a}\right)^2} = (n-1)!! a^n. \qquad (2)$$

For odd n the result is zero. With the boundary condition of vanishing Φ at infinity, the solution to the Poisson problem is

$$\Phi(\vec{r}) = \frac{1}{4\pi\varepsilon_0} \int d^3r' \frac{\rho(\vec{r}')}{|\vec{r}-\vec{r}'|}. \tag{3}$$

For convenience, we consider coordinates that have been rotated to the frame where σ is diagonal via an orthogonal transformation. If $\rho(x,y,z)$ is as given above, we have to evaluate integrals of monomials times Gaussians of the form

$$I_{l,m,n}(\vec{r}) = \int d^3r' \frac{x'^l y'^m z'^n \exp\left(-\frac{1}{2}\left(\frac{x'^2}{a^2} + \frac{y'^2}{b^2} + \frac{z'^2}{c^2}\right)\right)}{\sqrt{(x-x')^2 + (y-y')^2 + (z-z')^2}}. \tag{4}$$

We note that by introducing the linear term $\alpha x' + \beta y' + \gamma z'$ into the argument of the exponential, differentiating with respect to α, β, and γ, completing the square, making a linear change of variables and taking the appropriate limit, we have

$$I_{l,m,n}(\vec{r}) = \lim_{\alpha,\beta,\gamma \to 0} \frac{\partial^l}{\partial\alpha^l} \frac{\partial^m}{\partial\beta^m} \frac{\partial^n}{\partial\gamma^n} \left(\exp\left(\frac{(\alpha^2 a^2 + \beta^2 b^2 + \gamma^2 c^2)}{2}\right) G(x,y,z,\alpha,\beta,\gamma) \right) \tag{5}$$

where,

$$G(x,y,z,\alpha,\beta,\gamma) = \int d^3r' \frac{\exp\left(-\frac{1}{2}\left(\frac{x'^2}{a^2} + \frac{y'^2}{b^2} + \frac{z'^2}{c^2}\right)\right)}{\sqrt{(x-x'-\alpha a^2)^2 + (y-y'-\beta b^2)^2 + (z-z'-\gamma c^2)^2}}. \tag{6}$$

Since the square root kernel of this integral is symmetric in x and x' etc., the derivatives of $G(x,y,z,\alpha,\beta,\gamma)$ with respect to α, β and γ are simply related to those with respect to x, y, and z, namely

$$\frac{\partial G}{\partial \alpha} = -a^2 \frac{\partial G}{\partial x} \tag{7}$$

In the limit of α, β, γ going to zero, these are the derivatives of the potential for the simple Gaussian distribution

$$G_0(x,y,z) = \int d^3r' \frac{\exp\left(-\frac{1}{2}\left(\frac{x'^2}{a^2}+\frac{y'^2}{b^2}+\frac{z'^2}{c^2}\right)\right)}{\sqrt{(x-x')^2+(y-y')^2+(z-z')^2}}. \quad (8)$$

The required derivatives of G_0 may be automatically calculated by the automatic differentiation routines of a Taylor mapping code. The extension to treating a distribution function that is a Taylor or Hermite series times a Gaussian is accomplished by forming the appropriate weighted sums of these derivatives. The appropriate sums for a particular monomial integral $I_{l,m,n}$ are found from (6). For example, the normalized (a=1) 1D forms are

$$\lim_{\alpha \to 0} \frac{\partial^n\left(e^{\alpha^2/2} G(x,\alpha)\right)}{\partial \alpha^n} = \sum_{m=0}^{n} \binom{n}{m} \begin{Bmatrix} (n-m-1)!!, & n-m \text{ even} \\ 0, & n-m \text{ odd} \end{Bmatrix} (-1)^m \frac{\partial^m G_0}{\partial x^m}. \quad (9)$$

Having produced the $I_{l,m,n}$ for an arbitrary monomial times Gaussian distribution, constructing Φ for the more general charge distribution of the type of Eq. 1 is accomplished by a straightforward summation of terms. Noting that each differentiation of a Taylor series reduces its' order by one, the initial computation of G_0 and the linear combinations of Eq. 9 must be carried out to twice the order of accuracy desired. Similar remarks apply to the second moments. Of course, this order doubling can be done only where required.

Formulas (3) through (9) can be verified to apply to numerical computation of the electric field components, virtually without modification. Above first order, it is more efficient to compute the integral of the potential rather than those for the three components of the electric field. Since the accuracy of the fields produced by differentiation is reduced by one order, one can always compute the potential to one higher order and differentiate.

Pushing the particle distribution

The main result of the Taylor map code is the non linear map $\mathcal{M}(t)$, six truncated multivariate Taylor series, one for each independent coordinate, that take initial coordinates to final coordinates $\zeta(t)=\mathcal{M}(t)\zeta(0)$. As before, ζ represents the 6 coordinates of phase space. The particle distribution function evolves as

$$f(\zeta,t) = f(\mathcal{M}^{-1}(t)\zeta,0), \quad (10)$$

where $\mathcal{M}^{-1}(t)$ is the inverse map. It is not hard to show that moments of the distribution evolve similarly. For example, a second moment evolves as

$$\sigma_{XX}(t) = <x^2>(t) = \int d^3x \, d^3p \, (\mathcal{M}(t)x)^2 f(x,p,0). \tag{11}$$

As many have observed, the bulk of the space charge force is fairly well represented by models that depend on the second moments of the distribution alone. In recognition of this notion we might make a choice for the distribution function by demanding that the coefficients of the Gaussian's argument are exactly these second moments. Clearly, other choices could be made.

In a lattice without misalignments, one can assume

$$f(\varsigma) = A_0 \exp\left(-\frac{1}{2}\sum_{i,j} \sigma^{-1}_{i,j} \varsigma_i \varsigma_j - g(\varsigma)\right), \tag{12}$$

where $g(\zeta)$ contains terms of third order and higher. The argument of the exponential is convected in time using the inverse map and the quadratic part of the argument is transformed via the linear portion of this map. The terms of cubic order and up in the argument are exponentiated to get the Taylor series multiple.

An Example Problem

Here, we present some numerical results from a simple transverse FODO, longitudinal FOFO lattice, supporting a 3D bunched beam. All the focusing forces are idealized as discrete linear kicks, applied at distinct periodic locations along the lattice. For vanishing space charge forces, the linearly matched beam Twiss parameters are easily found for this lattice and, to linear order, any function of the quadratic form calculated from these Twiss parameters and given emittance is a matched periodic solution. The main non linearities in this sample problem arise from space charge forces.

We model a relativistic, 10 MeV, electron beam in this lattice. Transverse focusing coefficients are arranged to give a zero current transverse phase advance per cell of 72 degrees per FODO cell and 20 degrees longitudinally. Figure 1 shows the depression of the transverse and longitudinal phase advances with beam current compared to that found with the linear Trace3D code and with that found with Topkark's ray tracing code. The distribution function for this case was constrained to be a simple Gaussian and the space charge depressed phase advance was calculated from the linear part of the one lattice period map.

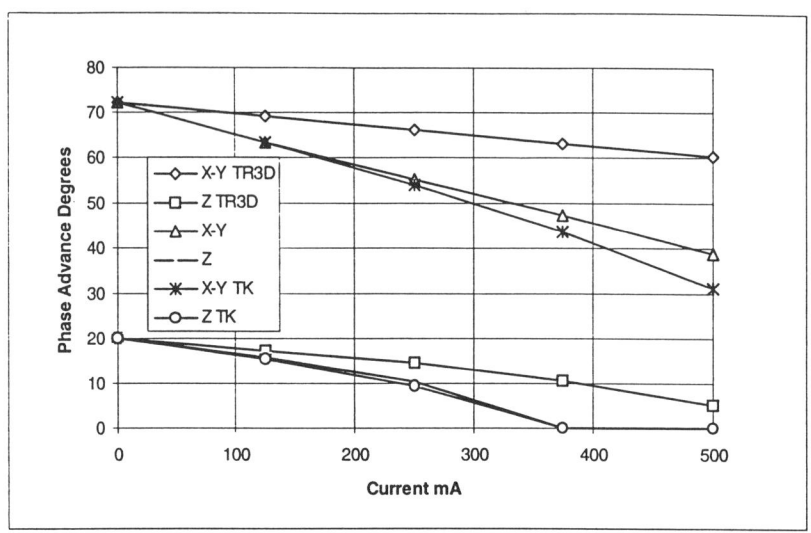

Fig. 1 - Depression of phase advance with current

The greater depression of the tune in the Gaussian cases is due to the stronger radial field near the bunch center which is about three times greater than the uniform bunch Trace 3D model.

In conclusion, the technique of this paper may be useful in studying the space charge forces of general particle distributions with applicability to several interesting problems including halo formation. Remaining work includes studies convergence of the method and benchmarking against other methods of treating space charge problems.

References

[1] Martin Berz, The Description of Particle Accelerators Using High-Order Perturbation Theory on Maps, AIP Conference Proceedings 184, Physics of Particle Accelerators Volume I, American Institute of Physics, New York 1989.

[2] E. Forest et al., Normal Form Methods for Complicated Periodic Systems, Particle Accelerators, 1989, Vol 24, pp 91-107.

[3] Lie Algebraic Treatment of Space Charge, Robert D. Ryne, Ph.D. dissertation, Department of Physics, University of Maryland (1987)

[4] D. L. Bruhwiler and M. F. Reusch, "High Order Optics with Space Charge: The Topkark code," Computational Accelerator Physics, AIP Conf. Proc. 297 (1993) 524.

A Tracking Code for Injection and Acceleration Studies in Synchrotrons *

E. Lessner, K. Symon[†]

Argonne National Laboratory
9700 Cass Ave., Argonne, IL 60439

Abstract. CAPTURE_SPC is a Monte-Carlo-based tracking program that simulates the injection and acceleration processes in proton synchrotrons. The time evolution of a distribution of charged particles is implemented by a symplectic, second-order-accurate integration algorithm. The recurrence relations follow a time-stepping leap-frog method. The time-step can be varied optionally to reduce computer time. Space-charge forces are calculated by binning the phase-projected particle distribution. The statistical fluctuations introduced by the binning process are reduced by presmoothing the data by the cloud-in-cell method and by filtering. Both the bin size and amount of filtering can be varied during the acceleration cycle so that the bunch fine structure is retained while the short wavelength noise is attenuated. The initial coordinates of each macro particle together with its time of injection are retained throughout the calculations. This information is useful in determining low-loss injection schemes.

INTRODUCTION

CAPTURE_SPC is a Monte-Carlo-based computer program developed for tracking simulation studies of the injection and acceleration processes in high intensity proton synchrotrons.

Charges in the bunch are represented by a distribution of macro particles whose time evolution follows a symplectic integration algorithm. The integrating time-step can be chosen so that it follows closely the dynamical process, or it can be made larger to reduce execution time, provided it is kept smaller than any relevant period of the motion. The initial particle phase-space configuration can be generated internally, according to some specified distribution, or can be read as input.

---

*Work supported by U. S. Department of Energy, Office of Basic Energy Sciences under Contract No. W-31-109-ENG-38.
[†]Also Dept. of Physics, University of Wisconsin-Madison, Madison, WI 53706

One of the main concerns in high intensity proton machines is the level of activation resulting from beam losses. The program provides information on when the losses occur and the dynamical conditions related to the rf voltage and space-charge at the time of loss. By pointer-bookeeping, the program provides also the initial phase-space coordinates and time of injection of the lost particles. This information is useful for reformulating the injection scheme to mitigate losses.

The code was tested extensively [1]. In the following, the algorithm used in the program and some of its capabilities are discussed.

TRACKING ALGORITHM

The equations of motion of a charged particle in a synchrotron describe the time-evolution of a particle subject to an accelerating force that depends on the rf phase and to forces that act on the particle. These equations can be modified to describe the motion of particles whose energies remain close to the energy of a synchronous particle, defined as that particle which remains in the equilibrium orbit determined by the magnetic field B(t). The equations so obtained are derivable from a Hamiltonian and suitable for rapid numerical integration by the leap-frog algorithm [2]. The coordinates of each particle, W and Φ, are advanced by a time-step τ according to:

$$W_{n+1/2} = W_{n-1/2} + \frac{eV_n\tau}{2\pi}(sin\Phi_n - sin\Phi_{s,n}) + \frac{e^2 g_0}{4\pi\varepsilon_0}\frac{h^2\tau}{R\gamma_{s,n}^2}(\frac{d\lambda(\Phi)}{d\Phi})_n, \quad (1)$$

$$\Phi_{n+1} = \Phi_n + h\tau(\frac{\eta_s\omega_s^2 W}{\beta_s^2 E_s})_{n+1/2} + \Phi_{s,n+1/2} - \Phi_{s,n-1/2}, \quad (2)$$

where $W = (E - E_s)/\omega_s$, E is the energy, the subscript s refers to the synchronous particle, and ω is the angular frequency. Φ denotes the phase of the gap voltage at the time the the particle crosses the accelerating cavity, V is the total rf gap voltage, h is the harmonic number, $\eta = (\gamma_{tr}^2)^{-1} - (\gamma^2)^{-1}$, and the subscript n indicates that a quantity is evaluated at a time $t = n\tau$. These equations are accurate to second order since the derivatives are calculated at the midpoint of each time interval. Moreover, they have the correct long-term behavior, because the phase-space area element $dWd\Phi$ is conserved to first order in W.

The last term in Eq. (1) is the electric field contribution from a continuous beam of cross-sectional radius a, moving in a smooth cylindrical pipe of radius b, and for which the wavelength associated with changes in the charge density is much greater than b/γ. $\lambda(\Phi)$ denotes the number of particles per unit rf-phase, and g_0 is a capacitive geometrical factor given by $g_0 = 1 + 2ln(b/a)$.

The numerical simulation of the injection, capture, and acceleration processes follows an ensemble of macro particles whose initial phase-space coordinates can be user-specified and read as input data, or generated by the program. In the latter case, the initial phase coordinates are derived from a random uniform distribution, whereby a stratified sampling in Φ is used to avoid clumping. The particle energies can be derived from either a cosine or a Gaussian distribution.

The time-step should be kept small whenever greater accuracy is needed, and the dynamical changes followed closely, e. g. , during injection and capture. It should be kept much smaller than the synchrotron period or any relevant period of the motion. Experience shows that a time-step less than 1/30 of the synchrotron period is quite adequate. The computer time required to track a fixed number of particles was shown to be 60% longer than the time to track the same number of particles during the same interval, with a time-step five times larger. With the larger time-step, the peak-current values are overestimated by 0.4%.

By using a system of pointers, the initial coordinates and time of injection of the lost particles can be determined. This knowledge is useful in exploring injection schemes and rf programs to minimize losses. Figure 1 shows an example of the initial phase-space coordinates of the particles uncaptured a few turns after the end of injection. The rf voltage is raised adiabatically. The loss pattern shown is typical of that obtained with adiabatic captures.

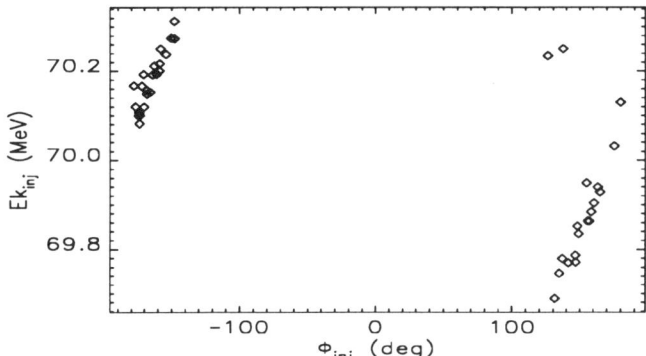

Figure 1: Phase-space coordinates of uncaptured particles, represented by the squares, soon after injection completion. The injected beam occupies uniformly a rectangle defined by $\Delta E = \pm 0.35$ MeV and $\Delta \Phi = \pm 180°$.

The stable phase-space area is determined with and without space-charge contributions. In cases when there are particles outside the rf bucket, as shown in Figure 2-(b), the space-charge forces shift the unstable fixed points. Then, one of the turning points of the motion is calculated by searching the rightmost (below transition and $sin\Phi_s > 0$) intersection of the instantaneous total energy gain per turn, $\Delta E - \Delta E_s$, with the zero axis. In this example, 10,000 macro particles are used to represent the 1.0×10^{14} particles of a beam injected at 400 MeV. Due to the space-charges, the unstable fixed point is slightly displaced from 178.3° to 177.6°. The conjugated turning point is obtained from the usual expression relating the two extreme phases, including the space-charge potentials at both phases.

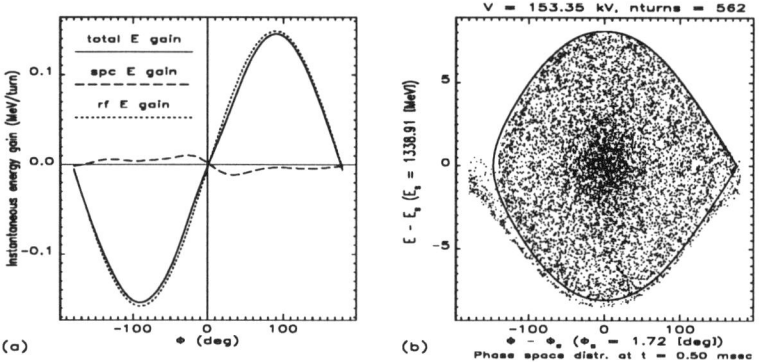

Figure 2: (a) Plot of the instantaneous energy gain per turn with the axes superimposed. (b) Bucket and phase-space distribution of the beam whose energy gain per turn is shown in (a).

The space-charge fields are calculated by using binning and data smoothing techniques that alleviate the statistical fluctuations introduced by the relatively small number of macro particles used in the simulations [3].

The macro particles are binned first into an appropriate number of bins. The number of bins must be chosen so that the average number of macro particles per bin is not too small and the bin length not too large so as to reveal the bunch structure. The projected phase distribution is binned such that, for each particle, a weighted contribution is assigned to the two closest grid points, according to how far the particle is from these points. The data is then fast-Fourier-transformed and convolved with a $sink/k$ kernel, where k is a harmonic number. This smoothens the binned density and provides a finer grid from which the fields can be obtained by interpolation. Finally, the Fourier components are multiplied by a Lanczos convergence factor whose cutoff harmonic number can be varied. As the bunch length shortens during the cycle, higher cutoff frequencies should be used. The filtering process eliminates the ringing at the ends of the bunch, but has the disadvantage of overestimating the density at the ends of the bunch and underestimating its peak. Figure 3 shows the linear particle density distributions obtained after application of the same kernel interpolation but with cutoff harmonics of 9 and 17. The coarsely-binned data, represented by the squares, is obtained by using 32 bins. The interpolated data is obtained with 512 bins. As shown in the figure, the cutoff at the 9th harmonic underestimates the peak of the

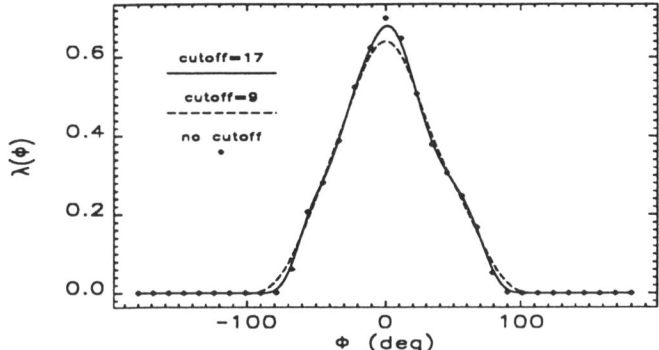

Figure 3: Linear particle density distribution obtained by using different filter cutoff frequencies.

distribution by 8% when compared with the peak of the distribution obtained from the 32-binned unfiltered data. The same cutoff overestimates the density at the ends of the bunch by 10%.

Input and Output Information

The program requires an input data file containing the time variation of the rf voltage to be tested. If desired, the variation of g_0 during the cycle can also be given as an input file. Two preprocessors can be used with CAPTURE_SPC. One preprocessor calculates the rf voltage program based on the desired rf bucket area. In this way, the bucket area can be tailored from injection through extraction to meet the design requirements of the accelerator. The reduction of the bucket area due to space-charge is not included in this calculation. The program allows the requirement for a fixed or ramped bucket area, linear or otherwise, through injection, and tailoring of the rf voltage at extraction to adjust the bunch length and momentum spread to meet the requirements for beam stability. The other preprocessor calculates the ratio of chamber size to average beam size as a function of beam energy. This allows comparisons of the effect of the space-charge impedance on the machine performance as a result of different injection phase-space painting schemes and different types of vacuum chambers.

Output files containing the energy and phase coordinates of the surviving particles, together with the separatrix coordinates, can be obtained at any specified time. The values of the energy and phase of the synchronous particle, the rf voltage, and number of turns at this time are also given, as can be seen in Figure 2-(b). As mentioned, information about the initial configuration of the uncaptured particles is available. Files containing the space-charge potential as a function of rf phase and the instantanous energy gain per turn, as depicted in Figure 2-(a), are also provided. The variation of the bucket area, bucket height, peak rf and space-charge potentials, and bunch harmonics is provided as a function of time.

FUTURE DEVELOPMENTS

Presently, only the image forces due to the space-charge fields account for the interactions of the beam with its environment through the capacitive factor g_0. Preliminary tests of the resistive-wall effects have been performed. These calculations did not include transient effects. Inclusion of general wake field contributions from elements in the ring will be implemented in a future version of the program. The algorithm will include corrections to the impedance calculations due to wake field decay times larger than the time it takes for beam parameters changes to occur. A detailed description of the algorithm is given elsewhere in these Proceedings [4].

ACKNOWLEDGEMENTS

The two preprocessors used with the program were designed and written by K. Harkay (ANL).

References

[1] Y. Cho, E. Lessner and K. Symon, "Injection and Capture Simulations for a High Intensity Proton Synchrotron," *Proc. of the European Particle Accelerator Conference*, London, p. 1228 (1994).

[2] K. R. Symon, "Synchrotron Motion with Space-Charge," ANL Report NSA-94-3 (April 4, 1994).

[3] E. Lessner, Y. Cho, K. Harkay, K. Symon, "Longitudinal Tracking Studies for a High Intensity Proton Synchrotron," *AIP Conference Proc. 377*, p. 375 (1996).

[4] K. R. Symon, "Longitudinal Wake Field Corrections in Circular Machines," *these Proceedings*.

The Dynamics of Space Charge Compensation

Reinard Becker

Institut für Angewandte Physik der Johann Wolfgang Goethe - Universität
D-60054 Frankfurt, Germany

Abstract. The classical operation of EBIS (electron beam ion source) avoided to use confinement times close to the compensation time. However the results of EBIT (electron beam ion trap) as well as of EBIS sources operating with a gas mixture - purposely or due to a too high residual gas pressure - have shown, that advantage may be taken from confinement times longer than the compensation time. By numerical simulations it has been found that compensation by thermal ions does not lead to a complete loss of radial confinement. Instead, the compensated electron beam acts as a radial energy filter for the ions produced: Hot ions are lost, cold ions become confined to the beam even better.

By a dynamical simulation of space charge compensation, the time dependence of important features can be studied and interpreted with respect to experimental results, like electron-ion-overlap, ion cooling by newly born ions and by evaporated ions, and the loss of thermal ions until compensation is reached.

INTRODUCTION

The operation of EBIS and EBIT is governed in different ways by space charge compensation of the electron beam by the produced ions. The naive picture that space charge compensation must be avoided, because the radial trapping action of the electron beam for ions would be lost, has ruled classical EBIS operation. On the other hand, in EBIT it is possible to trap ions for hours, well above any compensation time related to the residual gas pressure. This has been explained by "evaporative cooling" [1], where newly born low charged and cold ions take away the heat from highly charged hot ions and leave the trap, while highly charged ions can be trapped further inspite of the heat transferred to them in small angle Coulomb collisions [2] of the ionizing electrons.

This steady state operation involves a thermal distribution of the trapped ions and a loss mechanism of the trap. While these losses occur in axial direction by low barrier potentials in EBIT [3] as well as in the "leaky" dc operation of EBIS [4], establishing space charge compensation by raising the confinement time beyond the compensation time has shown the mechanism of "evaporative cooling" taking place in EBIS by losses in radial direction [5,6]. For understanding the radial distribution of thermal ions and its impact on stepwise ionisation, the Poisson equation has been solved in a self-consistent way, including both, thermal electrons and thermal ions

[7]. Also the set of differential equations, describing the time evolution of charge states has been augmented by a thermal loss term [8]. As a result, the steady state charge distribution, which is typical for EBIT operation, but has been shown for EBIS as well [5], may be simulated unambiguously. In this paper, the dynamic evolution of space charge compensation is investigated by numerical simulation.

SPACE CHARGE COMPENSATION OF ELECTRON BEAMS

The radial Poisson equation

$$\Delta U(r) = \frac{\partial^2 U}{\partial r^2} + \frac{1}{r}\frac{\partial U}{\partial r} = -\frac{\rho_e + \rho_i}{\varepsilon_o} \tag{1}$$

where U stands for the potential, ε_o denotes the dielectric constant, and ρ_e and ρ_i give the space charge of electron beam and compensating ions, being defined by:

$$\rho_e = \frac{I_e}{\pi r_e^2 \sqrt{\frac{2e}{m} U_0}}, \tag{2}$$

where I_e is the beam current, r_e is the beam radius, and U_0 the potential difference on the axis to the cathode

and $\quad \rho_i = -f_c \rho_e \exp\left\{-\frac{U - U_o}{U_i}\right\}, \tag{3}$

where f_c is the degree of central compensation and $U_i = kT_i/e$ is a voltage related to the ion temperature T_i, has been solved [7] for various ratios of

$$\mu = \frac{U(r_e) - U(r_o)}{U_i}, \tag{4}$$

which measures the potential difference in the uniform (uncompensated) beam in units of the temperature of the compensating ions. For any given temperature T_i, there is one distinguished solution where the tube surrounding the beam contains per length the same amount of positive and negative charges. Only in this case, the electric field strength at the tube wall disappears. However, since the radial distributions of the beam electrons and the ions are different, there is electric field and potential variation between the beam center and the tube wall inspite of "full compensation".

NUMERICAL RECEPIES FOR THE DYNAMICAL SIMULATION OF SPACE CHARGE COMPENSATION

Dynamic simulations of the development of space charge compensation are more complex than static ones. Basically, the self consistent solver for the Poisson equation is operating all time with the addition that the ion space charge is augmented from time step to time step by an amount of

$$Q_{new} = -Q_{beam} * STEP, \qquad (5)$$

where STEP stands for the the integration time in units of the compensation time. A value of STEP=1E-4 has been found satisfactory in the beginning. However, as space charge compensation proceeds and the problem becomes more non-linear, the variable STEP should be reduced, in order to reduce the "kick" to the solution obtained so far by increasing the ionic space charge by a discrete step. A reasonable measure for decreasing STEP has been found in the ratio of lost to newly born ions. The amount of newly born ions per time interval is constant, however the amount of lost ions increases drastically in approaching full compensation - the ratio of both finally will become unity.

In order to overcome numerical instabilities due to the non-linear problem, a tridiagonal LSOR (Line Successive Over Relaxation) solver, similar to the one used in EGUN [9] and IGUN [10] is used, speeding up convergence. The LSOR solver can generate a "perfect" solution in one pass, however, since the space charge of the ions has to be adjusted according to eq. 3 (the change of the electronic space charge according to eq. 2 is neglected), a self-consistent solution needs an increasing number of iterations approaching compensation. The number of iterations needed typically varies from 5 in the beginning to 500 to the end per time step in order to observe less than 10^{-8} relative change of the potential difference between the axis and the tube. Additional cure for the non-linear problem is provided by an under-relaxation procedure for the space charge term.

RESULTS OF SIMULATIONS

All calculations have been performed for a ratio of 1:10 for the beam to tube radius. The most important results of these simulations are shown in Fig. 1, where the central degree of compensation f_c is plotted together with the ratio of electron-ion-overlap in dependence on the time steps until compensation occurs. The time scale is adjusted such that compensation would be reached for t=1 without losses. Due to the losses, however, f_c only asymptotically approaches 1. It has already been shown by static simulations [7] that under the assumptions made, f_c never can reach unity: "full" compensation with a vanishing electrical field on the inside surface of the tube surrounding the electron beam still has a value of $f_c < 1$. The overlap O between electrons and the thermal ions is calculated as

$$O = \frac{\int_0^{r_t} \rho_e \, \rho_i \, 2\pi r \, dr}{\int_0^{r_t} \rho_e \, 2\pi r \, dr \int_0^{r_t} \rho_i \, 2\pi r \, dr} , \text{ where } r_t \text{ is the tube radius} \qquad (6)$$

and shown as the almost constant curve in Fig. 1. The overlap is always between 70 - 85%, which does not change the naive picture of complete overlap significantly.

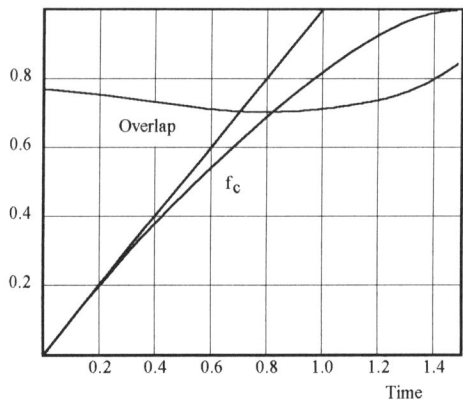

FIGURE 1. Degree of central compensation f_c and electron-ion overlap with time (the straight line shows, how compensation is reached without losses, scaling time)

In Fig. 2 the time dependences for the ion temperature U_i and for the energy of the newly born ions U_{new} are shown. Both are normalised to the potential depression ΔU in the uncompensated beam, because the starting values of 1/2 can be found by integration over the parabolic potential of the empty electron beam. After birth of the first ions, U_{new} is obtained by numerical integration through

$$U_{new} = \frac{\int_0^{r_e} U(r)\rho_e(r) 2\pi r \, dr}{\int_0^{r_e} \rho_e(r) 2\pi r \, dr}, \qquad (7)$$

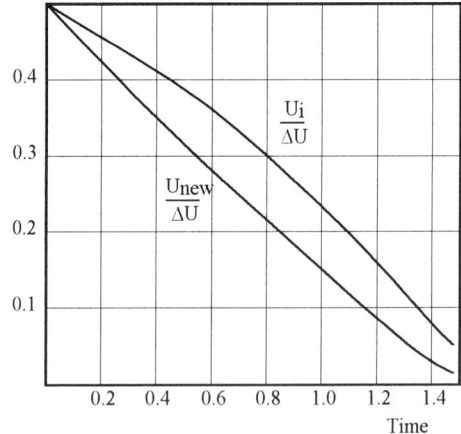

FIGURE 2. Energy of newly born and trapped ions compared to the potential depression ΔU in the empty beam with time

while the ion energy U_i is calculated from the thermal balance equation

$$U_{i,new} * Q_{i,new} = U_{i,old} * Q_{i,old} + U_{new} * Q_{new} - U_{loss} * Q_{loss} \qquad (8)$$

from time step to time step. Both, U_{new} and U_i decrease with time, U_{new} faster, which provides cooling for the trapped ions by newly born ones, especially in the beginning, where the amount of newly born ions Q_{new} compares well with the amount of trapped ions $Q_{i,new}$. To the end of the confinement time, cooling by the lost ions becomes predominant, as shown in Fig. 3. by the cooling rates of newly born and evaporated ions.

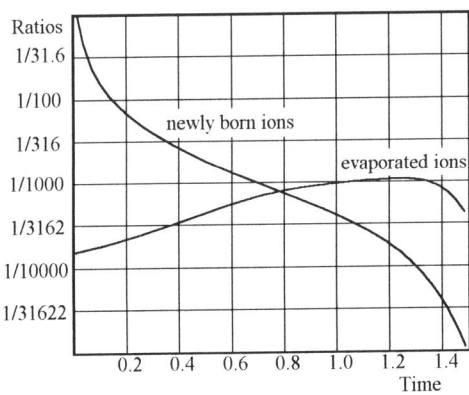

FIGURE 3. Heat of newly born ions and evaporated ones per time step, compared to heat of all trapped ions

In Fig. 4 the loss of ions is examined. The amount of lost ions may be obtained by integrating analytically the tail of the ion distribution function above the tube potential U_t, where no ions do exist in the numerical simulation:

$$Q_{loss} = \int_{U_t}^{\infty} \rho_i(U_0) \exp\left\{\frac{U - U_0}{U_i}\right\} dU, \qquad (9)$$

which gives $Q_{loss} = U_i * \rho_i(U_0) \exp\left\{\dfrac{U_t - U_0}{U_i}\right\}.$ (10)

In Fig. 4 this quantity is compared to the constant amount of newly born ions, showing how the loss to birth rate approaches unity. Also interesting is the sum of all lost ions as compared to the trap capacity, shown as the lower curve in Fig. 4: Only about 10% of all born ions are lost until compensation is established!

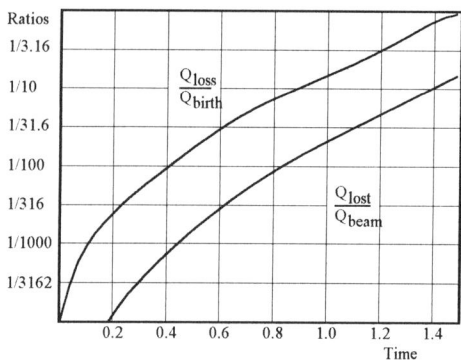

FIGURE 4. Loss of ions compared to newly born ones per time step and total amount of lost ions compared to the trap capacity

CONCLUSIONS

The dynamic evolution of space charge compensation has been simulated in order to understand basic properties during the confinement of ions in EBIS and EBIT devices. Assuming a Boltzmann energy distribution for the trapped ions, the radial functions of the potential and the space charge have been obtained in a self-consistent way by numerical simulations. From this, the time dependence of the central degree of compensation, the electron-ion-overlap, the cooling rates by newly born and evaporated ions, and the loss rates have been obtained and shown in figures. With progressing compensation during ion confinement, the temperature of the trapped ions is reduced considerably - in the beginning by newly born ions, towards the end by evaporated ions. The electron-ion-overlap is almost constant at 75%, and only about 10% of newly born ions will be evaporated off.

REFERENCES

1. B.M.Penetrante et al., Phys. Rev. **A43**, pp. 4873 - 4882, (1991)
2. R. Becker, Proc. IInd EBIS Workshop 1981, IPN2 Orsay and LNS Saclay, edts. J.Arianer, M.Olivier, pp. 185 - 196, (1981)
3. M.A.Levine et al., Physica Scripta **T22**, pp. 157 - 163, (1988)
4. R.Becker, M. Kleinod , and H.Klein, Nucl. Instrum. Methods **B24/25**, pp. 838 - 840,(1987)
5. R. Becker, M. Kleinod, H. Thomae, and E. D. Donets, Proc. HCI-92, Manhattan, edts. P.Richard, M.Stöckli, C.L.Cocke, and C.D.Lin, AIP Conf. Proc. **274**, pp. 686 - 689, (1993)
6. E. N. Beebe, Proc. IVth EBIS Workshop 1988, BNL, edt. A. Hershcovitch, AIP Conf. Proc. **188**, pp. 166 - 177, (1989)
7. R.Becker, Proc. Int. Symp. HIF, Frascati 1993, Nuovo Cimento **106A**, pp. 1613 - 1619, (1993)
8. R. Becker and M. Kleinod, Rev. Sci. Instrum. **65**, pp. 1063 - 1065, (1994)
9. W.B.Herrmannsfeldt, SLAC Report **331**, (1988)
10. R. Becker and W. B. Herrmannsfeldt, Rev. Sci. Instrum. **63**, pp. 2756 - 2758, (1992)

Generalized Time-Domain Method for Solution of Maxwell's Integral Equations

Mitsuo Hano

Department of Electrical and Electronic Engineering
Yamaguchi University, Ube 755, Japan

Abstract. This paper describes a generalized time-domain method to solve Maxwell's integral equations with a behavior of electric and magnetic fields in arbitrarily shaped electric devices. This formulation is based on a dual mesh system. A primary mesh mostly consists of tetrahedra which are used in a finite element method. A secondary mesh is built up of polyhedra whose edges connect circumcenters of adjacent primary cells. Updating equations of the electric and magnetic fields are obtained from Ampere's and Faraday's Law directly discretized on the respective mesh faces. This algorithm includes the Yee-Algorithm, and then it is fast as well as the Yee-Algorithm because it needs no procedure of field conversions at each time step.

I. INTRODUCTION

The traditional finite-difference time-domain (FDTD) method, based on the Yee-Algorithm[1], has been successfully applied to the analysis of many interesting problems. A regular grid of FDTD method, however, restricts modeling of curved surfaces. The surface must be approximated by highly refined staircase to reduce a discretization error. The contour-path FDTD method for modeling of curved surfaces was proposed[2]. This method is based on a regular orthogonal grid including grid locally deformed in the vicinity of the boundary surface. Updating equations of the electric and magnetic fields are obtained from Maxwell's integral equations applied to the deformed contour-path. But the modeling must often be done manually.

The formulations of the FDTD algorithm in generalized nonorthogonal coordinates were given[3,4]. These methods require an additional transformation from contravariant to covariant field components at every time step. A more robust techniques assuming a locally curvilinear coordinate system for each cell of an irregular structured grid are introduced[5].

More generalized techniques were recently presented[6,7]. These methods are based on a dual grid system. The secondary grid is built up of the closed polyhedra whose edges connect the centroids of adjacent primary cells. As a

© 1997 American Institute of Physics

normal vector to the primary cell face is not always parallel to a vector along the secondary cell edge. The flux density over the face is approximated by the local field values. This approximation must be evaluated at every time step.

On the other hand, a generalized time-domain method for a solution of Maxwell's equations in their integral forms has been proposed[8,9]. The formulation is based on the complementary nature of Delaunay tetrahedral and Voronoi polyhedral cells. The advantage of this formulation is that it is free from the field conversion at each time step, and it is based on an orthogonal coordinate system, so that the surface and line integrations of Faraday's and Ampere's Laws are very easily implemented.

In this paper, the generalized time-domain method is developed in Section II, and the reconstruction of primary cells is discussed in Section III.

II. GENERALIZED TIME-DOMAIN METHOD

The generalized time-domain method is based on a direct solution of Faraday's and Ampere's Laws in their integral form

$$\oint_{C_f} \boldsymbol{E} \cdot d\boldsymbol{l} = -\frac{\partial}{\partial t} \iint_{S_f} \boldsymbol{B} \cdot d\boldsymbol{s} \tag{1}$$

$$\oint_{C_a} \boldsymbol{H} \cdot d\boldsymbol{l} = \frac{\partial}{\partial t} \iint_{S_a} \boldsymbol{D} \cdot d\boldsymbol{s} + \iint_{S_a} \sigma \boldsymbol{E} \cdot d\boldsymbol{s} + \iint_{S_a} \boldsymbol{J} \cdot d\boldsymbol{s} \tag{2}$$

where $\boldsymbol{E}$ is the electric field, $\boldsymbol{H}$ is the magnetic field, $\boldsymbol{D}$ is the electric displacement and $\boldsymbol{B}$ is the magnetic flux density, respectively. S_f and S_a are the integral surfaces, and mutually link as a chain. C_f and C_a are the contours bounding S_f and S_a, respectively.

The electric and magnetic fields are discretized over a dual mesh structure formed by primary and secondary cells. The primary mesh is composed of relatively low order polyhedra, which must have circumcenters, distributed throughout the volume. The secondary mesh is built up of the relatively high order polyhedra whose edges connect the circumcenters of adjacent primary cells and consequently penetrate though the circumcenters of shared faces. Figure 1 illustrates adjoining primary cells and secondary cells. In Fig.1, six tetrahedra(solid and dash line) as primary cells and some portions of high order polyhedra(chain line) as secondary cells are shown. An electric field vector is located at the midpoint of each edge of the primary cells and a magnetic flux vector is located at the circumcenter of each primary cell in Fig.1.

Faraday's Law is approximated by discretizing surface integral over each primary cell face and line integral along each edge bounding the face. The fields are assumed to be constant over their respective faces and along each

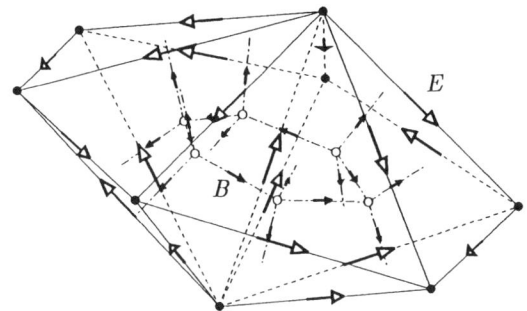

FIGURE 1. A dual mesh.

edge. The time derivative is then approximated using a central difference approximation. This leads to the discrete form of (1)

$$\sum_{p=1}^{N_i} E_{ip}^{n-\frac{1}{2}} \ell_{ip} = -\frac{B_i^n - B_i^{n-1}}{\Delta t} A_i \tag{3}$$

where Δt is the time step interval and the superscripts indicate the time index, B_i ia the normal component of the magnetic flux to the i-th face of the primary cell and A_i is the i-th face area, E_{ip} is the tangential component of the electric field along the p-th edge bounding the i-th face and ℓ_{ip} is the p-th edge length, and N_i is the number of edges bounding the i-th face.

On the other hand, Ampere's Law is approximated by discretizing surface integral over each secondary cell face and line integral along each edge bounding the face. The discrete form of (2) is obtained as

$$\sum_{q=1}^{N_j} \frac{B_{jq}^n \ell_{jq}}{\mu_{av,jq}} = \varepsilon_{av,j} \frac{E_j^{n+\frac{1}{2}} - E_j^{n-\frac{1}{2}}}{\Delta t} A_j + \sigma_{av,j} \frac{E_j^{n+\frac{1}{2}} + E_j^{n-\frac{1}{2}}}{2} A_j + I_j^n \tag{4}$$

where A_j is the j-th face area of the secondary cell, ℓ_{jq} is the length of the q-th edge bounding the j-th face of the secondary cell, N_j is the number of edges bounding the j-th face, I_j is the total current flowing along the j-th edge of the primary cell, and $\varepsilon_{av,j}$, $\sigma_{av,j}$ and $\mu_{av,jq}$ are derived in Appendix.

The field solution is obtained from (3) and (4) resulting in the explicit time stepping algorithm

$$B_i^n = B_i^{n-1} - \frac{\Delta t}{A_i} \sum_p E_{ip}^{n-\frac{1}{2}} \ell_{ip} \tag{5}$$

$$E_j^{n+\frac{1}{2}} = \frac{\varepsilon_{av,j} - \sigma_{av,j}\Delta t/2}{\varepsilon_{av,j} + \sigma_{av,j}\Delta t/2} E_j^{n-\frac{1}{2}} + \frac{\Delta t}{(\varepsilon_{av,j} + \sigma_{av,j}\Delta t/2) A_j} \left(\sum_{q=1}^{N_j} \frac{B_{jq}^n \ell_{jq}}{\mu_{av,jq}} - I_j^n \right) \tag{6}$$

In the above dual mesh system, each edge of the primary cell always intersects the corresponding face of the secondary cell at the circumcenter of the face with a right angle and each edge of the secondary cell always intersects the corresponding face of the primary cell at the circumcenter of the face with a right angle. Subsequently, the advantage of this technique is that a normal component of the electric displacement to the secondary cell face and a tangential component of the magnetic field along the secondary cell edge are directly estimated from a tangential component of the electric field along the primary cell edge and a normal component of the magnetic flux to the primary cell face, by using isotropic constitutive relationships, respectively. This nature maintains a fast computation as well as the Yee-Algorithm[1]. The transformation from contravariant to covariant field components in the Nonorthogonal Curvilinear FDTD Algorithm[3,4] and the interpolation of the electric displacement and the magnetic flux density in the Generalized Yee-Algorithm[6,7], whose method were developed to reduce a discretization error in the curved surface geometry, must be numerically calculated at every time step. These calculations require a significant load for computation. The disadvantage of the generalized time-domain method is that each primary cell must own a circumcenter. Fortunately, any tetrahedral has always a circumcenter. If the volume of interest is wholly subdivided by tetrahedra, the present method can be always applied to calculate the electromagnetic fields varying in time.

III. RECONSTRUCTION OF PRIMARY CELLS

Each face area and edge length of the primary cell are not always zero since the volume of the cell is not equal to zero. But the face area and edge length of the secondary cell become zero on rare occasions owning to a misarrangement of primary cells. In this case, two or more circumcenters of the primary cells are located at the same position. Since ℓ_{jq} is equal to zero, B_{jq} never contribute the updating of E_j in (6). Furthermore, if A_j in (6) is equal to zero, E_j never be updated. If A_j becomes zero as a result of the degeneration of circumcenter, E_j must be rejected from this system. The rejection procedure of useless unknowns identifies a reconstruction of primary cells which own circumcenters locating at the same position. Examples of the reconstruction are shown in Fig.2. This technique is applied to the regular structured cell including six tetrahedra as shown in Fig.2(b), and then Eqs.(5) and (6) identify the Yee-Algorithm.

IV. CONCLUSION

In this paper, the generalized time-domain method has been developed. The algorithm is an explicit time-marching method based on the discretization of

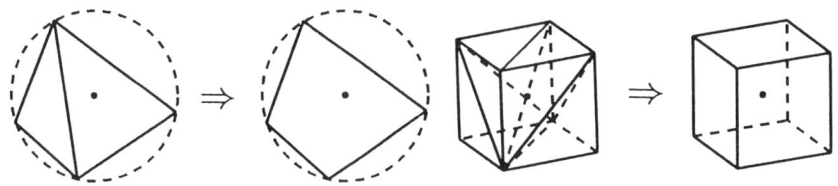

(a) Two dimensional case   (b) Three dimensional case
FIGURE 2.   Reconstruction of primary cells.

Maxwell's integral equations using the complementary nature of Delaunay tetrahedral and Voronoi polyhedral cells. This algorithm is fast as well as the Yee-Algorithm. It is anticipated that the generalized time-domain method will be an efficient tool for the analysis of high-frequency electromagnetic devices.

APPENDIX

In this Appendix, the average permittivity, conductivity and permeability in (6) are introduced. It is assumed that the permittivity, conductivity and permeability are constant throughout each primary cell. Consider a face on the secondary cell associating with the primaly cell edge j as shown in Fig.3. Each edge of the face connects the circumcenters of adjacent primary cells and is normal to the interface of boundary between them because of the nature of the circumcenter. If a normal component of the electric field $\boldsymbol{E}_j$ to the secondary cell face is assumed to be constant over the face, surface integral terms of an electric displacement and an eddy current in (2) are expressed as

$$\int_{S_a} \varepsilon \boldsymbol{E}_j \cdot d\boldsymbol{s} = \left(\sum_{q=1}^{N_j} \varepsilon_{jq} A_{jq}/A_j \right) E_j A_j \tag{7}$$

$$\int_{S_a} \sigma \boldsymbol{E}_j \cdot d\boldsymbol{s} = \left(\sum_{q=1}^{N_j} \sigma_{jq} A_{jq}/A_j \right) E_j A_j \tag{8}$$

where $A_j (= \sum_{q=1}^{N_j} A_{jq})$ is the total face area of the secondary cell face. The average permittivity and conductivity are introduced as

$$\varepsilon_{av,j} = \sum_{q=1}^{N_j} \varepsilon_{jq} A_{jq}/A_j \tag{9}$$

$$\sigma_{av,j} = \sum_{q=1}^{N_j} \sigma_{jq} A_{jq}/A_j \tag{10}$$

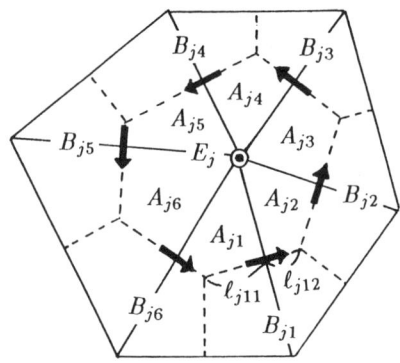

FIGURE 3. Secondary cell face.

where N_j is the number of the adjacent primary cells to the electric field E_j.

On the other hand, if a normal component of the magnetic flux density $\boldsymbol{B}_{jq}$ to the primary cell face is assumed to be constant along the secondary cell edge, a line integral term of a magnetic field in (2) is expressed as

$$\oint_{C_a} \boldsymbol{H} \cdot d\boldsymbol{\ell} = \sum_{q=1}^{N_j} \left(\sum_{m=1}^{2} (\ell_{jqm}/\mu_{jqm})/\ell_{jq} \right) B_{jq} \ell_{jq} \qquad (11)$$

where $\ell_{jq} (= \sum_{m=1}^{2} \ell_{jqm})$ is the total edge length of the secondary cell edge. An average permeability is introduced as

$$\mu_{av,jq} = \ell_{jq} / \sum_{m=1}^{2} (\ell_{jqm}/\mu_{jqm}) \qquad (12)$$

REFERENCES

1. K. S. Yee, *IEEE Trans. Antennas Propagat.*, **AP-14**, 302-307 (1966).
2. T. G. Jurgens, A. Taflov, K. Umashankar and T. G. Moore, *IEEE Trans. Antennas Propagat.*, **40**, 357-365 (1992).
3. R. Holland, *IEEE Trans. on Nuclear Science*, **NS-30**, 4589-4591 (1983).
4. M. A. Fusco, M. V. Smith and L. W. Gordon, *IEEE Trans. Antennas Propagat.*, **39**, 1463-1471 (1991).
5. P. H. Harms, J. F. Lee and R. Mittra, *IEEE Trans. on Microwave Theory and Techniques*, **40**, 741-746 (1992).
6. N. K. Madsen, *J. Comput. Phys.*, **119**, 34-45 (1995).
7. S. D. Gedney, F. S. Lansing and D. L. Rascoe, *IEEE Trans. on Microwave Theory and Techniques*, **44**, 1393-1400 (1996).
8. A. J. Butler, "Time Domain Solutions of Two- and Three- Dimensional Transient Electromagnetic Fields (Two- Dimensional)" Ph.D Thesis, Carnegie-Mellon University (1990).
9. M. Hano and T. Itoh, *IEEE Trans. on Magnetics*, **32**, 946-949 (1996).

Space-Charge-Dominated Beam Dynamics Simulations Using the Massively Parallel Processors (MPPs) of the Cray T3D

Hongxiu Liu

Thomas Jefferson National Accelerator Facility
12000 Jefferson Avenue, Newport News, VA 23606, USA

Abstract. Computer simulations using the multi-particle code PARMELA with a three-dimensional point-by-point space charge algorithm have turned out to be very helpful in supporting injector commissioning and operations at Thomas Jefferson National Accelerator Facility (Jefferson Lab, formerly called CEBAF). However, this algorithm, which defines a typical N^2 problem in CPU time scaling, is very time-consuming when N, the number of macro-particles, is large. Therefore, it is attractive to use massively parallel processors (MPPs) to speed up the simulations. Motivated by this, we modified the space charge subroutine for using the MPPs of the Cray T3D. The techniques used to parallelize and optimize the code on the T3D are discussed in this paper. The performance of the code on the T3D is examined in comparison with a Parallel Vector Processing supercomputer of the Cray C90 and an HP 735/125 high-end workstation.

I. INTRODUCTION

Massively Parallel Processing (MPP) is of common interest for numerically intensive industrial and scientific calculations (1–3). It may provide a new approach to simulating three-dimensional space-charge-dominated beam dynamics with its larger amounts of CPU time and memory. Space charge simulation has turned out to be very helpful in supporting injector commissioning and operations at Thomas Jefferson National Accelerator Facility. It also is important for our future injector design and development.

Jefferson Lab's 4 GeV superconducting electron beam accelerator is providing new opportunities for pursuing new knowledge of internal structures of matter. It consists of a 45 MeV electron injector, two sections of north and south 400 MeV linear accelerators, and five recirculation passes. The injector design is unique in that it can deliver three CW high-intensity electron beams simultaneously to the main accelerator, which accelerates and finally separates the three beams for injection into three nuclear physics experiment halls. The quality of the beams is determined largely by the front end of the machine, i.e., the injector, where space charge is a dominating factor affecting beam generation, transport and bunching.

Space charge refers to the repulsive Coulomb interacting forces among the

© 1997 American Institute of Physics

charged particles that tend to blow up the beams. Its effects on the dynamics of a beam are so complicated, both in theoretical treatment and in experimental practice, that computer simulation often is the most effective shortcut in delivering the first and most accurate answers to the problems of concern. For example, the space charge effects on bunching of electrons in the Jefferson Lab's injector were clarified and corrected for a bunch length setting in use through computer simulation (4); beam transmission through this injector has been improved by a factor of two with the help of computer simulations.

The code we have been using for space-charge-dominated beam dynamics simulations is PARMELA (5) that has been maintained and modified by the author to meet various special needs for simulating electron injectors. Its space charge subroutine is based on a point-by-point method (6, 7), and has been used extensively in two contexts: designing a photoemission gun injector test stand (8–11) for high-power industrial free electron laser applications, and supporting the Jefferson Lab main accelerator's injector commissioning and operations (4). The algorithm was benchmarked (12) with the simulation results from the PIC code ISIS (13).

A longest space charge run we once had consumed a CPU time of 8 days on an HP 735/125 workstation with 8000 macro-particles that simulated beam transport and bunching through the entire FEL injector (11). Apparently, it is necessary to seek for a faster approach to running space charge jobs. Motivated by this, we have parallelized the code for using the MPPs of the Cray T3D. The techniques used to parallelize the code on the T3D are discussed in this paper. The performance of the code on the T3D is analyzed in detail.

II. PROCESSES IN A COMPLETE PARMELA RUN

A PARMELA input deck is composed of various elements as listed in Table 1. The elements chopper (#6), backb (#34), alpham (#36), poisson (#37), bfield (#38), ! (#39), nbend (#40), kicker (#41) and b_earth (#42) were added by the author for modeling chopper systems (14), backbombardment in microwave guns (15), α-magnets (16–18), finite-length solenoids, indexed-field bends, beam orbital corrections and the earth fields. The "!" element is used for inserting comments.

TABLE 1. Elements constituting a PARMELA input deck

(drift,1)	(solenoid, 2)	(quad, 3)	(bend, 4)	(buncher, 5)	(chopper, 6)
(cell, 7)	(tank, 8)	(trwave, 9)	(coil,10)	(run,11)	(input,12)
(output,13)	(title,14)	(scheff,15)	(zout,16)	(adjust,17)	(start,18)
(restart,19)	(continue,20)	(save, 21)	(end, 22)	(limit, 23)	(errors, 24)
(change, 25)	(rotate, 26)	(sbload, 27)	(cfield, 28)	(dpout, 29)	(cathode, 30)
(design, 31)	(pipe, 32)	(foclal, 33)	(backb, 34)	(wiggller, 35)	(alpham, 36)
(poisson,37)	(bfield, 38)	(!, 39)	(nbend, 40)	(kicker, 41)	(b_earth, 42)

The processes in a complete run are threaded in Fig. 1. After initialization, an electron beam is generated with N macro-particles. The six coordinates of the particles are stored in a two-dimensional array named cord. Then, the subroutine pardyn is called to initiate the processes of solving particle dynamics equations. Each process contains two major loops, one on time steps, and the other on particles. The space charge subroutine scheff is called once on all the particles, with the resultant momentum increments superimposed to those from external elements to advance each particle during the present time step. When all the processes come to the end of a beamline successfully, the program exits the subroutine pardyn and saves the coordinates of the particles before ending execution.

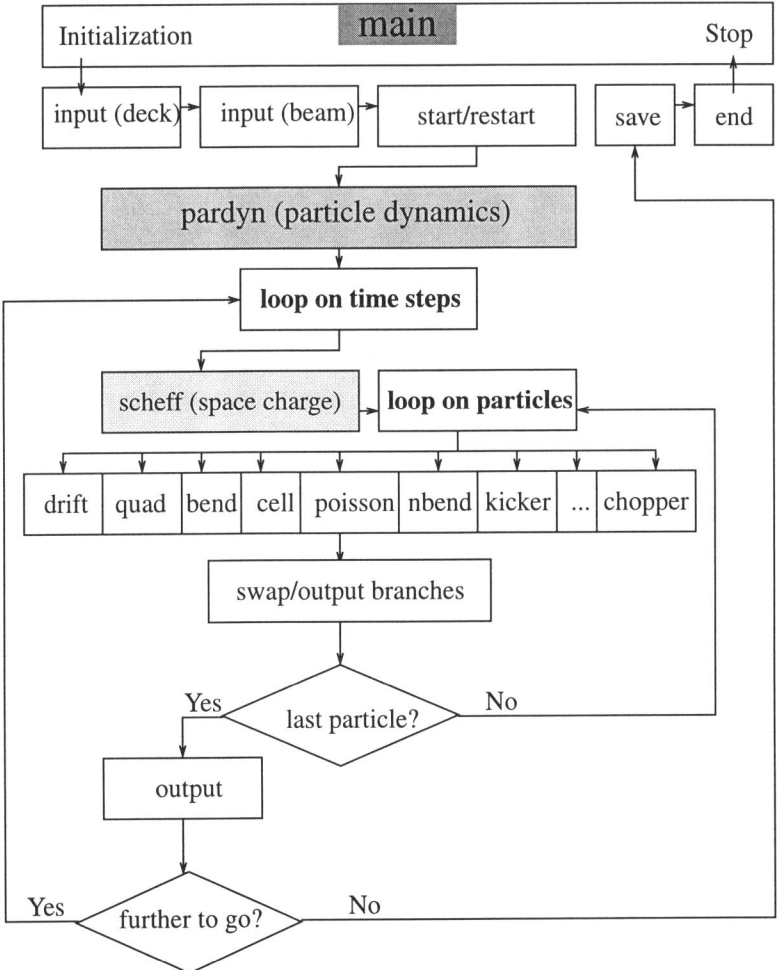

Figure 1. Processes in a complete PARMELA run.

III. POINT-BY-POINT SPACE CHARGE ALGORITHM

In the point-by-point space charge algorithm that we have been using (6, 7), there are two major loops, the *i*-loop and the *j*-loop, to calculate the electric ($\vec{E}$) and magnetic ($\vec{B}$) fields produced by a moving charge Q

$$\vec{E}_{ij} = Q\gamma_j \vec{r}_{ij}/[\vec{r}_{ij} \cdot \vec{r}_{ij} + (\vec{r}_{ij} \cdot \overrightarrow{(\gamma\beta)}_j)^2]^{3/2}, \qquad \vec{B}_{ij} = \vec{\beta}_j \times \vec{E}_{ij}, \qquad (1)$$

where *j* denotes the particle applying electromagnetic forces to the *i*th particle. The *i*-loop specifies the source particles, and the *j*-loop calculates and sums up the E and B fields from all other particles for the *i*th particle. A final loop applies the impulses obtained from the previous two loops to each particle. As is seen, the CPU time spent executing this algorithm is proportional to N^2, where N is the number of macro-particles, which defines a typical N^2 problem in CPU time scaling. It becomes more and more time-consuming as N goes up rapidly.

IV. PARALLELIZATION ON THE T3D

The Cray T3D is an MPP supercomputer with 256 processing elements (PEs) (19–21). On the T3D, four MPP programming methods (22) are available: data sharing, work sharing, message passing, and explicit shared memory. In order to minimize the frequency of data exchange among the PEs, the arrays that contain the net momentum increments for each particle, dbgx, dbgy and dbgz, instead of cord, are chosen as the shared ones. The cyclic data distribution mechanism is used as follows

```
CDIR$ GEOMETRY G(:BLOCK(1))
    real dbgx(imaa), dbgy(imaa), dbgz(imaa)
CDIR$ SHARED (G) :: dbgx, dbgy, dbgz
```

where imaa = 16384 specifies the maximum number of particles that can be loaded in a run, which must be a power of 2 on the T3D.

Work sharing is achieved by adding the following DOSHARED directive

```
CDIR$ DOSHARED (i) ON dbgx(i)
```

prior to the *i*-loop. This way, the *j*-loop is not parallelized since it is a nested loop inside the *i*-loop. The last loop applying the impulses to all the particles is split into two, and a SHMEM subroutine shmem_put (23) is used for the Master PE to collect all other particles from all other PEs. Finally, the master task copies the arrays cord and gam across to the other tasks or PEs.

The loop on particles in pardyn (see Fig. 1) has been parallelized as well. The major problem encountered was to deal with output and particle loss/swap. This requires extra caution for correct communication among all the nodes; otherwise, particles would walk into wrong PEs, leading to wrong results. Output is controlled using the MPP intrinsic function MY_PE() which designates a master PE region. When a call to output is made, the Master PE collects all other particles it does not have in its own memory from all other PEs, and then conducts all statistic computations. LOCKs are used to count the particles to determine whether all the alive particles have passed the last exit and the program execution should stop.

V. PERFORMANCE

In this section, we examine the code performance on the T3D in comparison with an HP 735/125 workstation and the Cray C90. The major hardware/software configurations of each computing vehicle are listed in Table 2.

TABLE 2. Comparison of configurations on different machines

Configuration	HP 735/125	Cray C90	Cray T3D
CPU Type	PA-RISC 7150	Custom-made	DEC EV4 α (21064)
Number of CPU(s)	1	16	256
Precision for real (bits)	32	64	64
Memory	64 MB	268 Mw	256 x 8 Mw
Clock Period (MHz)	125	240	150
MFLOPS/CPU	201 (SPECfp92)	960 (peak)	150 (peak)
FP Computation	IEEE + sqrt	–	IEEE
Primary Cache	256KBI+KBD off-chip	Fast memory and vector registers with no cache (25)	8 KB on-chip
Secondary Cache	none		none

We choose an input, shown in Fig. 2, that does a real job of simulating beam quality degradation due to space charge effects. This input is set up to reflect all the aspects that parallelization must deal with. The emittance values of 7.380, 30.081, 47.710, and 62.840 in the *x*-direction are used to check that the same results are obtained at the same exits despite of any changes and/or modifications to the code.

```
TITLE - mpp test (cap96_1.in)-
RUN /IRUN=1 /IP=1 /FREQ=1497. MHZ /Z0=-3.5 CM /W0= 0.511 MEV /LTYPE=1
OUTPUT 6
DRIFT /L=2.1 /APER=2.54 /IOUT=1
DRIFT /L=6.7 /APER=2.54 /IOUT=1
DRIFT /L=6.6 /APER=2.54 /IOUT=1
DRIFT /L=8.9 /APER=2.54 /IOUT=1
ZOUT
INPUT 16 /NP=999 10.0 0.0 0 10.0 0.0 0 31.8 0.0 0 0 0 0 0 0 1.0
SCHEFF /BEAMI=-1025.E8 /RMESH=2.5 /ZMESH=8.0 /NR=10 /NZ=30 0  0. 0  1.5  0 /POINT=3. /SS=1.0 0
START /WTO=0. /DWT=7.2 /NSTEPS=999999 /NSC=1 /NOUT=40
END
```

Figure 2. An input deck with 10^3 particles for testing the performance of the code.

The quantitative approach to measuring the performance of a computer differs from workstation to supercomputer. Workstations are benchmarked using "SPECfp" with which the MFLOPS number is a geometric mean from a number of application runs, while supercomputers are rated using "peak performance," which is sort of "the speed of light" that the manufacturers guarantee that nobody can exceed. In this paper, we simply use the following formula to calculate the MOPS contained in the input shown in Fig. 2

$$MOPS = 70 \times M \times N^2, \qquad (2)$$

where 70 is the total number of operations per step per particle contained in the inner loop, M is the total time steps executed from start to finish for the job, and N is the number of macro-particles. With the input shown in Fig. 2, $M = 90$, $N = 1000$, so it needs 6300 MOPS total. This number will be used as the same measure in the following context to calculate MFLOPS numbers for all the platforms.

The baseline performance of the code on a single node is shown in Table 3. For each platform, different compile flags have been tested to make sure that the FORTRAN compiling optimizer built with each computer has been used to the largest possible extent. It is seen that an optimization functionality may make a huge difference in speeding up the CPU time that this input may take. On the C90, the inner loop is vectorized and the outer loop is autotasked (parallelized), making it ideal for running this algorithm. The wall-clock time for this input on the C90 is about 7 seconds only with a usage of ~ 50% of 16 CPUs.

TABLE 3. Baseline performance on a single node

	$(\Delta t)_{cpu}$ (s)	MFLOPS (% of peak)	Compile Flags
HP	412	15 (7.6 of SPECfp)	f77 -R8 +e +E1
	134	47 (23 of SPECfp)	f77 -V -R8 +OP -WP, -o=4 +e + E1
C90	52	121 (12)	cf77 -Zp -Wd"-du -l tmp.l" -c
	61	103 (10)	cf77 -Wd"-du -l tmp.l" -c
	148	43 (4.3)	cf77 -O vector0 -c
	561	11 (1.1)	cf77 -O scalar0 -O vector0 -c
T3D	417	15 (10)	cf77 -Wl"-lfi" -c
	389	16 (11)	cf77 -Wf"-o unroll" -Wl"-lfi" -c
	388	16 (11)	cf77 -Wf"-o aggress" -Wl"-lfi" -c
	392	16 (11)	cf77 -Wf"-o unroll -o aggress" -Wl"-lfi" -c
	390	16 (11)	cf77 -Wf"-o unroll -o noieeedivide" -Wl"-lfi" -c
	389	16 (11)	cf77 -Wf"-o unroll " -Wl"-lfi -D rdahead=on" -c

The difference in CPU time between default optimization and maximized optimization is 1.2 on the C90, 1.1 on the T3D, and 3.1 on the HP. Aligning cache boundaries was tried on the T3D, but seemed not very helpful, possibly because the number of variables involved in a single loop is much larger than the cache size, and cached data cannot be reused effectively. The single node performance is 12% of peak on the C90, and 11% of peak on the T3D. Using the single node CPU time spent on each platform, we see that a single C90 CPU is 2.6 times faster than the HP, the HP is 2.9 times faster than a single T3D CPU, and a single C90 CPU is 7.5 times faster than a single T3D CPU, which is close to the peak performance ratio (6.7) of the C90 to the T3D.

The optimized performance, shown in Table 4, refers to the performance when the code is manually optimized to take full advantage of the architectural features of each machine. Based on the observations that pipelining and cache reuse are effective only when as few branches as possible are contained in a loop (2), the code has been rewritten on the C90 with the original two loops split into three parts: one is for calculating shielding factors for each particle in advance, one is to calculate interacting forces without image charge, and the third one includes image charge calculations. Due to limitations on the single node memory size, the shielding factors are not pre-calculated on the HP and T3D. After the algorithm has been recoded, we found that the code is 1.6 times faster than it was on the HP, 3 times faster on the C90, but only 1.2 times faster on the T3D.

TABLE 4. Optimized performance on a single node

	$(\Delta t)_{cpu}$ (s)	MFLOPS (% of peak)	Compile Flags
HP	345	18 (9 of SPECfp)	f77 -R8 +e +E1
	85	74 (37 of SPECfp)	f77 -V -R8 +OP -WP, -o=4 +e + E1
C90	22	286 (29)	cf77 -Zp -Wd"-du -I tmp.l" -c
	23	274 (27)	cf77 -Wd"-du -I tmp.l" -c
	164	38 (3.8)	cf77 -O vector0 -c
	589	11 (1.1)	cf77 -O scalar0 -O vector0 -c
T3D	319	20 (13)	cf77 -Wl"-lfi" -c
	318	20 (13)	cf77 -Wf"-o unroll" -Wl"-lfi" -c
	342	18 (12)	cf77 -Wf"-o aggress" -Wl"-lfi" -c
	326	19 (13)	cf77 -Wf"-o unroll -o aggress" -Wl"-lfi" -c
	321	20 (13)	cf77 -Wf"-o unroll -o noieeedivide" -Wl"-lfi" -c
	318	20 (13)	cf77 -Wf"-o unroll " -Wl"-lfi -D rdahead=on" -c

It is noticed that functional calculations of sqrt are significantly slower on the T3D than on the C90 and on the HP. On the T3D/T3E the standard "libm" "sqrt()" is implemented in Alpha assembly language, and the routine has about two dozen floating point adds/multiplies; the latest sqrt() on the T3D takes about 129 clock periods, and on the T3E about 67 clock periods (24). The C90 vector sqrt function requires 21 floating point operations, 8 multiplies, 10 adds, and 3 reciprocal approximations plus some overhead to set up the vectors; for vectorized code with arrays on the order of 1 million elements, sqrt achieves around 700 MFLOPS, roughly 1 sqrt result per each 7.2 clock ticks (25). It is noticed that the DEC's Alpha architecture CPUs have no sqrt FP computation, unlike MIPS, PA-RISC, PowerPC and SPARC (26).

Table 5 shows the partition of CPU time among different parts of the program executed with one PE and 32 PEs respectively, based on apprentice analysis on the T3D (27). The measured elapsed time was 748 seconds with 1 PE and 35.5 seconds with 32 PEs in contrast with 318 seconds with 1 PE and 20 seconds with 32 PEs when compiled with no apprentice. Therefore, the program was slowed down by a factor of 1.78 with 1 PE, and a factor of 2.35 with 32 PEs by apprentice. The MFLOPS numbers are 8.7 for program and 14.7 for scheff2 with 1 PE, and 137 for program and 384 for scheff2 with 32 PEs, according to apprentice. However, these MFLOPS numbers from apprentice are less accurate, since sqrt has not been instrumented in counting MOPS, but its time is counted in calculating MFLOPS.

TABLE 5. Partition of CPU time among different parts of the program on the T3D

Parts	Δt in seconds	
	NPES = 1	NPES = 32
program/parmela/pardyn	368/0.04/3.21	755/26.9/19.5
scheff2/_sqrt/drift	218/145/0.668	268/145/0.677
shmem_broadcast/_barrier	0.008/0.003	294/79.7
shmem_put/set_lock/clear_lock	0.200/0/0006/0.0006	1.14/1.82/0.06

It is seen from Table 5 that sqrt is responsible for 39% of the CPU time in the case of one CPU run. According to Amdahl's law, the program is slowed down by a factor 1.5 with the assumption that 24/70 = 34% of MOPS attributable to sqrt and that it is $(24 \times 129 \times 240)/(21 \times 7.2 \times 150) = 33$ times slower than on the C90. In addition, overhead resulting from communications among multiple PEs has increased significantly from the case of 1 PE to the case of 32 PEs, as indicated by large amounts of CPU time with 32 PEs from shmem_broadcast and _barrier shown in the table.

The performance of the code vs. the number of nodes on the T3D is shown in Table 6, with space charge on and off. It is seen that the performance is linearly scalable for time-consuming space charge runs. The T3D starts to outperform the C90 at 32 processors. In the near future, optimization and use of the code for our space-charge-dominated beam dynamics simulations will be switched to the NERSC's Cray T3E which is about six times faster than the T3D.

TABLE 6. Performance vs. number of nodes on the T3D

Number of nodes	Space charge on		Space charge off	
	$(\Delta t)_{cpu}$ (s)	MFLOPS	$(\Delta t)_{cpu}$ (s) (N_P = 1000)	$(\Delta t)_{cpu}$ (s) (N_P = 16384)
1	318	20	3.95	63.3
2	162	39	3.36	38.0
4	84	75	3.30	21.2
8	46	137	3.78	13.2
16	28	225	4.56	9.85
32	20	315	5.50	9.07

ACKNOWLEDGMENTS

I thank C. Sinclair and J. Bisognano for their support and S. Corneliussen for editing the manuscript. The access to the T3D and C90 was provided by NERSC/LLNL. This work was supported by DOE contract no. DE-AC05-84ER40150.

REFERENCES

1. A. Koniges and K. Lind, Comput. Phys., **9**, 399 (1995).
2. V. Decyk, S. Karmesin, A. Boer and P. Liewer, Comput. Phys., **10**, 290 (1995).
3. R. Ryne and S. Habib, *LHC95*, 15–21 October 1995, CERN, Geneva.
4. H. Liu and J. Bisognano, to be published.
5. Originally from K. Crandall and L. Young at Los Alamos.
6. K. McDonald, *IEEE Trans. Electron Devices* **ED-35**, 2052 (1988).
7. H. Liu, *Computational Accelerator Phys.*, AIP Conf. Proc. No. 297, 508(1994).
8. H. Liu et al., *Nucl. Instr. Meth.* **A 358**, 475 (1995).
9. H. Liu et al., *Proc. of 1995 Particle Accelerator Conf.*, **2**, 942 (1995).
10. H. Liu, *Microbunches Workshop*, AIP Conf. Proc. No. 367, 56 (1995).
11. *CEBAF FEL Conceptual Design Report*, Chapter 4, 1995, unpublished.
12. H. Liu, *CEBAF Tech. Notes*, TN# 94-040, 1994.
13. M. Jones and B. Carlsten, *Proc. 1987 IEEE Particle Accelerator Conf.*, p. 1319.
14. H. Liu, *Computational Accelerator Phys.*, AIP Conf. Proc. No. 297, 385 (1994).
15. H. Liu, *Nucl. Instr. Meth.* **A 302**, 535 (1991).
16. H. Enge, *Rev. Sci. Instr.* **34**, 385 (1963).
17. H. Liu, *Nucl. Instr. Meth.* **A 294**, 365 (1990).
18. H. Liu, *J. of Electronics* **13**, 293 (1991).
19. *An introduction to the T3D at LLNL*, NERSC document, July 1995.
20. *Cray T3D system architecture overview*, Revision 1.B, March, 1993, CRI.
21. *Cray T3D software 1.0 release overview*, CRI manual HR-04033 Ro-5215 1.0, 1993, CRI.
22. *Cray MPP FORTRAN reference manual*, SR-2504 6.2.2, 1995, CRI.
23. *SHMEM Technical Note for FORTRAN*, Revision 2.3, October 25, 1994, CRI.
24. T. Welcome, private communication.
25. M. Stewart, private communication.
26. D. Bhandarkar, *Alpha Implementations and Architecture*, Digital Press, 1996.
27. *Introducing the MPP Apprentice tool*, IN-2511 1.2, 1994, CRI.

PARTICLE TRACKING AND BEAM TRANSPORT

Optimization of Dynamic Aperture for the KEKB B-Factory

K. Oide, H. Koiso, and K. Ohmi

KEK, National Laboratory for High Energy Physics
Oho, Tsukuba, Ibaraki 305, Japan

Abstract. A few methods are introduced to optimize the dynamic aperture of the KEKB B-Factory. The basic idea is to fit the linear optics around orbits with finite amplitudes in the desired acceptance. Chromaticity correction was done by this method for orbits with finite momentum offsets, and sufficient momentum acceptance, ±2.5%, was achieved. This method was extended to the transverse direction, then resulted in significant increase of the dynamic aperture. A survey on the relationship between a Taylor map and the dynamic aperture was also done.

Machine Overview

KEKB[1] is an asymmetric double-ring e^+e^- collider with a design luminosity of $10^{34}\text{cm}^{-2}\text{s}^{-1}$. It consists of two rings, HER and LER, whose parameters are listed in Table 1. A high luminosity collider inevitably requires high beam currents ($I = 2.6/1.1$ A for LER/HER) and small β_y^* ($= 1$ cm for both rings) at the interaction point(IP). The latter causes a high chromaticity at the final quadrupoles around IP, and limits the dynamic aperture of the ring. In

TABLE 1. Machine Parameters of KEKB.

		LER	HER	
Beam Energy	E	3.5	8.0	GeV
Luminosity	$\mathcal{L}$	\multicolumn{2}{c}{1.0×10^{34}}	$\text{cm}^{-2}\text{s}^{-1}$
Luminosity Reduction Factor	$R_\mathcal{L}$	\multicolumn{2}{c}{0.845}	
Half crossing angle	θ_x	\multicolumn{2}{c}{11}	mrad
Tune shifts	ξ_x/ξ_y	\multicolumn{2}{c}{0.039/0.052}	
Tune shift reductions	$R_{\xi x}/R_{\xi y}$	\multicolumn{2}{c}{0.737/0.885}	
Beta functions	β_x^*/β_y^*	\multicolumn{2}{c}{0.33/0.01}	m
Beam current	I	2.6	1.1	A
Particles/bunch	N	3.3×10^{10}	1.4×10^{10}	
Emittances	$\varepsilon_x/\varepsilon_y$	\multicolumn{2}{c}{$1.8 \times 10^{-8}/4.3 \times 10^{-10}$}	m
Bunch length	σ_z	\multicolumn{2}{c}{4}	mm
Momentum spread	σ_δ	7.7×10^{-4}	7.8×10^{-4}	
Synchrotron tune	ν_s	\multicolumn{2}{c}{0.01~0.02}	
Momentum compaction factor	α_p	\multicolumn{2}{c}{$1 \times 10^{-4} \sim 2 \times 10^{-4}$}	
Betatron tunes	ν_x/ν_y	45.52/45.08	46.52/46.08	
Circumference	C	\multicolumn{2}{c}{3016}	m

© 1997 American Institute of Physics

the case of KEKB, the dynamic aperture must be enough large to accept the injected beam. The momentum acceptance is also important to keep the Touschek lifetime sufficiently long, for LER in particular. In this sense LER's dynamic aperture is more critical than HER's, therefore we mainly discuss LER's dynamic aperture hereafter.

Evolution of the Unit Cell

The design of a unit cell in the arc of KEKB has evolved starting from a conventional interleaved $\pi/3$-cell, as being summarized in Fig. 1. The final design, the non-interleaved 2.5π-cell, has unique characteristics in efficient sextupole allocation, large transverse aperture, tunability in the momentum compaction factor and the horizontal emittance[2]. Its chromatic performance is greatly enhanced by combining with the local chromaticity correction in the IP section. The local chromaticity correction is applied only in vertical plane with a pair of sextupoles at each side of IP. It is necessary only for LER because of the Touschek lifetime. In this paper the dynamic aperture was estimated by particle tracking for 1,000 turns with synchrotron motion, and the initial ratio of the transverse actions was fixed as $J_x/J_y = 9$. Figure 2 shows the comparison of dynamic apertures of those schemes for LER.

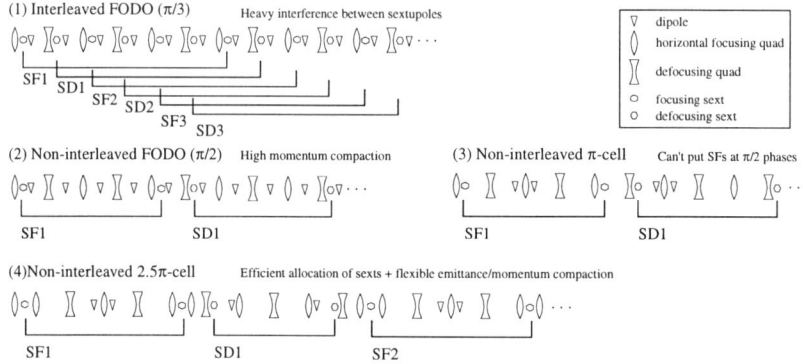

FIGURE 1: Evolution in the structure of the unit-cell.

Finite-Bandwidth Chromaticity Correction

Chromaticity correction of a function $f(\delta)$ by making its derivatives at the origin zero up to the m-th order, i.e.,

$$f^{(k)}(0) = 0, \qquad k = 1, .., m \qquad (1)$$

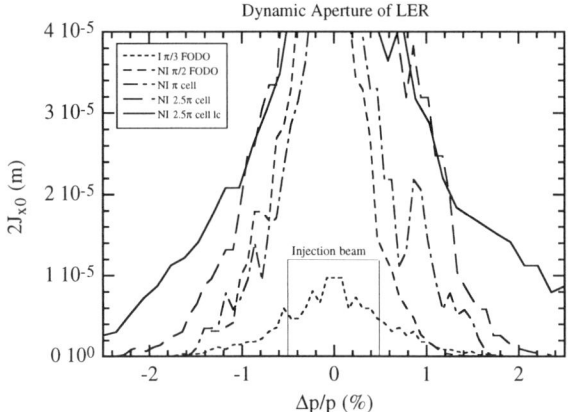

FIGURE 2: Comparison of dynamic apertures for various schemes.

is inappropriate unless the contributions from the higher orders are small. Always one has to calculate the function $f(\delta)$ in the desired bandwidth $-\Delta < \delta < \Delta$ to confirm the applied chromaticity correction is sufficient. Then it is more practical to optimize $f(\delta)$ directly so that

$$||f(\delta) - f(0)|| \to 0, \qquad -\Delta < \delta < \Delta , \qquad (2)$$

than using the derivatives. This is done numerically by choosing n points in the range $-\Delta < \delta < \Delta$. The number of points n must be big enough so that $f(\delta)$ is smooth between these points. At KEKB, chromaticities in total tunes, $\alpha^*_{x,y}$, $\beta^*_{x,y}$, R-matrix(x-y coupling matrix) at IP, $\alpha_{x,y}$ and $\beta_{x,y}$ at the middle of the rf section are corrected in this way[3]. Typically $\Delta = 3\%$, $n = 30$ are chosen, and more than 50 sextupole families participate the optimization.

Finite-Amplitude Matching

The finite-amplitude matching (FAM) is an extension of the finite-bandwidth method of the chromaticity correction to the transverse dimension. Consider one-turn orbits starting at a particular location in a ring with finite-amplitudes in the transverse dimensions. Although such an orbit never closes in one turn, the transfer matrix around the orbit becomes equal to the design matrix M_0 around the closed orbit if the machine is completely linear. Then it should not be a bad strategy to match the one-turn matrices Ms around the finite-amplitude orbits to the design matrix M_0. The FAM method has a practical merit, because the computation of the orbit and the transfer matrix around it is quite fast and suitable to an optimization process which requires many

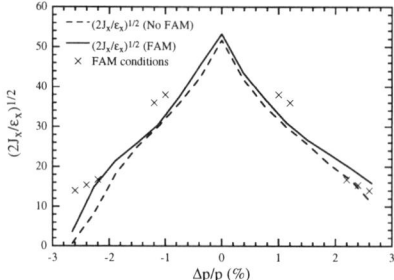

FIGURE 3: Effect of the finite-amplitude matching(FAM) method. The dashed and solid lines show the dynamic apertures, averaged over 100 samples, before and after FAM, respectively. The FAM orbits are shown by markers.

iterations.

In the case of KEKB, the FAM method is applied to enlarge the momentum bandwidth, which is in particular necessary to increase the Touschek lifetime. The initial conditions were

$$(x, p_x, y, p_y, z, \delta) = \begin{cases} (x_k \cos \phi_x, x_k \sin \phi_x, 0, 0, 0, \delta_k) \\ (0, 0, y_k \cos \phi_y, y_k \sin \phi_y, 0, \delta_k) \end{cases}, \quad (3)$$

where the phases are $\phi_{x,y} = (0, 2\pi/3, 4\pi/3)$ and $y_k = x_k/3$, which corresponds to the ratio of amplitudes of the injection beam. In this case x_k ranges from $14\sigma_x$ to $38\sigma_x$, and $\delta_k = (\pm 2.6, \pm 2.4, \pm 2.2, \pm 1.2, \pm 1.0)\%$.

In order to investigate the effect of FAM, we randomly distributed the strengths of sextupoles and then applied the finite-bandwidth chromaticity correction and the FAM correction. Figure 3 compares the dynamic apertures, averaged of 100 samples, before and after FAM. Figure 4 shows the distribution

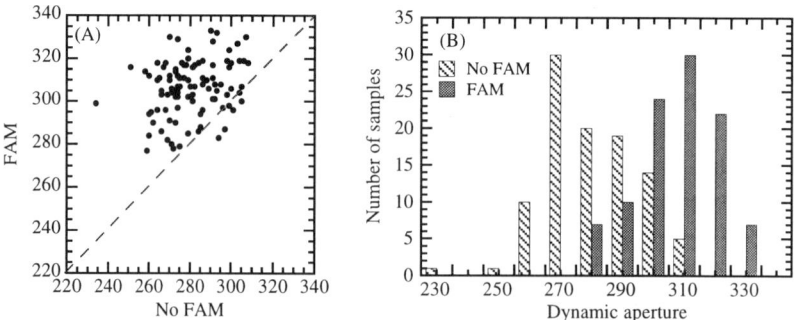

FIGURE 4: (A): Comparison of the dynamic aperture before/after FAM for 100 samples of sextupole settings. (B): Distribution of the dynamic apertures.

of the dynamic apertures, before and after FAM. The dynamic aperture A_i for the i-th sample is defined as $A_i = a \int \sqrt{2J_{x0}} d\delta$ with an arbitrary constant a. Although there are small number of exceptions, FAM improved the dynamic aperture significantly.

Taylor Map and Dynamic Aperture

For each sample of sextupole settings obtained above, we calculated a Taylor map in 6 dimensions up to the 6th order in Hamiltonian. The nonlinear Hamiltonian was expressed in the "resonant basis", $a_{x,y,z}^\pm = (X, Y, Z \pm i p_{X,Y,Z})/\sqrt{2}$. An example of the result is shown in Fig. 5, where the strength of each term is normalized at a typical size of the dynamic aperture, $2(J_{x1}, J_{y1}, J_{z1}) = (28^2 \varepsilon_x, 28^2 \varepsilon_x/9, 20^2 \varepsilon_z)$. The Taylor map was calculated by a C++ routine coded by K. Ohmi, and accessed from SAD[4] by a SADScript programming language.

First we examined a correlation r_{ik} of the coefficients c_{ik} of the Taylor map and the dynamic aperture A_i, over the 100 samples. The best correlation in the amplitudes $|c_{ik}|$ was about -0.6 for a few coefficients, but entire distribution was broad (Fig. 6). The average of the distribution shifts somewhat negative, which simply implies "smaller nonlinearity gives larger aperture." Although there are many coefficients with anti-correlations (better aperture corresponds to larger magnitude), the correlations tend to be more negative for coefficients with larger strengths, as shown in Fig. 6.

Besides such a general tendency, no singular coefficient which dominantly determines the aperture was seen by this simple analysis. This does not,

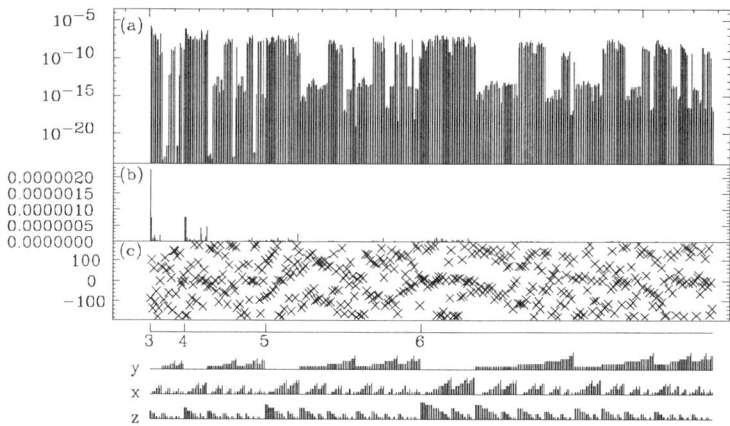

FIGURE 5: An example of the result of a Taylor Map. (a,b): Normalized amplitude (log,linear), (c): phase (degree). The chart at the bottom shows the orders in (X, p_X), (Y, p_Y), (Z, p_Z) of each term.

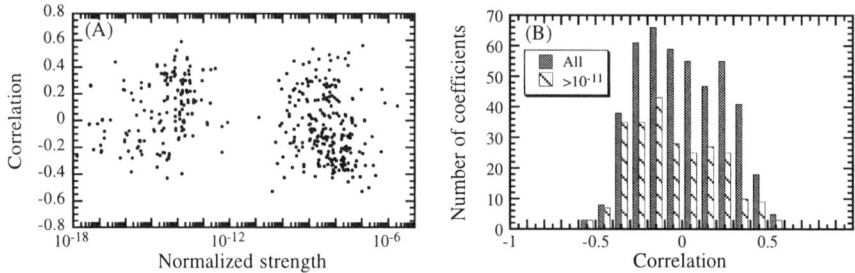

FIGURE 6: (A): Correlations with dynamic aperture versus the normalized strength is shown for coefficients of the Hamiltonian up to the 6th order. The strength was averaged over 100 samples. (B): Distribution of correlations for all coefficients and those with normalized strength bigger than 10^{-11}.

however, eliminate a possibility that the dynamic aperture is determined by a factor that is not very sensitive to the sextupoles.

Summary

A few experiences were made through the optimization of the dynamic aperture of KEKB: (1) The non-interleaved 2.5π-cell has the best results, together with the local chromaticity correction. (2) The finite-bandwidth chromaticity correction and the finite-amplitude matching were practically effective to increase the dynamic aperture. (3) The relation between coefficients of Taylor map and the dynamic aperture did not become clear by the simple analysis in this paper.

ACKNOWLEDGMENTS

The authors thank K. Hirata, S-I. Kurokawa, K. Satoh, and all KEKB stuffs for help in the design of the lattice and the development of the simulation codes.

REFERENCES

[1] *KEKB B–Factory Design Report*, KEK Report 95–7(1995).
[2] H. Koiso and K. Oide, Proc. 1995 Part. Accel. Conf. at Dallas, **TAG11**, 2780(1995).
[3] K. Oide and H. Koiso, Phys. Rev. **E47**, 2010(1993).
[4] Information on SAD is available at
http://www–acc–theory.kek.jp/SAD/sad.html

Differential Algebras with Remainder and Rigorous Proofs of Long-Term Stability

Martin Berz [1]

Department of Physics and Astronomy and
National Superconducting Cyclotron Laboratory
Michigan State University, East Lansing, MI 48824

Abstract. It is shown how in addition to determining Taylor maps of general optical systems, it is possible to obtain rigorous interval bounds for the remainder term of the n-th order Taylor expansion. To this end, the three elementary operations of addition, multiplication, and differentiation in the Differential Algebraic approach are augmented by suitable interval operations in such a way that a remainder bound of the sum, product, and derivative is obtained from the Taylor polynomial and remainder bound of the operands.

The method can be used to obtain bounds for the accuracy with which a Taylor map represents the true map of the particle optical system. In a more general sense, it is also useful for a variety of other numerical problems, including rigorous global optimization of highly complex functions. Combined with methods to obtain pseudo-invariants of repetitive motion and extensions of the Lyapunov- and Nekhoroshev stability theory, the latter can be used to guarantee stability for storage rings and other weakly nonlinear systems.

INTRODUCTION

The study of single particle dynamics in accelerators, storage rings, spectrographs and beamlines requires the determination and analysis of the functional dependence

$$\vec{z}_f = \mathcal{M}(\vec{z}_i, \vec{\delta})$$

describing how the final coordinates $\vec{z}_f$ of a system depend on the initial coordinates $\vec{z}_i$ as well as a set of system parameters $\vec{\delta}$. Because of the weakly nonlinear character of the motion, the problem is particularly amenable to a perturbative treatment, usually in the form of a Taylor expansion. Previous

---

[1] Supported in part by the US Department of Energy and the Alfred P. Sloan Foundation.

perturbative techniques like those employed in older codes [1] [2] [3] attack the problem in an analytical, order-by-order way. Because of the tremendous complexity of the resulting formulas, these attempts have been limited to order three; only a custom formula manipulator [4] could succeed to produce a fifth order code [5].

The differential algebraic methods [6] [7] have overcome this problem by providing a morphism from operations of a space of functions to objects that can be handled on a computer. Almost all map codes and tool libraries that have been developed since the introduction of the differential algebraic method employ this approach as the fundamental mechanism to handle and manipulate maps [8] [9] [10] [11] [12] [13] [14].

In the following we will show how it is possible to amend this approach in such a way that it is not only able to handle the perturbative tasks to any order, but also to provide rigorous estimates for the errors that are made up to a given order. To this end, it is first necessary to develop an understanding of the algebraic and topological structure of the various types of function spaces. After this, we will provide a new morphism from these function algebras to corresponding adjoint algebras, and in particular, we will discuss the concept of differential algebras with remainders (RDA).

ALGEBRAS OF FUNCTIONS

The detailed study of functions and their properties, like the transfer maps, requires the study of function spaces and their properties. The basic properties of functions are algebraic in nature, and are connected to addition, multiplication, and multiplication with scalars. Demanding merely addition and scalar multiplication leads to the concept of a **Vector Space**; including multiplication leads to the concept of an **Algebra**.

An algebra consists of a set A, two operations denoted "+" and "·", as well as a field K connected to A via a scalar multiplication "·". With the operations "+" and "·", it forms a vector space over K, and under the operations "+" and "·", it forms a ring. Mixed products are assumed to satisfy $(sa) \cdot (tb) = (st)(a \cdot b)$.

The next question is that of convergence of sequences of functions, for example those that appear when trying to introduce exponentials of functions via the exponential series. The study of these issues require the presence of a **Norm** to measure distances and benefit from **Cauchy-completeness**, which is important to study convergence properties. A normed, Cauchy-complete algebra is called **Banach**.

Often it is important to study functions on the function spaces, which historically often have been referred to as **operators** or **functionals**. In their role as functions, they can be added and multiplied with scalars and form a vector space. In many applications it is important to study the **commu-**

tator $[O_1, O_2] = O_1 O_2 - O_2 O_1$ of such operators O_1 and O_2, which can be viewed as an "multiplicative" operation that turns out to be antisymmetric and non-associative; instead it satisfies a Jacoby identity. This behavior matches of what is summarized in the structure of the **Lie Algebra**, which is a modification of the concept of the algebra in that the condition of associativity of multiplication is dropped and replaced by the Jacoby identity $(a \cdot (b \cdot c)) + (b \cdot (c \cdot a)) + (c \cdot (a \cdot b)) = 0$, and the multiplication is assumed to be antisymmetric, i.e. $a \cdot b = -b \cdot a$. Note that by virtue of the fact that conventional associativity does not hold, **a Lie algebra is not an algebra.**

In the study of series like $\exp(tL_{\vec{f}})$, which is one of the important objects studied in nonlinear dynamics, questions of convergence are important; a detailed study of the issue reveals that among other things what is needed is in fact again Cauchy-completeness, and hence a **Banach Lie algebra.**

In many cases, including most of what is done in nonlinear dynamics, functions are not only defined by certain algebraic properties, but rather through some processes involving differentiation, like in differential equations. Hence the concept of the Algebra, based on two arithmetic operations, is too small, and a larger structure is needed. This is achieved in the case of the **Differential Algebra**, which is an algebra with a derivative-like operation ∂, formally called a derivation, that is linear in $+$ and $\cdot$ and that satisfies $\partial(a \cdot b) = (\partial a) \cdot b + a \cdot (\partial b)$. It is enlightening to observe that the presence of derivations ∂ in a natural way allows to construct a **Lie Algebra from a differential algebra** by utilizing conventional addition as well as the commutator on the space of derivations.

ADJOINTS TO FUNCTION ALGEBRAS

Given the exceedingly large differential algebra of infinitely often differentiable functions, for any practical computation it is now important to move to a smaller space containing all the necessary information of interest. In the simplest case, such "information of interest" could consist of just some function values at specified points. In the case of the DA method [6], this is achieved via the equivalence relation T, which extracts from a function f its Taylor expansion F of a given order around a given reference point:

$$f \xrightarrow{T} F$$

Let us assume now we are given a particular binary operation $*$ in the space of functions. The question now becomes whether to this operation, we can find a matching operation $\circledast$ such that for all functions f, g, we have $T(f * g) = T(f) \circledast T(g)$; or in modern terminology, we want to find $\circledast$ in such a way that the following diagram commutes:

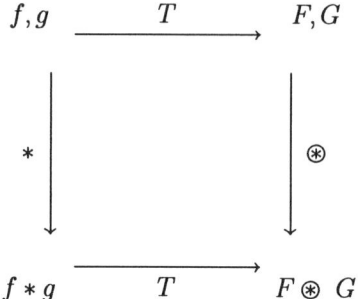

Similarly, we want to have that for a unary operation ∧, we can find an adjoined unary operation $\overline{\wedge}$ such that for any f, we have that $T(\wedge f) = \overline{\wedge}\, T(f)$. In this case, the algebraic structure of the original set induces a similar structure on the new set based on the adjoint operations. In the case of the transition T from the function f to the Taylor series F, for example, this is achieved via the differential algebra of truncated power series described in [6] [7].

The question is now whether this approach can be extended to not only provide a Taylor series to order n, but also a bound for the remainder. Altogether, it is necessary to craft adjoint operations $\oplus, \odot$ and $\partial_\bigcirc$ to $+, \cdot$ and ∂ such that the following diagrams commute:

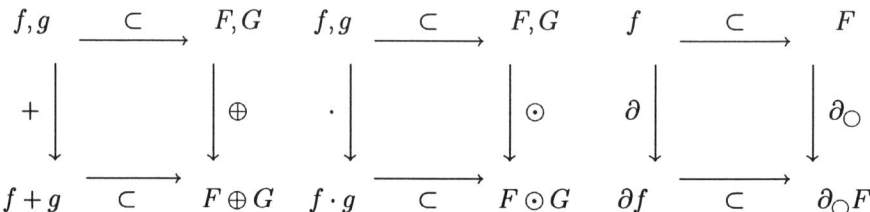

The next section will address how this can be achieved.

DIFFERENTIAL ALGEBRAS WITH REMAINDER

In order to study how to introduce the operations $\oplus, \odot$ and $\partial_\bigcirc$, we begin with the definition of a Taylor model. Let f be $C^{(n+1)}$ on $D_f \subset R^v$, and $\vec{B} = [a_1, b_1] \times ... \times [a_v, b_v] \subset D_f$ an interval box containing the point $\vec{x}_0$. Let T be the nth order Taylor polynomial of f around the point $\vec{x}_0$. We call the interval I an nth order **Remainder Bound** of f on $\vec{B}$ if

$$f(\vec{x}) - T(\vec{x}) \in I \text{ for all } \vec{x} \in \vec{B}.$$

We call the pair (T, I) an nth order **Taylor Model** of f. As a first example, we see that low-order polynomials have trivial remainder bounds; since every polynomial of order not exceeding n agrees with its nth order Taylor polynomial, the optimal remainder bound in this case is the interval $[0, 0]$.

For practical purposes, it is important that if the original interval box $\vec{B}$ decreases in size, then according to the Lagrange remainder formula, the optimal remainder bound will decrease in size with a power of $n + 1$ and hence will become small quickly. In particular, this entails that the knowledge of a good Taylor model of a function on an interval box $\vec{B}$ allows a rather accurate estimate of the range of the function.

In the following sections, we will develop the elementary operations on the Differential Algebra with Remainders (RDA).

A Addition on RDA

Let (T_f, I_f) and (T_g, I_g) be nth order Taylor models of the functions f and g on the interval box $\vec{B}$. Clearly, the Taylor polynomial of $(f + g)$ is simply $T_f + T_g$; on the other hand, we know that on $\vec{B}$, $f(\vec{x}) \in T_f(\vec{x}) + I_f$ and $g(\vec{x}) \in T_g(\vec{x}) + I_g$. Thus obviously,

$$(f + g)(\vec{x}) \in (T_f + T_g)(\vec{x}) + (I_f + I_g) \text{ for all } \vec{x} \in \vec{B},$$

and so $(T_f + T_g, I_f + I_g)$ is a Taylor model for $(f + g)$ on $\vec{B}$. Furthermore, the interval $I_{f+g} = I_f + I_g$ is the sharpest remainder bound that can be guaranteed from the knowledge of the Taylor models for f and g. For practical purposes, it is also important to note that if I_f, I_g are "fine of order $\vec{B}^{n+1}$", i.e. their size scales with the size of $\vec{B}$ to the $(n+1)$st power, so is I_{f+g}. In the same way we see that $(T_f - T_g, I_f - I_g)$ is a Taylor model for $(f - g)$.

B Multiplication on RDA

In order to study multiplication, let (T_f, I_f) and (T_g, I_g) be nth order Taylor models of the functions f and g on the interval box $\vec{B}$. As pointed out before, the Taylor polynomial $T_{f \cdot g}$ of $f \cdot g$ can then be obtained by multiplication of T_f and T_g and subtracting the polynomial $\bar{T}_{f \cdot g}$ consisting of the terms whose order exceeds n. For any $\vec{x} \in \vec{B}$, there are values $e_f \in I_f$ and $e_g \in I_g$ such that $f(\vec{x}) = T_f(\vec{x}) + e_f$ and $g(\vec{x}) = T_g(\vec{x}) + e_g$. So we obtain

$$(f \cdot g)(\vec{x}) = (T_f(\vec{x}) + e_f) \cdot (T_g(\vec{x}) + e_g)$$
$$= T_f(\vec{x}) \cdot T_g(\vec{x}) + T_f(\vec{x}) \cdot e_g + T_g(\vec{x}) \cdot e_f + e_f \cdot e_g$$
$$= T_{f \cdot g}(\vec{x}) + \{\bar{T}_{f \cdot g}(\vec{x}) + T_f(\vec{x}) \cdot e_g + T_g(\vec{x}) \cdot e_f + e_f \cdot e_g\}.$$

The first term is the Taylor polynomial of $f \cdot g$. The term in curly brackets describes the behavior of the error; it is a polynomial in the $v + 2$ variables

$(\vec{x}, e_f, e_g) \in \vec{B} \times I_f \times I_g$ and is denoted by $E(\vec{x}, e_f, e_g)$. Since no knowledge about the correlation between $\vec{x}$ and e_f, e_g is contained in the Taylor models of f and g besides the fact that $e_f \in I_f$, $e_g \in I_g$, the sharpest possible remainder bound of $f \cdot g$ that can be obtained is given by the extrema of $E(\vec{x}, e_f, e_g)$ on the interval box $\vec{B} \times I_f \times I_g$. Thus the determination of $I_{f \cdot g}$ is reduced to a polynomial bounding problem; for details, refer to [15].

C Intrinsic Functions on RDA

Let (T_f, I_f) be an nth order Taylor model for f on $\vec{B}$, and let g be an elementary function appearing in a computer environment like exp or log. The goal is to find a Taylor model for $g \circ f$. While there appears to be no fully universal strategy that works for any function g, in most cases it is possible to follow an approach similar to what we develop here for the exponential function $g(x) = \exp(x)$. Writing $T_f(\vec{x}) = c + \bar{T}_f(\vec{x})$, we obtain

$$\exp(T_f(\vec{x}) + I_f) = \exp(c) \cdot \exp(\bar{T}_f(\vec{x}) + I_f).$$

For the second exponential, we use the Taylor formula of exp around the origin with Lagrange remainder, i.e. $\exp(x) = \sum_{\nu=0}^{n} x^\nu / \nu! + x^{n+1} \exp(\xi)/(n+1)!$, where ξ is between 0 and x. In our case, $\xi \in \bar{T}_f(\vec{B}) + I_f$, and so we obtain the following inclusion:

$$\exp(\bar{T}_f(\vec{x}) + I_f) \subset \sum_{\nu=0}^{n} \frac{(\bar{T}_f(\vec{x}) + I_f)^\nu}{\nu!} + (\bar{T}_f(\vec{B}) + I_f)^{(n+1)} \cdot \frac{\exp(\bar{T}_f(\vec{B}) + I_f)}{(n+1)!}.$$

The polynomial in the resulting Taylor model on the right is the Taylor polynomial of $\exp(f)$. Apparently, sharper inclusions are possible if there are dedicated methods for the compuation of polynomials of Taylor models. In a

Besides providing the operations $\oplus$ and $\odot$ as well as intrinsic functions for Taylor models such that the corresponding diagrams commute, there are a variety of other operations that have to be ported to the Taylor models; these include $\partial_\bigcirc$ and several derived operations, including composition of maps. For reasons of space, we have to restrict ourselves here to a referral to more detailed papers about the matter [15] [16].

RIGOROUS TREATMENT OF LONG-TERM STABILITY

As advertised above, the method of Taylor models can be used to obtain fully rigorous bounds for the long term stability of motion. To this end, one utilizes a method outlined in [17] and in various modifications frequently used in nonlinear dynamics: One obtains a family of pseudo-invariants of the motion and uses a simple geometric argument that links the quality of invariance to a prediction of stability [18]. To this end, the following three steps are necessary:

1. Decide on a family of pseudo-invariants
2. Compute the transfer map with remainder bound for the system
3. Determine a bound for the maximum deviation of invariance

Items number 2 and 3 require the use of RDA methods to be carried out rigorously; item number 1 is independent of the RDA approach and can use invariants based on normal form methods [7], interpolation [19], or other approaches including hybrids of the two previous ones.

Currently, a full implementation of the RDA method is under development. In the special case of normal form invariants, a substantial simplification is possible that allows an efficient bounding of the deviation from invariance. It does not require RDA, but merely the approach of Interval Chains [20] [18]. While it is not able to solve the problem number 2 above and is not applicable to general families of pseudo-invariants, it allows to obtain first experience about the quality of invariants that is to be expected from a full RDA implementation.

REFERENCES

1. Wollnik, H., Hartmann, B., and Berz, M., Principles behind GIOS and COSY. *AIP Conference Proceedings* **177**, 74 (1988).
2. Brown, K. L., The ion optical program TRANSPORT. Technical Report, SLAC **91** (1979).
3. Dragt, A. J., Healy, L. M., Neri, F., and Ryne, R., MARYLIE 3.0 - a program for nonlinear analysis of accelerators and beamlines. *IEEE Transactions on Nuclear Science*, **NS-3,5**, 2311 (1985).
4. Berz, M., and Wollnik, H., The program HAMILTON for the analytic solution of the equations of motion in particle optical systems through fifth order. *Nuclear Instruments and Methods*, **A258**, 364 (1987).
5. Berz, M., Hofmann, H. C., and Wollnik, H., COSY 5.0, the fifth order code for corpuscular optical systems. *Nuclear Instruments and Methods*, **A258**, 402 (1987).
6. Berz, M., Differential algebraic formulation of normal form theory. In *Berz, M., Martin, S., and Ziegler, K., editors, Proc. Nonlinear Effects in Accelerators*, 77, IOP Publishing (1992).
7. Berz, M., High-order computation and normal form analysis of repetitive systems. In *Month, M., editor, Physics of Particle Accelerators*, AIP **249**, 456, American Institute of Physics (1991).
8. Makino, K., and Berz, M., COSY INFINITY version 7, *these proceedings*.

9. Berz, M., COSY INFINITY Version 7 reference manual. Technical Report **MSUCL-977**, National Superconducting Cyclotron Laboratory, Michigan State University, East Lansing, MI 48824 (1995).
10. Michelotti, L., MXYZTPLK: A practical, user friendly c++ implementation of differential algebra. Technical report, Fermilab (1990).
11. van Zeijts, J., and Neri, F., The arbitrary order design code TLIE 1.0. In *Berz, M., Martin, S., and Ziegler, K., editors, Proceedings Workshop on Nonlinear Effects in Accelerators*, IOP Publishing (1993).
12. Yan, Y., and Yan, C.-Y., ZLIB, a numerical library for differential algebra. Technical Report, SSCL **300** (1990).
13. Davis, W. G., Douglas, S. R., Pusch, G. D., and Lee-Whiting, G. E., The Chalk River differential algebra code DACYC and the role of differential and Lie algebras in understanding the orbit dynamics in cyclotrons. In *Berz, M., Martin, S., and Ziegler, K., editors, Proceedings Workshop on Nonlinear Effects in Accelerators*, IOP Publishing (1993).
14. Iselin, F. C., The classic project, *these proceedings*.
15. Makino, K., and Berz, M., Remainder Differential Algebras and their applications. In *Berz, M., Bischof, C., Corliss, G., and Griewank, A., editors, Computational Differentiation: Techniques, Applications, and Tools*. SIAM, Philadelphia, Penn. (1996).
16. Berz, M., and Hoffstätter, G., Computation and application of Taylor polynomials with interval remainder bounds. *Interval Computations*, submitted, (1995).
17. Warnock, R. L., and Ruth, R. D., Stability of orbits in nonlinear mechanics for finite but very long times. In *Nonlinear Problems in Future Accelerators*, 67-76, World Scientific, New York (1991). Also **SLAC-PUB-5304**.
18. Berz, M., and Hoffstätter, G., Exact bounds of the long term stability of weakly nonlinear systems applied to the design of large storage rings. *Interval Computations*, **2**, 68-89 (1994).
19. Warnock, R. L., Close approximation to invariant tori in nonlinear mechanics. *Physical Review Letters*, **66, 14**, 1803-1806 (1991).
20. Hoffstätter, G. H., *Rigorous bounds on survival times in circular accelerators and efficient computation of fringe–field transfer maps*. PhD thesis, Michigan State University, East Lansing, Michigan, USA (1994). Also DESY **94-242**.

Longitudinal Wake Field Corrections in Circular Machines *

Keith R. Symon

Argonne National Laboratory, 9700 South Cass Avenue, Argonne, IL 60439 USA
(permanent address: Dept. of Physics, University of Wisconsin-Madison,
1150 University Ave., Madison, WI 53706 USA.)

Abstract. In computations of longitudinal particle motions in accelerators and storage rings, the fields produced by the interactions of the beam with the cavity in which it circulates are usually calculated by multiplying Fourier components of the beam current by the appropriate impedances. This procedure neglects the slow variation with time of the Fourier coefficients and of the beam revolution frequency. When there are cavity elements with decay times that are comparable with or larger than the time during which changes in the beam parameters occur, these changes can not be neglected. Corrections for this effect have been worked out in terms of the response functions of elements in the ring. The result is expressed as a correction to the impedance which depends on the way in which the beam parameters are changing. A method is presented for correcting a numerical simulation by keeping track of the steady state and transient terms in the response of a cavity.

INTRODUCTION

In many calculations and simulations of longitudinal motion in particle accelerators we find the longitudinal electric field by first finding its Fourier transform and then inverting the transform. If we have the Fourier transform $\hat{E}(\omega)$ of the complete electric field history $E(t)$, then its inverse transform will of course give the correct field $E(t)$. However we do not know to begin with either the complete field history $E(t)$ or its transform $\hat{E}(\omega)$. That is one of the things we want to calculate, so it is available only at the end of the calculation. This problem has also been treated by J. Machlachlan[1].

---

*Work supported by the U.S. Department of Energy, Office of Basic Energy Sciences, under Contract No. W-31-109-ENG-38.

We will use the azimuthal coordinate θ to locate a point around the ring in a circular accelerator. In order to distinguish between rapidly varying quantities like θ and slowly varying quantities, we introduce a reference angle

$$\Theta = \int_0^t \Omega(t')dt', \qquad (1)$$

where $\Omega(t)$ is a slowly varying reference angular velocity which is close to the angular velocities of particles in the beam. It is intended that Θ shall maintain its position relative to the beam so that the relative phase

$$\phi = \theta - \Theta \qquad (2)$$

is slowly varying in comparison with the revolution time or the decay time of any circuit element around the ring.

In a circular machine, we usually know, at any time in the calculation, the linear particle density $\rho(\phi)$, which is slowly varying. We then proceed as follows. The linear charge density is $e\rho(\phi)$, where e is the charge on a particle. We write $\rho(\phi)$ in terms of its Fourier transform:

$$\rho(\phi) = \sum_{k=-\infty}^{\infty} \hat{\rho}_k e^{ik\phi}, \qquad (3)$$

and use the equation of continuity for the current to get

$$\hat{J}_k = e\Omega R \hat{\rho}_k, \qquad (4)$$

where $\hat{J}_k$ is the Fourier transform of the current $J(\phi)$ and $2\pi R$ is the circumference. We will express the azimuthal electric field in terms of the voltage per turn $V(\phi)$ delivered to a particle at ϕ. The Fourier transform of $V(\phi)$ may then be written:

$$\hat{V}_k = -Z(k\Omega)\hat{J}_k, \qquad (5)$$

where $Z(k\Omega)$ is the impedance at the frequency $k\Omega$ associated with the Fourier component k.

The above calculation of the voltage per turn neglects changes in $\rho(\phi)$ with time. If the ring contains any element which produces an extended wake field, it may be necessary to take into account such changes. The purpose of this paper is to find the correction to formula (5) which properly treats long lasting wake fields.

THE VOLTAGE ACROSS A RING ELEMENT

The voltage response function $K(t)$ of an element of the accelerator ring to a unit current impulse at $t = 0$ may be written in the form:

$$K(t) = -K_0 e^{-\gamma t} \sin(\omega t + \zeta), \qquad (6)$$

where the phase ζ is given by the condition that $\partial K/\partial t$ vanish at $t = 0$:

$$\tan \zeta = \frac{\omega}{\gamma}. \tag{7}$$

Equation(6) is written for the underdamped case, but the other cases may be handled in a similar way. The voltage across the element at any time t is

$$V(t) = \int_{-\infty}^{t} J(t')K(t-t')dt' = \int_{0}^{\infty} J(t-\tau)K(\tau)d\tau. \tag{8}$$

If we represent this element by a series RLC circuit with the voltage appearing across the capacitance, then we have the relations:

$$\begin{aligned}
\gamma &= \frac{R}{2L} = \frac{\omega_0}{2Q} \quad, \quad Q = \frac{\omega_0 L}{R} \quad, \quad \omega_0 = \frac{1}{\sqrt{LC}} \quad, \\
\omega &= \left(\frac{1}{LC} - \left(\frac{R}{2L}\right)^2\right)^{1/2} = \omega_0 \left(1 - \frac{1}{4Q^2}\right)^{1/2} \quad, \\
\tan \zeta &= \left(4Q^2 - 1\right)^{1/2} \quad, \quad K_0 \sin \zeta = \frac{1}{C} \quad, \quad K_0 = \frac{1}{C}\left(1 - \frac{1}{4Q^2}\right)^{-1/2}
\end{aligned} \tag{9}$$

An accelerating cavity can be represented as a set of modes, each represented by an RLC circuit as above. The complete ring can be represented as a sequence of elements, so that the total voltage per turn is the sum of the voltages of all the elements. Since the total charge density and current are sums over Fourier components as given by Eq.(3), we will consider the response of a single element to a single Fourier component of the current. Let the element j be located at the azimuth θ_j. Its response to the current

$$J_k = \hat{J}_k e^{ik\phi} = \hat{J}_k e^{ik\theta_j} e^{-ik\Theta(t)} \tag{10}$$

is given by Eq.(8):

$$V(t) = -\int_0^\infty \hat{J}_k(t-\tau) e^{ik\theta_j} e^{-ik\int_0^{t-\tau}\Omega(t')dt'} K_0 e^{-\gamma\tau} \sin(\omega\tau + \zeta)d\tau, \tag{11}$$

where we have explicitly indicated the possible time-dependence of $J(\phi)$ and Ω.

If we neglect these time-dependences, we get

$$V(t) = \hat{V}_k e^{ik\theta_j} e^{-ik\Omega t}, \tag{12}$$

where

$$\hat{V}_k = -\hat{J}_k Z(k\Omega), \tag{13}$$

and

$$Z(k\Omega) = \int_0^\infty e^{(-\gamma + ik\Omega)\tau} K_0 \sin(\omega\tau + \zeta)d\tau. \tag{14}$$

Equation (12) can be rewritten using Eq.(2) in the form:

$$V(\phi) = \hat{V}_k e^{ik\phi}, \quad (15)$$

where we have substituted $t = (\theta_j - \phi)/\Omega$, since the element j is located at θ_j. Equation (15) justifies the use of the notation $\hat{V}_k$ in Eq.(12).

The integral in Eq.(14) can be evaluated. It is not hard to show, using Eqs.(9), that

$$Z(k\Omega) = \frac{R - ik\Omega L}{1 - k^2\Omega^2 LC - iRk\Omega C}. \quad (16)$$

This is the correct formula for the impedance across a capacitance which is part of an RLC series circuit.

Now let us take into account the slow time-dependence of $J(\phi)$ and Ω. In order to write the result as a correction to Eq.(14), we substitute in Eq.(11):

$$\hat{J}_k(t-\tau)e^{-ik\Theta(t-\tau)} = \hat{J}_k(t)e^{-ik\Theta(t)}e^{ik\Omega(t)\tau}F(t,\tau), \quad (17)$$

where the function $F(t,\tau)$ approaches 1 as $\tau \to 0$ and is slowly varying in t. Equation (11) then again gives the generalization of Eq.(12):

$$V(t) = \hat{V}_k(t)e^{ik\theta_j}e^{-ik\int_0^t \Omega(t')dt'}, \quad (18)$$

where

$$\hat{V}_k(t) = -\hat{J}_k(t)Z_{\text{eff}}(k\Omega(t)) \quad (19)$$

is a slowly varying function of t and the effective impedance is

$$Z_{\text{eff}}(k\Omega) = \int_0^\infty F(t,\tau)e^{(-\gamma+ik\Omega(t))\tau}K_0\sin(\omega\tau + \zeta)d\tau. \quad (20)$$

This formula gives the correction to the impedance taking into account the slow time-dependence of $J(\phi)$ and Ω. The impedance is now also a slowly varying function of t. The correction factor given by Eq.(17) is:

$$F(t,\tau) = \frac{\hat{J}_k(t-\tau)}{\hat{J}_k(t)}e^{ik\int_0^\tau [\Omega(t-\tau')-\Omega(t)]d\tau'}. \quad (21)$$

Note that the exponent is of order τ^2.

IMPEDANCE CORRECTION – ANALYTIC CASE

The factors in Eq.(21) can be expanded in power series in τ:

$$\begin{aligned}
F(t,\tau) &= \frac{1}{\hat{J}_k(t)}\sum_0^\infty \frac{(-\tau)^n}{n!}\frac{\partial^n \hat{J}_k(t)}{\partial t^n} e^{-ik\sum_{n=2}^\infty \frac{(-\tau)^n}{n!}\frac{\partial^{n-1}\Omega(t)}{\partial t^{n-1}}} \\
&= \sum_{n=0}^\infty r_n(-\tau)^n. \quad (22)
\end{aligned}$$

The first few coefficients are

$$r_0 = 1 \quad , \quad r_1 = \frac{\dot{\hat{J}}_k}{\hat{J}_k} \quad , \quad r_2 = \frac{1}{2}\frac{\ddot{\hat{J}}_k}{\hat{J}_k} - \frac{1}{2}ik\dot{\Omega} \quad ,$$

$$r_3 = \frac{1}{6}\frac{\dddot{\hat{J}}_k}{\hat{J}_k} - \frac{1}{6}ik\ddot{\Omega} - \frac{1}{2}\frac{ik\dot{\Omega}\dot{\hat{J}}_k}{\hat{J}_k} \quad . \tag{23}$$

We substitute these results into Eq.(20) to obtain

$$Z_{\text{eff}}(k\Omega) = Z(k\Omega) + \sum_{n=1}^{\infty} r_n \frac{\partial^n Z(k\Omega)}{\partial \gamma^n}, \tag{24}$$

where $Z(k\Omega)$ is the uncorrected impedance given by Eq.(14). The sum in the second term is the correction due to the time dependence of $\rho(\phi)$ and $\Omega(t)$.

IMPEDANCE CORRECTION – NUMERICAL SIMULATION

In this section we will evaluate the impedance correction for a numerical simulation using the leap-frog algorithm to move the particles. A formula derived from Eqs.(20) and (21) is given in an ANL report [2]. We present here a simpler way to handle this problem by writing the voltage across a circuit element as the sum of a steady-state term plus a transient.

In a simulation using the leap-frog algorithm, the particle positions and the resulting density $\rho_n(\phi)$ are computed at the beginning of each time step at $t_n = n\Delta t$. We will use the subscripts n and $n+1/2$ to denote quantities evaluated at the corresponding times t_n and $t_{n+1/2} = (n+1/2)\Delta t$. The density $\rho_n(\phi)$ is taken to be constant during the time interval $t_{n-1/2} \leq t < t_{n+1/2}$. We will calculate from this density in a standard way, for example by using Fourier transforms, the steady state response $V_{\text{ssn}}(\phi) = V_{\text{ssn}}(t)$ of each circuit j during this n^{th} time interval, where ϕ and t are related by Eq.(2). We then write the total voltage as the steady state plus a transient:

$$V_n(t) = V_{\text{ssn}}(t) + V_{\text{trn}} e^{-\gamma(t-t_{n-1/2})} \sin[\omega(t - t_{n-1/2}) + \xi_{\text{trn}}], \tag{25}$$

where ω, γ, ζ (for element j) are given by Eqs.(7) and (9).

We now match the solution at the beginning of time step n to that at the end of time step $n-1$:

$$V_{n-1}(t_{n-1/2}) = V_{\text{ssn}}(t_{n-1/2}) + V_{\text{trn}} \sin \xi_{\text{trn}} \quad ,$$

$$\left.\frac{\partial V_{n-1}}{\partial t}\right)_{t=t_{n-1/2}} = \left.\frac{\partial V_{\text{ssn}}}{\partial t}\right)_{t=t_{n-1/2}} - \gamma V_{\text{trn}} \sin \xi_{\text{trn}} + \omega V_{\text{trn}} \cos \xi_{\text{trn}} \quad . \tag{26}$$

The solution of these equations for the transient during the nth time step is

$$V_{\text{trn}} = \left[A^2 + B^2\right]^{1/2} \quad , \quad \tan \xi_{\text{trn}} = \frac{A}{B} \quad , \tag{27}$$

where

$$A = V_{n-1}(t_{n-1/2}) - V_{\text{ssn}}(t_{n-1/2}) \quad , \quad B = \omega^{-1}\left[\left(\frac{\partial V_{n-1}}{\partial t}\right)_{t=t_{n-1/2}}\right.$$
$$\left. + \gamma V_{n-1}(t_{n-1/2}) - \gamma V_{\text{ssn}}(t_{n-1/2}) - \left(\frac{\partial V_{\text{ssn}}}{\partial t}\right)_{t=t_{n-1/2}}\right] \quad (28)$$

The reference angular velocity $\Omega(t) = \Omega_n$ is taken as constant during the interval $t_{n-1/2} \leq t < t_{n+1/2}$. Equation (1) then gives

$$\Theta(t) = \Omega_n(t - t_{n-1/2}) + \Theta_{n-1/2}, \quad (29)$$

where

$$\Theta_{n-1/2} = \sum_{n'=1}^{n-1} \Omega_{n'}\Delta t + \Omega_0 \frac{\Delta t}{2}. \quad (30)$$

According to Eqs.(2), (30) and (29) a particle with coordinate ϕ will arrive at the element j at times given by

$$\phi = \theta_j - \Theta_{n-1/2} + \Omega_n(t - t_{n-1/2}) + 2\pi\ell, \quad (31)$$

where ℓ is any integer which puts t in the interval $t_{n-1/2} \leq t < t_{n+1/2}$. The first such time is given by

$$t - t_{n-1/2} = \delta t = \frac{\phi - \theta_j - \Theta_{n-1/2} + 2\pi\ell}{\Omega_n}, \quad (32)$$

where ℓ is the smallest integer which makes δt non-negative. The particle at ϕ thus crosses the element j at the times $t = t_{n-1/2} + \delta t + 2\pi r/\Omega_n$. The integer r runs over the range $0 \leq r \leq r_1$, where

$$r_1 = [\text{ nearest integer } \leq (\Omega_n \delta t/2\pi) - 1]. \quad (33)$$

The total voltage increment given to the particle by the transient during the nth time interval is

$$\delta V = \sum_{r=0}^{r_1} V_{\text{trn}} e^{-\gamma[\delta t + (2\pi r/\Omega_n)]} \sin[\omega(\delta t + 2\pi r/\Omega_n) + \xi_{\text{trn}}]. \quad (34)$$

The energy of each particle at $t_{n+1/2}$ is obtained by adding to the energy at $t_{n-1/2}$ the increment $eV_{\text{ssn}}(\phi) + e\delta V$.

In the leap-frog algorithm, the new energies at time $t_{n+1/2}$ are then used to advance the particle phases from their values at time t_n to time t_{n+1}.

References

[1] J. A. MacLachlan, Fundamentals of Particle Tracking for the Longitudinal Projection of Beam Phase Space in Synchrotrons, FNAL Report FN-481, April 15, 1988.
[2] Keith Symon, Longitudinal Wake Field Corrections in Circular Machines, ANL Report NSA-96-1, August 14, 1996. [Available by request from the author.]

A C++ Implementation of the Differential-Algebraic Model for the TASCC Superconducting Cyclotron

S.R. Douglas, W.G. Davies, and G.E. Lee-Whiting

AECL Chalk River Laboratories
Chalk River ON, K0J 1J0

Abstract: A new differential-algebra (DA) computer model for the orbit dynamics of the TASCC superconducting cyclotron has been developed at Chalk River; it produces a Taylor series solution of Hamilton's equations to high order and high precision. Analytic expressions for the magnetic and RF fields are used. The equations of motion are integrated with solvers explicitly written for DA variables; linear and nonlinear stability analysis is made using Lie algebra, and normal forms. The code is written in C++, which allows the physics to be represented directly in the C++ language, including the DA variables, the vector potentials, and symplectic matrices. The code runs unchanged on both UNIX and MS-Windows. An analysis program, written in C++, for MS-Windows allows the visualization of beam trajectories and the results of the stability analysis. The features of object-oriented programming were necessary for a simple and successful implementation of this complex physical model. Examples of the object-oriented concepts applied here will be given to illustrate their power and utility.

INTRODUCTION

A new orbit dynamics code has been developed for the TASCC (1,2) superconducting cyclotron at Chalk River. The code uses techniques of differential algebra (DA) (3) to generate a Taylor series solution of Hamilton's equations about a central trajectory. Differential algebra provides a very accurate method (no finite differences, hence no truncation error) of obtaining aberration expansions. For DA to be used, the vector potential, A, appearing in the Hamiltonian must be in analytic form.

The calculation for the TASCC cyclotron uses a three-dimensional analytic model (4,5) of the magnetic vector potential, which satisfies Maxwell's equations exactly. The model consists of separate potentials for the superconducting coils, the saturated-iron poles, and 104 iron "trim rods", which are used instead of trim coils to make fine adjustments of the field. These potentials, accurately differentiable to high order, are computationally intensive even though considerable effort has been expended in optimizing performance. The radio-frequency accelerating field is represented in the Coulomb gauge by a time-dependent magnetic vector potential, expanded as a Fourier-Bessel series, and also satisfies Maxwell's equations to the required order.

Since Hamilton's equations form a Lie algebra, powerful techniques (2) can be used to analyse and interpret results of the calculation. DA also provides accurate tools for transforming the Taylor series into the Lie representation.

The C++ programming language provides powerful, elegant tools for efficiently implementing DA and the cyclotron model.

CALCULATIONAL MODEL AND TYPES OF SOLUTION

What does the program calculate? Solving Hamilton's equations as parametrized by angle produces the trajectory of an ion through many turns of the cyclotron. If the integrator is symplectic, or at least very accurate, phase space volumes will not shrink or grow during a hundred turns in the cyclotron. One key to further information is to use automatic differentiation via differential algebra (DA) to do a sensitivity analysis about a central trajectory (3). All algorithms, including the numerical integrator, formulas for vector potentials, and Hamilton's equations, must be replaced by their DA equivalents. The numerical solution of Hamilton's equations uses either a DA Runge-Kutta integrator or a DA Bulirsch-Stoer integrator. Symplectic integrators have also been considered. The DA integrations of Hamilton's equations produce maps (linear, quadratic, cubic, etc, to arbitrary order) of a Taylor expansion about the central trajectory, which is also found.

Linear maps are elements of a linear symplectic group and can be analysed for beam stability. The full nonlinear map can be rephrased in the language of Lie groups and used to study hidden symmetries in the beam dynamics or to study nonlinear resonances by normal form analysis. Single-turn, multi-turn, and partial-turn maps can be determined, both with and without the RF field. Maps also provide phase space tracking by iterating ions through a map and producing a Poincare section in phase space. All of these approaches improve understanding of beam stability.

Automatic differentiation using DA can also be targetted at sensitivity studies of parameters in the field models, thereby investigating dynamics issues such as the effect of vertical deflections or the contribution of field components to nonlinear resonance behaviour. In summary, the reformulation of beam dynamics using DA greatly extends the range of beam analysis. The drawbacks are complexity of formulas, increased computer time, and the need for analytic expressions for A. The next section explores these drawbacks from the perspective of object-oriented programming.

COMPUTER ISSUES

General Object-Oriented Design

Step 1 in an object-oriented (OO) design is to talk to the expert in the language of the problem: beam dynamics -- ions, fields, Hamilton's equations, phase space, nonlinear resonances, maps, and so on. The discussion scrupulously avoids any software issues and instead focusses on concepts that are fundamental to beam dynamics. They will form permanent building blocks of any software to describe it. The expert explains the relationship between key concepts, which will form the basis of the design.

Step 2 repeats step 1, but with a shift in emphasis. With the expert's help, structural relationships of concepts and service relationships between concepts are noted. A structural framework for the program gradually forms and the service properties of each key concept are revealed.

Step 3 considers building a "new" computer language tailored to beam dynamics. What features of this new language would make it ideal? Obviously, these features are the key concepts and their service and structural relationships. Code would be written using variables or "objects" that appear indivisibly such as a Hamiltonian, a Vector Potential, an Ion, a Map, and so on. With the classing mechanism in C++, each of these concepts is expressed as a "class", which is a factory for creating objects to represent specific examples of a concept. Helper concepts may also be useful, such as a place to store points for scatter plots.

Step 4 writes the software using these objects as needed and implements the solution with the structure and services natural to beam dynamics. A class is written to provide properties and actions for each key concept. A specific object of a class in an OO program responds when asked with data or actions. The level of abstraction and the interface of response is strictly controlled for a class, thereby hiding messy details and leaving flexibility for changes. Testing, validation, and verification takes place on a class by class basis.

Specific Object-Oriented Implementation

Here we describe the implementation of the beam dynamics calculation for the TASCC cyclotron using OO programming in C++. Class names are written as capitalized words with intervening spaces removed, e.g., PhaseSpacePoint.

A simple example of the class mechanism is the Hamiltonian class. An object of a Hamiltonian class describes the physical system. As part of its data each Hamiltonian object has an Ion object and an ElectroMagneticPotential object, two class abstractions at a level appropriate to the Hamiltonian concept. In turn, an Ion object has a Charge object, a Mass object, and a PhaseSpacePoint object of initial conditions. The Ion object will reply to questions concerning mass, position, charge, energy, and momentum as well as know how to be written to or read from disk. An ElectroMagneticPotential object has a list of VectorPotential objects where any number of field models may contribute to the potential A. The ElectroMagneticPotential object provides vector potential components and derivatives as needed by Hamilton's equations.

A simple example of an inheritance hierarchy is the VectorPotential class. As noted, a VectorPotential object represents part of the total electromagnetic potential of the field models. An ElectroMagneticPotential object is expecting to deal with each listed VectorPotential object in the same way using the same interface. However, although the questions asked of each VectorPotential object may be the same, the details of how the components of A are calculated certainly are not. To provide a common interface but a customized implementation, each vector potential in the field model is derived from a base VectorPotential class using public inheritance. The base class provides the common interface, but each derived class, such as CoilPotential, customize the required actions and store any special parameter data such as currents in the superconducting coils. By separating objects of the VectorPotential within an ElectroMagneticPotential class, parts of the field model are easily excluded without disturbing the solution structure.

How can new vector potentials be added to the model without changing the underlying solution structure? The answer to this question illustrates a key feature of OO programming. The interface actions of the base VectorPotential class, such as values and derivatives with respect to canonical variables, are declared to be "virtual", which allows specific actions of a derived class to be slotted into the formulas for Hamilton's equations at run time. No need to recompile or modify the solution structure to accomplish this; the C++ compiler handles it explicitly, even for derived VectorPotential classes created long after compilation. This makes the code very flexible and supports easy modifications to the overall model.

Another idea of OO programming with C++ that was used extensively is operator overloading. The C++ language allows the reuse of any standard arithmetic symbols (such as +, -, *, and /) for any class for which the symbols have sensible meanings as defined by the programmer. Furthermore, standard mathematical functions such as sqrt() or exp() can be overloaded for these classes. Analytical formulas for the vector potentials of coils, poles, and trim rods are complicated enough without the added complexity of DA evaluation. By overloading all arithmetic and mathematical functions for the DA class, the formulas for the vector potentials remain unchanged, although nonlinear maps are produced during integration instead of only the central trajectory. Also, operators relating the Map class, the PhaseSpacePoint class, and the SymplecticMatrix class were overloaded so that standard formulas in Lie groups and algebras would have their usual meanings. Controlling the complexity of formulas was vital to the successful completion of this project.

For a calculation of many turns of an ion through the cyclotron, millions of DA objects are created and destroyed dynamically on the heap. Profiling the code showed that this is a serious bottleneck to fast execution. C++ permits the programmer to take control over memory allocation on a class by class basis. Specifically, the "new" and "delete" operators for the DA class were overloaded to provide alternate memory allocation instead of the heap. At the start of a simulation, a large number of DA objects are acquired from the heap and placed in a stack hidden within the DA class itself. The stack is declared static, which means it is the same and available to every DA object. The DA constructors and destructor and the DA new and delete operators then acquire pre-allocated storage transparently from the stack memory pool. Execution time reduced by an order of magnitude. This is an excellent example of the power of C++ to allow the programmer to take control of details whenever it becomes necessary, while still maintaining simplicity of coding style and of interfaces for the rest of the code.

Parts of the model for A that span different vector potential classes need to evaluate the same function with the same argument. Recalculation is wasteful, but breaking the simplicity of a class hierarchy by global variables makes code inflexible and unwieldy. The solution is "memoization", a concept borrowed from the AI community. Results of the member function in question are cached in a short static list in its base class. A new function evaluation scans this list for the function argument and returns the result if known or calculates and caches it if unknown. The speedup is substantial.

Container classes, such as the stack just mentioned, are common tools of OO programming. Containers allow the storage, retrieval, and manipulation of objects without placing restriction on the number or exact type of the object. For classes derived from base classes that use the virtual function mechanism, containers holding objects of these derived classes can provide customized actions in a uniform way. The list of VectorPotential objects in an ElectroMagneticPotential object is one example. Another is the storage and manipulation of phase space points for scatter plots.

CONCLUSIONS

This paper describes some advantages of using OO features of the C++ language in writing a program for the orbit dynamics of the TASCC superconducting cyclotron. Differential algebra allows the accurate study of high-order, nonlinear effects arising from magnetic and RF potentials within the Hamiltonian. Storing the coefficients of a multi-dimensional Taylor series, a DA object is complicated, but the combined power of classes with operator overloading makes it appear like a simple scalar. The need for accurate analytic vector potentials makes the model calculationally intensive. The use of "memory pools" and "memoization" speed up the computations substantially. Classes for phase space, maps, and symplectic matrices help simplify the use of DA in the Lie algebraic formalism.

The beam dynamics calculation runs on both MS-Windows/DOS and UNIX(SGI). A graphical user interface is provided for the MS-Windows version, although it is shut off by conditional compilation for the UNIX version.

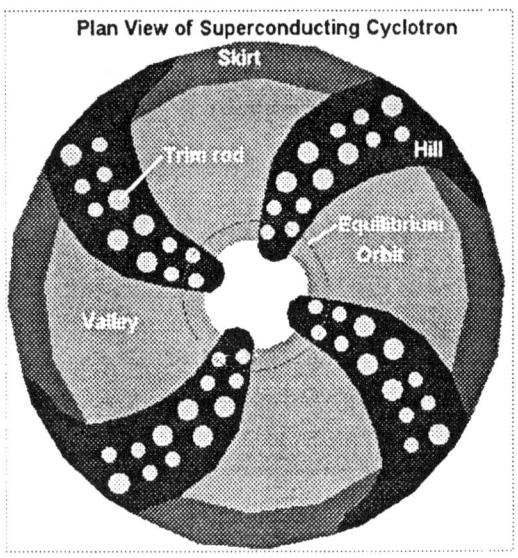

FIGURE 1. Plan view of the TASCC cyclotron. The magnetic field model generated the features shown. Also shown is an inner orbit about which are computed the DA maps.

ACKNOWLEDGEMENTS

The authors thank G. Pusch, D. Dunford, and A. Leung for their efforts on this project and thank H. Schmeing and J. Hardy for their support.

REFERENCES

1. Davies, W.G., Douglas, S.R., Pusch, G.D., and Lee-Whiting, G.E., "The Chalk River Differential Algebra Code DACYC and the Role of Differential and Lie Algebras in Understanding the Orbit Dynamics of Cyclotrons", Inst. Phys. Conf. Ser 131(1993)147.
2. Davies, W.G., "Advantages of Using Differential-Algebra and Lie-Algebra in the Orbit Dynamics of Cyclotrons", Proc. 13 International Conf. on Cyclotrons and their Applications, World Scientific, Singapore, (1992) 372.
3. Berz, M., "Differential Algebraic Description of Beam Dynamics to Very High Order", Part. Accel 24(1989)109
4. Lee-Whiting, G.E., and Davies, W.G., "Analytic Representation of Cyclotron Magnetic Field", 13th Inter. Conf. on Magnet Tech. Victoria, IEEE Trans. Magn. 30(1994)1663.
5. Davies, W.G., Lee-Whiting, G.E., Douglas, S.R., and Pusch, G.D., "Three-Dimensional Analytic Model of the Magnetic Field of the Chalk River Superconducting Cyclotron", 13th Inter. Conf. on Magnet Tech. Victoria, IEEE Trans. Magn. 30(1994)1839.

MXYZPTLK and BEAMLINE: Status and Future

L. Michelotti
Fermi National Accelerator Laboratory[†]
P. O. Box 500, Batavia, IL 60510, USA

Abstract

MXYZPTLK and **BEAMLINE** resulted from early attempts to write C++ class libraries for accelerator analysis and modeling. We will describe a few features of their more recent, and possibly final, versions.

MXYZPTLK

The first version of **MXYZPTLK** (pronounced $z\bar{\imath}' \cdot tl$), was finished by 1990. [1, 2] It was originally intended to demonstrate how one might go about implementing the ideas of automatic differentiation and differential algebra using object-oriented programming (OOP), in particular, using C++ . Even back then, several languages were available for doing OOP: ADA, Smalltalk, Objective-C, Eiffel; even LISP could be manipulated to make it seem object oriented. (And there is now an "object-oriented LISP," although naively this seems to contradict OOP's strong typing requirements.) However, my primary goal was, and remains, to make the interface simple and transparent so that the syntax would be as natural as possible. The principal reason for choosing C++ even though it was at that time *very* immature, was based mostly on its syntactical style. Because of it, I felt that scientists be more attracted to this language not so violently opposed to learning it.[*]

[†] Operated by the Universities Research Association,Inc, under contract with the U.S. Department of Energy

[*] In this context, operator overloading was an especially necessary requirement. It is interesting that JAVA expressly forbids it.

MXYZPTLK: Environments

A few points of conceptual difference between **MXYZPTLK** and many other differential algebra packages emerged early. To begin with, the classes (objects) are **Jet**s, not simply truncated power series. Mathematically, a "Jet" is an equivalence class of functions, the equivalence being based on equality of all derivatives, up to some order, at some point in the functions' domain, what we will call its "reference point." Implemented as a C++ class, the Jet is more than a truncated polynomial; it is a truncated polynomial *and* an "environment." Each Jet carries around a pointer to its environment, which contains data like (a) the number of variables in the Jet, (b) the separation between "phase space" and "control" variables, (c) the maximum degree of its accurately computed coefficients – very useful information after derivatives are taken – and (d) its reference point.

Why bother carrying the reference point? There are two reasons for carrying this extra information around. A Jet variable is supposed to model a function, or more precisely, an equivalence class of functions. Not carrying information about that point means the representation is not complete. For example, if one tries to multiply two Jets whose reference points differ, the results are meaningless unless the reference point information is explicitly taken into account. (And even than it is incorrect if the function being modelled is not legitimately a polynomial.) The second reason is related to the first. Let g be the Jet attached to point $a \in R$ containing the function $x \mapsto e^x$; let h be the Jet attached to point $b \in R$ containing $x \mapsto e^x$; let f be the Jet attached to a point $c \in R^2$ containing $(x, y) \mapsto e^{x+y}$. Upon evaluation, one would like to write simply $e^w = f(w) = g(w) = h(\rho, w - \rho)$, at least to the appropriate level of approximation. On the other hand, if these were just polynomials, one would have to write, $e^w = f(w - a) = g(w - b) = h(\rho - c_1, w - \rho - c_2)$. This is not only looks awkward; more importantly, it forces the programmer to perform extra manipulations *that are more appropriately handled by the object itself*. In turn, this increases the likelihood of error, especially when Jets are passed to subroutines and modules disconnected from their initial declarations.

Another difference worth mentioning is that in **MXYZPTLK** each Jet explicitly tracks its most accurate order through operations. In this way, one is assured of not operating on meaningless data (after derivatives are taken, for example).

The motivation for both of these features (and others) was to relieve the user of having to "remember" details that can be handled by the object itself, which decreases the opportunities for programming errors. Avoiding making such unnecessary demands on the programmer is part of the "user-friendly" goals of **MXYZPTLK**.

One of the major limitations of earlier versions of **MXYZPTLK** was the fact that there was one and only one environment for the entire calculation, defined by a static initialization function, **Jet::Setup**. Thus, in order to do both a 12th order calculation in six variables and a 1st order calculation in one variable in the same program, one was forced to use the environment of the former to perform the latter. This deficiency has been corrected in more recent versions. Multiple environments are possible. They are specified by sandwiching declarations of **coord** variables – i.e., the coordinates of the calculation – between **Jet::BeginEnvironment** and **Jet::EndEnvironment** statements. The argument of **Jet::BeginEnvironment** specifies the order associated with the environment, while arguments of the **coord** declarations fix the reference point. The **coord**s themselves serve as the starting objects for all calculations within the environment.

An environment is itself an object with functionality. The **Jet::EndEnvironment** statement actually returns a pointer to the environment created. If that pointer is stored, one can access the environment's **coord**s at any time or query the environment for other information about itself. In addition, **coord** objects can themselves be changed in the program using an overloaded **coord::operator=**.[†]

MXYZPTLK: JetVector, Map, and LieOperator

A confusing "feature" in former versions of **MXYZPTLK** was that the **LieOperator** object had more functionality than was appropriate to it. This has been fixed by building separate **Map** and **LieOperator** objects upon the base class of **JetVector**. **JetVector** handles the behavior appropriate to a vector space over the reals, or, in fact, over Jets. To this is added functionality appropriate to **Map**s and **LieOperator**s without mixing: class **Map** acts like a multi-variable function, while class **LieOperator** acts like a derivation.

These classes are extended into the complex domain with **JetCVector**, **MapC**, and **CLieOperator** objects. Conversion constructors and operators easily handle transitions from the double based objects to their complex based counterparts.

BEAMLINE

The first **BEAMLINE** library was finished a year after **MXYZPTLK** [3] and has been upgraded continually since then. The intent was to provide C++ objects for modelling the components of an accelerator. Integration with the **MXYZPTLK** library, for the purpose of calculating maps, was part and parcel of this from the beginning. The **BEAMLINE** library contains classes

---

[†]However, this does have the effect of changing the reference point of the environment, altering the interpretation of all the Jets already associated with that environment.

for modelling magnetic and electrostatic elements and editing functions for building them into accelerator models. Virtual **.propagate** functions allow for tracking or construction of maps, when linked with the **MXYZPTLK** library.

BEAMLINE: Calculator objects and Barnacles

One approach for doing accelerator calculations which we have found very useful is to design a class which does the calculation and attaches its results to elements upon request. The base class for all such information is called **Barnacle**; essentially, it acts as a sort of Post-It note. Each beamline element contains a **BarnacleList** object which collects these notes for later access.

The classes which are derived from **Barnacle** usually contain information left behind by a **Calculator** object, such as the local dispersion or lattice functions. But the information really can be anything, such as hooks for accessing data from a control console. [4] Classes derived from Barnacle will tag their information with an ASCII string upon attaching the Barnacle. It can then be accessed any time later by a module visiting that beamline element by invoking the appropriate string. Thus, many different Barnacles can be attached to the same element without confusion.[‡]

For example, the EdwardsTeng Calculator was the original motivation for the invention of the Barnacle. Among the parameters associated with the EdwardsTeng functor, or any other Calculator, are user-defined "criterion" functions, used to determine to which beamline elements the information Barnacles get attached. In principle, the criterion can be set up in any desired manner. To make use of this, it is essential that elements of the **BEAMLINE** library possess RTTI-like functions (see below).

Notice that there is some similarity between a Calculator object and what is now called a Visitor, enough that the former could be translated into the latter with little difficulty.

BEAMLINE: RTTI-like functions

We do not have the time to consider all the features that have been added to **BEAMLINE** objects over the years: member functions that edit beamlines, splitting elements, being able to insert thin elements into other elements, and so on. Worth mentioning in passing, however, are virtual member functions — such as **.Clone**, **.Name**, and **.Type** — which enable a basic level of emulating RTTI-like behavior. All are pure virtual, inlined functions. The **.Clone** function return a new beamline element – which may, in fact, be a beamline –

[‡]This is different from a "dictionary," in that many Barnacles can have the same key. In such cases, the ordering is important.

that is a copy of the element calling it; the **.Name** function returns the name of the object, if it has one, and has been especially useful for the interactive graphics of the beamline model; the **.Type** function returns a string which is the particular type of the beamline element. These, and others, have been essential in setting up the criterion functions for Calculator objects.

MXYZPTLK/BEAMLINE: Current and future usage

We have been using the **MXYZPTLK/BEAMLINE** libraries for some time to do off-line calculations of Fermilab's accelerators' properties. These include the usual repertoire: closed orbits; tunes; dispersion, and other "lattice functions"; tune shifts and footprints. Most recently, features have been added to the **BEAMLINE** objects so that models can be connected directly to our accelerator control systems. These "on-line" models can read parameters directly from and output results of calculations directly to control consoles. A new beamline element, the **TclMonitor**, provides a TCL/TK graphic interface to the screen that can sample and control tracking calculations at any position (or positions) in the accelerator. [4] Other recent applications have included such things as: (a) studying the influence of rolled quads on beam optics in the CDF and D0 interaction regions [5]; (b) calculating beambeam tuneshifts exactly for Run II, using the correct, coupled "clothed" orbits; (c) modelling extraction in a fixed target configuration of the Tevatron and associated beamlines; (d) usage as base classes for the development of more realistic hardware modelling objects, to be used in a C++ version of Nikolai Mokhov's MARS code [6], which calculates energy deposition due to particles scattered out of the beam[§]; (e) and studying matching from uncoupled to coupled lattices [7].

For the immediate future, **MXYZPTLK** and **BEAMLINE** will continue to be used in the ongoing development of applications already mentioned, especially in conjunction with the online model. In addition, both libraries are more or less in a continual state of upgrade. For example, the highest order retained for each variable need not be the same for all variables. (One will be able to specify this with a second argument to the **coord** constructor.) The old "dlist" structure will give way to STL containers and iterators. "Slot" objects are going to be introduced to completely separate geometry from the beamline element. This last, in particular, is an idea fostered by the **CLASSIC** collaboration. [8] Ultimately, the **BEAMLINE** interface will be rewritten to comply with **CLASSIC** header files. I hope that the **CLASSIC** collaboration will see fit to preserve some of the more useful **MXYZPTLK/BEAMLINE**

[§]This work is being done by Mokhov's graduate student, Oleg Krivosheev features, especially those related to "user-friendliness." How much ultimately survives, either in software or ideas, remains to be seen.

References

[1] Leo Michelotti. MXYZPTLK: A C++ hacker's implementation of automatic differentiation. In G. Corliss and A. Griewank, editors, *Automatic Differentiation of Algorithms: Theory, Implementation, and Application.* SIAM, 1991. Philadelphia, PA.

[2] Leo Michelotti. MXYZPTLK: A practical, user-friendly C++ implementation of differential algebra: User's guide. Fermi Note FN-535, Fermilab, January 31, 1990.

[3] Leo Michelotti. MXYZPTLK and BEAMLINE: C++ objects for beam physics. In *Advanced Beam Dynamics Workshop on Effects of Errors in Accelerators, their Diagnosis and Correction. (Corpus Christi, Texas. October 3-8, 1991).* American Institute of Physics, 1992. Conference Proceedings No.255.

[4] J. A. Holt, A. Braun, L. Michelotti, and M. Martens. Accelerator physics computing in a control system enviroment. This Conference.

[5] J. A. Holt, M. A. Martens, L. Michelotti, and G. Goderre. Calculating luminosity for a coupled Tevatron lattice. In *Proceedings of the 1995 IEEE Particle Accelerator Conference. Dallas, TX.*, 1995.

[6] N. V. Mokhov. The MARS code system user's guide, version 13(95). Technical report, Fermilab, 1995. Fermilab-FN-628.

[7] Jean-Francois Ostiguy, Leo Michelotti, and James A. Holt. Automatic differentiation and lattice function matching. This Conference.

[8] F. Christoph Iselin. The CLASSIC project. This Conference.

TRACE 3-D Code Improvements*

W. P. Lysenko, D. P. Rusthoi, K. C. D. Chan,

AOT-1, MS H808, Los Alamos National Lab., Los Alamos, NM 87545, USA

G. H. Gillespie and B. W. Hill

G. H. Gillespie Associates, Inc., P.O. Box 2961, Del Mar, CA 92014, USA

Abstract. TRACE 3-D is an interactive beam-transport code for bunched beams that includes accelerating elements and linear space-charge forces. It has been integrated with an improved GUI (graphic user interface) based on the Shell for Particle Accelerator Related Codes. Recent modifications to the code include centroid tracking and an improved beam description consisting of a set of beam slices, each having its own 6D centroid and sigma matrix. This allows us to study some nonlinear effects, such as wakefields, that are related to the variation of the beam bunch along the longitudinal direction.

INTRODUCTION

TRACE 3-D (1) is an interactive beam-transport code for bunched beams. It includes accelerating elements and linear space-charge forces. It is useful for designing linear accelerators and transport lines because of its ease of use and its ability to do matching and other types of optimizations. TRACE 3-D describes a beam by a 6 × 6 sigma matrix, which is the collection of second moments of the phase-space distribution. Since the code computes the evolution of the distribution directly instead of dealing with the single-particle motion, it is a natural environment for matching computations.

Recently, we added two new capabilities to TRACE 3-D. We now track centroids in addition to sigma matrices. This allows us to study offsets in beams and optical elements. Also, in a more substantial change, we can now represent the beam bunch as a collection of slices that interact with the external forces and each other. This allows us to study certain nonlinear effects like short-time (single bunch) wakefields. The changes to the code were made

---

*Work supported by US Department of Energy, Defense Programs, TTI/SBI Cooperative Research and Development program.

in manner that was very compatible with the existing code structure, existing input data files, and the GUI (graphic user interface) of the MacTrace™ version (5, 6). Bunch slicing is turned on by setting the existing variable IBS, which controls the generation of the initial beam, to a value of 2.

Simulation of wakefield effects has been reviewed by Chan (2). Our approach follows that of the LTRACK code (3), which was originated by Chao and Cooper (4). What is new here is that we do not assume that all the particles have velocity c. Our code can take into account, for example, bunch lengthening caused by space-charge interactions between the bunch slices.

CENTROIDS

Initial beam offsets are specified in a new array in the data file. Element offsets are introduced by a new rotate/translate element that replaces the old rotate element. Any offset automatically turns on centroid tracking. Centroids are plotted on the usual graphs (profiles and phase-space). The quadrupole element also has been expanded to include built-in rotations and offsets that can take on random values. The centroid capability is useful for aperture studies and is also required for the wakefield feature, described next.

BUNCH SLICING AND WAKEFIELDS

When we use the new wakefield feature, the beam bunch is divided into a user-defined number of slices longitudinally. We describe *each slice* by its 6D centroid and its 6 × 6 sigma matrix. The code generates this collection of slices from the usual input Courant-Snyder parameters and emittances.

Generating the Initial Beam

We assume a uniformly filled, upright ellipsoid in (x,y,z) with $-z_{max} \le z \le +z_{max}$. We make $2N+1$ equal-length slices, labeled from $-N$ (head) to $+N$ (tail). Slice number 0 is centered at $z = 0$. Let z_i be the z value at the upstream face of slice i and define z_{N+1} to be $-z_{max}$. The number of particles in a slice of length dz at position z is proportional to $(1 - (z/z_{max})^2)\,dz$. With this distribution, the z-centroid of slice i is given by

$$<z>_i = \frac{6z_i^2 - 3z_i^4/z_{max}^2 - 6z_{i+1}^2 + 3z_{i+1}^4/z_{max}^2}{12z_i - 4z_i^3/z_{max}^2 - 12z_{i+1} + 4z_{i+1}^3/z_{max}^2} \qquad (1)$$

and the fraction of particles in slice i is given by

$$N_i = \frac{1}{4}\left(3\frac{z_i}{z_{max}} - \left(\frac{z_i}{z_{max}}\right)^3 - 3\frac{z_{i+1}}{z_{max}} + \left(\frac{z_{i+1}}{z_{max}}\right)^3\right). \qquad (2)$$

To estimate the z' centroids $<z'>_i$, we assume the ratio $<z'>_i/<z>_i$ is the same as z'_e/z_m, where z_m is the maximum value of z for the z-z' ellipse and z'_e is the value of z' at $z = z_m$. Since $z_m = \sqrt{\beta_z \epsilon_z}$ and $z'_e = -\alpha_z \sqrt{\epsilon_z/\beta_z}$, we have

$$<z'>_i = -\frac{\alpha_z}{\beta_z} <z>_i \tag{3}$$

for the value of the z' centroid of slice i. When we compute the slice sigma matrices, we adjust the x and y emittances at each slice according to

$$\epsilon_{x,y} \to \left[1 - \left(\frac{<z>_i}{z_{max}}\right)^2\right] \frac{\epsilon_{x,y}}{f}, \tag{4}$$

where the factor f is the ratio of average to maximum emittance value

$$f = \sum_{i=-N}^{N} \left[1 - \left(\frac{<z>_i}{z_{max}}\right)^2\right] N_i. \tag{5}$$

The σ_{55} and σ_{66} for the i-th slice are estimated from

$$\sigma_{55i} = (\Delta z/2)^2, \tag{6}$$

$$\sigma_{66i} = \left[\frac{\epsilon_z}{\beta_z} - \left(\frac{<z>_i}{\beta_z}\right)^2\right] \frac{1}{e}, \tag{7}$$

where e is the ratio of average to maximum σ_{66} value

$$e = \frac{\beta_z}{\epsilon_z} \sum_{-N}^{N} \left[\frac{\epsilon_z}{\beta_z} - \left(\frac{<z>_i}{\beta_z}\right)^2\right] \tag{8}$$

Recombining the Slices

Whenever the codes needs to generate output or compute space charge, we combine the slices into a single centroid and sigma matrix. Then we use existing mechanisms in the code to generate output or compute space charge. Let u represent (x, x', y, y', z, z'). The overall centroid components are given by

$$<u> = \sum_{i=-N}^{N} N_i <u>_i. \tag{9}$$

The matrix elements σ_{ij} for the overall sigma matrix are defined by

$$\sigma_{ij}/5 = <(u_i - <u_i>)(u_j - <u_j>)> = <u_i u_j> - <u_i><u_j>. \tag{10}$$

The matrix elements for the k-th slice are given by

$$\sigma_{ijk}/5 = <u_i u_j>_k - <u_i>_k <u_j>_k. \tag{11}$$

The overall sigma matrix can be determined to be

$$\frac{\sigma_{ij}}{5} = \sum_{k=-N}^{N} N_k \left(\frac{\sigma_{ijk}}{5} + <u_i>_k <u_j>_k \right) - \left(\sum_{k=-N}^{N} N_k <u_i>_k \right) \left(\sum_{k=-N}^{N} N_k <u_j>_k \right). \tag{12}$$

Transporting the Beam

The bunch centroid is transported as a single particle using the R matrix at each step in the simulation. Space charge does not contribute to this motion. For a sliced-beam simulation, the slice centroids are transported similarly except that they are also acted upon by space charge forces, as described later.

In a sliced-beam simulation, the sigma matrix for each slice is transported similarly to the bunch sigma matrix in the original code. The new wakefield elements have a new feature, however, in that each slice has a different R matrix that acts on it, which depends on the slices ahead.

Space Charge

When we compute space charge in a wakefield run, we combine the slice centroids and sigma matrices into an overall sigma matrix, which we use to compute the space-charge R matrix. This R matrix then transports all the slice sigma matrices.

The overall centroid of the beam is not affected by space charge. For a wakefield simulation, however, we have a collection of slice centroids and these *are* affected by space charge. To handle this, we subtract the overall centroid from the slice centroids to get a collection of relative centroids. We then transport these using the space-charge R matrix.

The overall sigma matrix is checked for correlations to determine if the spatial ellipsoid is tilted, just as in the usual no-wakefield situation. If there is a correlation (this can be caused by a bend, rotation, or wakefield), we rotate into a coordinate system in which the overall spatial ellipsoid is upright.

Wakefield Elements

Three new elements have been added to TRACE 3-D to model the monopole, dipole, and quadrupole wakefields. These elements are inserted immediately following the element that is responsible for generating the wakefield. The monopole wakefield changes the energies of the bunch slices, the dipole wakefield cause deflections of the transverse centroids of the slices, while the quadrupole wakefield affects the sigma matrices of the slices in addition to the energy and transverse centroids. The three wakefield multipoles are expressed in terms of a wake function strength per unit length, $W_o(s)$, $W_1(s)$, and $W_2(s)$.

TABLE I. Comparison of output beams for test problem with no wakefields.

Mode	α_x	β_x	ϵ_x	α_y	β_y	ϵ_y	α_z	β_z	ϵ_z
IBS=0 (no slicing)	4.49	1.28	6.00	10.9	1.72	6.00	33.6	47.3	40.0
IBS=2 (with slicing)	4.49	1.28	6.00	10.9	1.72	6.00	33.1	46.5	39.8
IBS=0 (0 mA)	3.51	.955	6.00	5.70	.844	6.00	15.5	22.7	40.0

Each wakefield acts over the length L of the element responsible for generating the wakefield. The wakefield functions are modeled as second degree polynomials:

$$LW_o(s) = p_o(1) + p_o(2)s + p_o(3)s^2 \tag{13}$$

$$LW_1(s) = p_1(1) + p_1(2)s + p_1(3)s^2 \tag{14}$$

$$LW_2(s) = p_2(1) + p_2(2)s + p_2(3)s^2 \tag{15}$$

The three coefficients for a given multipole, $p_m(1)$, $p_m(2)$, and $p_m(3)$ are user inputs for the wakefield elements. Slice centroids and sigma matrices are transformed by the wakefield elements as in the LTRACK code (3).

INTEGRATION WITH THE GUI

The changes to TRACE 3-D described here have integrated into an improved version of the GUI based on the Shell for Particle Accelerator Codes, which is used in the MacTrace™ implementation of TRACE 3-D (5, 6). Of particular importance is the user defined element facility, which is how the new wakefield elements have been implemented. The TableBuilder application in the GUI allows the user to create custom Piece Windows, which are used to graphically construct beamlines containing the user-defined elements.

TEST RESULTS

Table I compares three TRACE 3- D runs for the same beamline and initial beam. There are no wakefields. The first line shows a normal run (no slicing). The second line shows a five-slice wakefield run with zero wakefields. The third line shows a normal no-slicing run with space charge turned off. We see that space charge causes a significant bunch lengthening (increase in β_z) and that the bunch-slicing run correctly simulates this behavior. Generally, bunch slicing is valid whenever the longitudinal temperature is not too high. This representation of the distribution breaks down if the velocity spread is large enough to cause the slices to overlap significantly during the simulation.

Figure 1 shows part of the graphical output of a TRACE 3-D run with a quadrupole wakefield. Phase-space ellipses are shown for all five slices. Transverse ellipses near the tail are rotated relative to those near the head. This

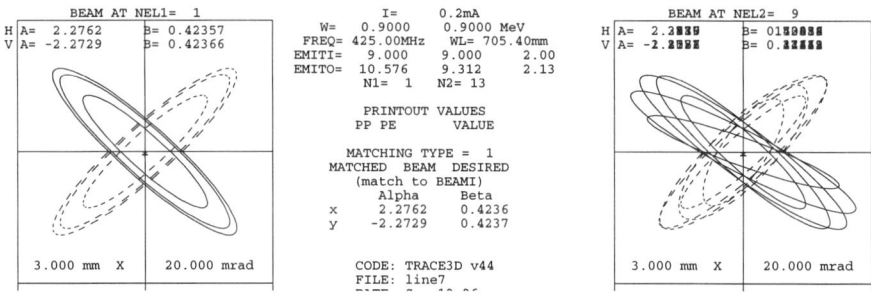

FIGURE 1. Quadrupole wakefield example. Phase-space ellipses for each slice are shown. Initial beam is at left, final is at right. Solid curves are for x, dashed are for y.

accounts for the emittance growth. The wakefield function for this example case peaks at about 1.5 mm, which is approximately the bunch length.

CONCLUSION

We have demonstrated the flexibility in the moment approach in computational beam dynamics by introducing a significant new capability to the TRACE 3-D code. This is especially significant because of the ease of use of this code, particularly in the MacTrace™ implementation.

REFERENCES

1. Crandall, K. and Rusthoi, D., *TRACE 3-D Documentation*, Los Alamos National Laboratory Report LA-UR-90-4146, 1990.

2. Chan, K. C. D., "Computer Codes for Wakefield Analysis in rf-Based Free-Electron Laser," in *Proceedings of the Beijing FEL Seminar*, Singapore, World Scientific Publishing Co. Pte. Ltd., 1989, pp. 172–192.

3. Chan, K. C. D. and Cooper, R. C., "LTRACK—Beam-Transport Calculation Including Wakefield Effects," in *AIP Conference Proceedings* **177**, New York, American Institute of Physics, 1988, p.37–44.

4. Chao, A. W. and Cooper, R. K., "Transverse Quadrupole Wake Field Effects in High Intensity Linacs," *Particle Accelerators* **13**, 1–12, (1983).

5. Gillespie, G. H., "The Shell for Particle Accelerator Related Codes (SPARC)—A Unique Graphical User Interface," in *AIP Conference Proceedings* **297**, New York, American Institute of Physics, 1993, pp. 576–583.

6. Gillespie, G. H. and Hill, B. W., "A Graphical User Interface for TRACE 3-D Incorporating Some Expert System Type Features," in *1992 Linear Accelerator Conference Proceedings (Ottawa)*, AECL-10728, 1992, pp. 787–789.

COSY INFINITY Version 7

Kyoko Makino and Martin Berz

*Department of Physics and Astronomy and
National Superconducting Cyclotron Laboratory
Michigan State University, East Lansing, MI 48824*

Abstract. An overview over the features of version 7 of the code COSY INFINITY is given. Currently distributed to about 160 registered users, the code allows the computation and manipulation of maps of arbitrary order for arbitrary arrangements of fields. Besides the conventional analysis tools including various techniques for symplectic tracking, normal form methods, hardware and reconstructive aberration correction, and achromat design, several new features are presented. These include efficient methods to treat fringe field, a variety of methods to directly use measured field data for the computation of maps, as well as the computation and analysis of spin dynamics. Furthermore, we give references regarding the practical use of the new remainder differential algebraic methods, which among other things allow fully rigorous estimates of stability times of nonlinear repetitive motion.

INTRODUCTION

This paper provides background of the code COSY INFINITY and its language, and examples of some of the new features in COSY INFINITY [1]. These new features include the SYSCA fast fringe field methods [2–5], the computation of maps from arbitrary measured fields, the computation and analysis of spin dynamics [6,7] which is getting more important due to the increasing number of studies connected to the acceleration of polarized particles, as well as the tools used for the design of fifth order achromats [8–10], and the new remainder differential algebraic method [11–14]. Besides these features based on new techniques, there are also a variety of other new tools of a more technical nature. These are connected to standard problems of accelerator design, to interactive graphics, as well as to several improvements of existing tools. A good overview over the key features of the code can be found in the extensive demo that is part of the COSY shipment. Information about COSY can also be obtained at www.beamtheory.nscl.msu.edu/cosy.

THE CODE COSY

COSY INFINITY is a code for the simulation, analysis and design of particle optical systems. Since the first official version in 1989, a total of seven releases with an increasing number of features have been provided, and currently there are a total of about 160 registered users as shown in Figure 1. COSY is based on differential algebraic methods, which are described in detail elsewhere [15–17], and which lately have been widely used also in most of the other newly emerging codes.

For the sake of portability, the code is based on standard FORTRAN 77, which is still the most widespread language on the computers used for accelerator design and simulation. Since the code employs its own programming language which is object oriented and has the flavor of PASCAL, COSY's programmers and users are free from FORTRAN. The COSY language has very simple syntax which makes programming easy, and its elements are as follows:

```
BEGIN ;      END ;       { Begins and ends program          }
INCLUDE ;    SAVE ;      { Includes and saves compiled code }
VARIABLE ;               { Declares a local variable        }
```

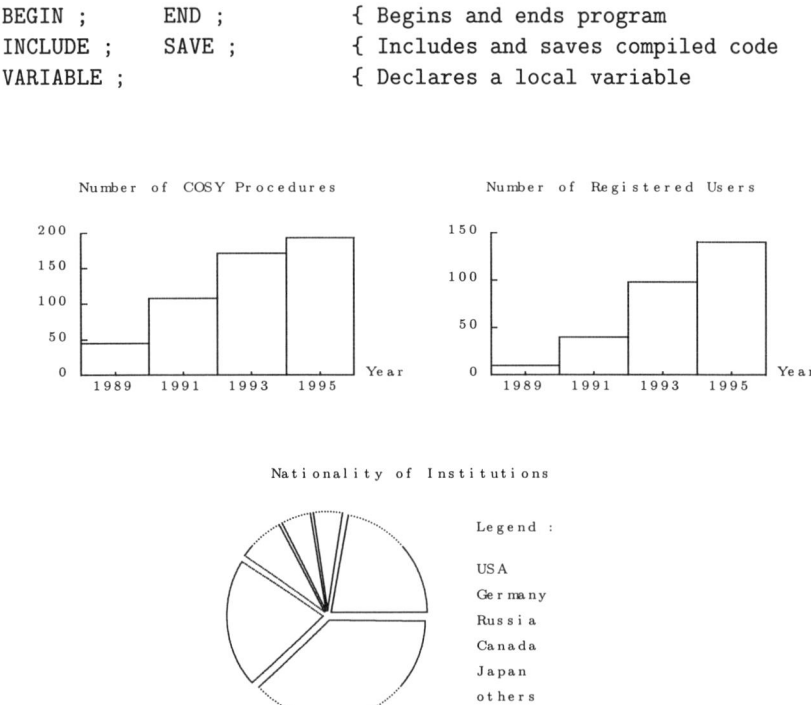

FIGURE 1. Some statistics about COSY INFINITY (picture generated with COSY's graphics environment and part of the COSY Demo).

```
PROCEDURE ;       ENDPROCEDURE ;   { Declares a local procedure        }
FUNCTION ;        ENDFUNCTION ;    { Declares a local function         }
< assignments > ;                  { Sets value of variable            }
< procedure calls > ;              { Calls previously defined procedure }
IF ;              ENDIF ;          { Executed once if argument is true }
WHILE ;           ENDWHILE ;       { Executed while argument is true   }
LOOP ;            ENDLOOP ;        { Stepping argument                 }
FIT ;             ENDFIT ;         { varying arguments to fit conditions }
```

Except for the last one, the flow control statements are rather standard. The unique FIT block is used for efficient utilization of COSY's various optimizers. The ENDFIT statement contains the number of the optimizer to be used and the objective quantities, as well as the tolerance and the maximum number of iterations allowed. The statements in the block are executed over and over again, where for each pass the optimization algorithm changes the values of the variables listed in the FIT statement and attempts to minimize the objective quantity.

The compiler for this language comprises about 6,000 of the approximately 25,000 lines of COSY's FORTRAN code. It has a completely rigorous syntax and error analysis and is comparable in speed to other compilers. The compiled code is stored in a binary file, which can either be saved for inclusion in later code or executed directly. The object oriented features of the code allow a direct use of the differential algebraic operations contained in the 16,000 line DA package.

THE FAST FRINGE FIELD CALCULATION

COSY has several options to take account of the fringe field. One is to compute the exact fringe field through DA integration of the equations of motion. In this case, the accuracy is limited only by the accuracy of the numerical integrator, which is adjusted via automatic step size control. While this method provides a very detailed fringe-field calculation, the computational expense becomes quite high. One of the other approaches to compute fringe-field in COSY is the SYSCA method. It is based on a combination of geometric scaling in TRANSPORT coordinates and symplectic rigidity scaling [2–5].

It uses parameter dependent symplectic representations of fringe-field maps stored on files to approximate the fringe field via symplectic scaling. This method computes fringe fields with a very high accuracy, yet its computational expense is two or three orders of magnitude less than direct integration and in the same terrain as that of main field only calculation. While the reference file can be produced by a user in COSY according to the detailed shape of the fringe fields at hand, one standard reference file comes as part of the COSY shipment. The SYSCA fringe-field mode is especially helpful in the final design stages of a realistic system after approximate parameters of the elements have been obtained by neglecting fringe fields or with other rough fringe-field calculation.

THE MEASURED FIELD DATA ELEMENT

One of new elements in COSY INFINITY, the measured field data element, allows the computation of the map of any magnetic field supplied by midplane measured data. The measured data has to be supplied at equi-distant grid points in cartesian coordinates.

The necessary interpolation method has to guarantee the differentiability to fit with the differential algebraic computation. Because of this, the evaluation of the field strength in the element in COSY is done by Gaussian interpolation, which is a powerful special case of a wavelet transform:

$$B_y(x,z) = \sum_{i_x, i_z} \mathrm{BY}(i_x, i_z) \frac{1}{\pi S^2} \exp\left[-\frac{(x-x_{i_x})^2}{\triangle x^2 S^2} - \frac{(z-z_{i_z})^2}{\triangle z^2 S^2}\right],$$

where $\mathrm{BY}(i_x, i_z)$ are the supplied measured data. Since the gaussian function falls off quickly, the time consuming summation over all the gaussians can be replaced by the summation of only the neighboring gaussians, which is in the vein of other wavelet transforms and greatly improves efficiency.

The method is used extensively in the simulation of the S800 Spectrograph at NSCL/MSU [18], construction of which has just been completed, as well as the various spectrographs at CEBAF.

SPIN DYNAMICS

In version 7, features to analyze spin motion have been added in COSY [6,7]. The classical equation of motion for spin has the form

$$\frac{d\vec{s}}{dt} = \vec{w} \times \vec{s};$$

for details, see [7]. The solution is a linear orthogonal transformation depending on orbital variables, thus

$$\vec{s}_f = \hat{A}(\vec{z}) \cdot \vec{s}_i, \quad \text{where} \quad \hat{A}(\vec{z}) \in SO(3).$$

The motion of a particle with spin can be described as a nine dimensional motion, neglecting spin-orbit coupling, as

$$\begin{pmatrix} \vec{z} \\ \vec{s} \end{pmatrix}' = \vec{F}(\vec{z}, \vec{s}, s) = \begin{pmatrix} \vec{f}(\vec{z}, s) \\ \hat{W}(\vec{z}, s) \cdot \vec{s} \end{pmatrix}$$

$$\begin{pmatrix} \vec{z}_f \\ \vec{s}_f \end{pmatrix} = \vec{M}(\vec{z}_i, \vec{s}_i, s) = \begin{pmatrix} \mathcal{M}(\vec{z}_i, s) \\ \hat{A}(\vec{z}_i, s) \cdot \vec{s} \end{pmatrix}$$

To reduce dimensionality and utilize linearity, it is advantageous to set up the equation of motion for $\hat{A}$. Insertion yields the equation of motion for the 3×3 spin matrix depending on only the six orbital variables:

$$\hat{A}'(\vec{z}, s) = \hat{W}(\vec{z}, s) \cdot \hat{A}(\vec{z}, s).$$

The COSY spin features have been used extensively at DESY and a variety of other places.

OTHER FEATURES

A recent application of COSY was the design of higher order achromats, and recently some extensive work connected to fifth order achromats has been done [8–10].

The new remainder differential algebraic (RDA) method combines the rigor of interval computations and a reduction of blow-up due to its use of Taylor polynomials. For the purposes of beam physics, it allows the determination of rigorous bounds for the remainder term of Taylor maps, and combined with methods to determine approximate invariants of the motion, it can be used for guaranteed estimates of stability times in circular accelerators [11–14].

REFERENCES

1. Berz, M., COSY INFINITY Version 7 reference manual. Technical Report **MSUCL-977revised**, National Superconducting Cyclotron Laboratory, Michigan State University, East Lansing, MI 48824 (1996).
2. Hoffstätter, G., and Berz, M., Efficient computation of fringe fields using symplectic scaling. In *Third Computational Accelerator Physics Conference*, AIP Conference Proceedings **297**, 467 (1993).
3. Hoffstätter, G., and Berz, M., Accurate and fast computaton of high–order fringe field maps via symplectic scaling. *Nuclear Instruments and Methods*, **352** (1994).
4. Hoffstätter, G. H., *Rigorous bounds on survival times in circular accelerators and efficient computation of fringe-field transfer maps*. PhD thesis, Michigan State University, East Lansing, Michigan, USA (1994). also DESY **94-242**.
5. Hoffstätter, G. H., and Berz, M., Symplectic scaling of transfer maps including fringe fields. *Physical Review E*, **54** (1996).
6. Berz, M., Differential algebraic description and analysis of spin dynamics. In *Proceedings, SPIN94*, in press.

7. Balandin, V., and Berz, M., Computation and analysis of spin dynamics, *these proceedings*.
8. Wan, W., and Berz, M., Design of a fith order achromat. *Nuclear Instruments and Methods*, **352** (1994).
9. Wan, W., *Theory and Applications of Arbitrary-Order Achromats*. PhD thesis, Michigan State University, East Lansing, Michigan, USA (1995). also **MSUCL-976**.
10. Wan, W., and Berz, M., Analytical theory of arbitrary order achromats. *Physical Review E*, **54**, 2870 (1996).
11. Berz, M., and Hoffstätter, G., Computation and application of Taylor polynomials with interval remainder bounds. *Interval Computations*, submitted, (1994).
12. Berz, M., and Hoffstätter, G., Exact estimates of the long term stability of weakly nonlinear systems applied to the design of large storage rings. *Interval Computations*, **2**, 68-89 (1994).
13. Makino, K., and Berz, M., Remainder differential algebras and their applications. In *Berz, M., Bischof, C., Corliss, G., and Griewank, A., editors, Computational Differentiation: Techniques, Applications, and Tools*. SIAM, Philadelphia, Penn. (1996).
14. Berz, M., Differential algebras with remainder and rigorous proofs of long-term stability, *these proceedings*.
15. Berz, M., High-order computation and normal form analysis of repetitive systems. In *Month, M., editor, Physics of Particle Accelerators*, AIP **249**, 456, American Institute of Physics (1991).
16. Berz, M., Arbitary order description of arbitrary particle optical systems. *Nuclear Instruments and Methods*, **A298**, 426 (1990).
17. Berz, M., Differential algebraic description of beam dynamics to very high orders. *Particle Accelerators*, **24**, 109 (1989).
18. Berz, M., Joh, K., Nolen, J. A., Sherrill, B. M., and Zeller, A. F., Reconstructive correction of aberrations in nuclear particle spectrographs. *Physical Review C*, **47,2**, 537 (1993).

Zlib: a Numerical Library for Optimal Design of Truncated Power Series Algebra and Map Parameterization Routines

Yiton T. Yan

Stanford Linear Accelerator Center, Stanford University, Stanford, CA 94309

Abstract

A brief review of the Zlib development is given. Emphasized is the Zlib nerve system which uses the One-Step Index Pointers (OSIP's) for efficient computation and flexible use of the Truncated Power Series Algebra (TPSA). Also emphasized is the treatment of parameterized maps with an object-oriented language (e.g. C++). A parameterized map can be a Vector Power Series (Vps) or a Lie generator represented by an exponent of a Truncated Power Series (Tps) of which each coefficient is an object of truncated power series.

Introduction

Zlib Fortran version was developed in 1990 [1]. Its development was originated at fast computational speed for nonlinear analyses of high-order power-series maps of the Superconducting Super Collider (SSC) lattices. Since Supercomputers, such as Cray computers, were used, the algorithms used for manipulating truncated power series and Lie algebras were optimized for scalar, vector, and parallel computing. Of the most important part in achieving such optimization was the Zlib nerve system consisted of One-Step Index Pointers for optimized computation of TPSA routines such as multiplication, concatenation, partial derivative, Taylor map tracking, etc. Memories for the One-Step Index Pointers and necessary internal auxiliary arrays were dynamically allocated at the minimum required level per user's input for the maximum order and number of variables. By the time of the CAP93 Conference, there were more than 200 dynamically usable subroutines in Zlib Fortran version for TPSA and Lie algebraic mapping analysis [2].

In late 1993, upon termination of the SSC, about 20% of the Zlib Fortran subroutines were faithfully translated into C++ codes that form two fundamental classes of the TPSA [3]. These two classes were named ZSeries and ZMap which handles the algebra of truncated power series and Vector truncated power series respectively and have been included in Malitsky's Unified Library [4]. Recently at SLAC, aiming at further development for mapping analysis, the two classes ZSeries and ZMap have been rewritten and named as Tps and Vps and added or to be added upon them are many other classes for treating Lie algebras. Many of these classes are translated or to be translated from Zlib Fortran version developed in early 90's.

Truncated Power Series - Tps

Tps is an abbreviated name for the Truncated Power Series. A Tps truncated at an order of Ω can be mathematically written as [1] [5]

---

*Work supported by Department of Energy, contract DE-AC03-76SF00515.

© 1997 American Institute of Physics

$$U(\vec{z}) = \sum_{o=0}^{\Omega} u(\vec{k})z^{\vec{k}},$$

where, assuming n variables, $\vec{z}$ represents the variables labeled as $z_1, z_2, ...z_n$, $\vec{k}$ represents the power indices $(k_1, k_2, ...k_n)$ and so $z^{\vec{k}}$ represents $z_1^{k_1} z_2^{k_2}...z_n^{k_n}$, and $\sum_{o=0}^{\Omega}$ means summation over all possible monomials labeled by $\vec{k}$ with order given by $o = k_1 + k_2 + ... + k_n$ that is less than or equal to Ω. For an n-variable, Ω-order Tps, there are a total of $m(n, \Omega) = (n + \Omega)!/(n!\Omega!)$ monomials. In an optimized computation for the TPSA, the first step is to allocate minimum possible memory for storing the Tps coefficients. To achieve this goal, Zlib uses an integer sequence j's that starts from 0 to $m(n, \Omega) - 1$ (or from 1 to $m(n, \Omega)$ for the Fortran version) to label the the Tps coefficients, that is, there is a one-to-one correspondence between j's and $\vec{k}$'s. Such labeling is the same for all Tps's except that there may be order differences and so the label sequence starts with the lowest order (the 0th order) and then go on to the next order and so on. For example, for a 3-variable case, the corresponding labels between j's and $\vec{k}$'s up to third order are: $0 \equiv (0,0,0)$, $1 \equiv (1,0,0)$, $2 \equiv (0,1,0)$, $3 \equiv (0,0,1)$, $4 \equiv (2,0,0)$, $5 \equiv (1,1,0)$, $6 \equiv (1,0,1)$, $7 \equiv (0,2,0)$, $8 \equiv (0,1,1)$, $9 \equiv (0,0,2)$, $10 \equiv (3,0,0)$, $11 \equiv (2,1,0)$, $12 \equiv (2,0,1)$, $13 \equiv (1,2,0)$, $14 \equiv (1,1,1)$, $15 \equiv (1,0,2)$, $16 \equiv (0,3,0)$, $17 \equiv (0,2,1)$, $18 \equiv (0,1,2)$, $19 \equiv (0,0,3)$. In Zlib, the relation between j's and $\vec{k}$'s is governed by a simple formula which is only used to generate One-Step Index Pointers.

The Tps class in Zlib C++ version is designed to manipulate Tps represented by the above-described coefficients.

Vector truncated Power Series - Vps

Vps is an abbreviated name for the Vector truncated Power Series. A Vps truncated at an order of Ω can be mathematically written as [1] [5]

$$\vec{U}(\vec{z}) = \sum_{o=0}^{\Omega} \vec{u}(\vec{k})z^{\vec{k}}, \tag{1}$$

that is, each component of the Vps is a Tps represented by coefficients described in the last section. For example, the ith component would be represented by

$$U_i(\vec{z}) = \sum_{o=0}^{\Omega} u_i(\vec{k})z^{\vec{k}},$$

The Vps class in Zlib C++ version is designed to manipulate Vps described above.

One-Step Index Pointers

For optimized computation of the TPSA, besides efficient memory mangement, one would also like to achieve efficient calculation for each of the related algebras such as Tps multiplication, Vps concatenation, and Taylor map tracking, etc. The key is to have One-Step Index Pointers prepared only once for repeated use such that for any coefficient involved in a given calculation, it can be identified with a minimum index path. For example, let A and B be two Tps's (may be with different orders), such that Tps C = A * B. The task is to obtain all of the coefficients of C to a specified order derived from the orders of A and B and the preset cap order. Assuming the minimum and the maxmum orders for C derived are minimumOrderOfC and maximumOrderOfC, to obtain C, the code in Zlib would look like as follows.

```
for (order = minimumOrderOfC; order <= maximumOrderOfC; ++order) {
    lowOrder = MaximumOrderOf (order-maximumOrderOfB, minimumOrderOfA);
    highOrder = MinimumOrderOf (maximumOrderOfA, order-miniMumOrderOfB);
    for (j = monomialBegin[order]; j < monomial[order]; ++j) {
        for (i = ipBegin[j][lowOrder]; i < ipEnd[j][highOrder]; ++i) {
            C[j] += A[aOSIP[i]] * B[bOSIP[i]];
        }
    }
}
```

where except A, B, C (assumed cleared), all other variables are integers and are assumed to have been declared. Note that except lowOrder and highOder (obtained with negligible computer time), all other indices are obtained by assignment only (minimum index path) through One-Step Index Pointers (OSIP's). Note that, if a supercomputer is used, then the innest (i-) loop is vectorized through automatic gather while the j-loop is parallelized. This original Zlib One-Step Index Pointers scheme for Tps multiplication may be categorized as a "backward" scheme. Recently, Dragt seems to be interested in exploring a similar scheme which may be categorized as a "forward" scheme [6].

As another example of the OSIP's, let V be a Vps with a dimension of n (an n-dimensional Tps) reresenting a one-turn Taylor map, and z be a vector respenting the phase-space coordinates of a particle. Assuming no parameters, that is, in phase space, z has the same dimension as V, then a one-turn Taylor map tracking is to update the phase-space coordinates represented by z through evaluation of the Vps given by Eq. 1. The code in Zlib would look like as follows.

```
for (j = 1; j < monomial[orderOfV]; ++j) xx[j] = z[iOSIP[j]]*xx[jOSIP[j]];
for (i = 0; i < n; ++i) z[i] = V[i][0];
for (i = 0; i < n; ++i)
    for (j = 1; j < monomial[orderOfV]; ++j) z[i] += V[i][j]*xx[j];
}
```

where again iOSIP and jOSIP are One-Step Index Pointers and in the double loop, the inner one is vectorized while the outer one is parallelized. It was this fast computational process that allowed the fast Taylor-map tracking for the SSC to high orders (11- or 12-th order) with a computational speed two orders of magnitude faster than the conventional element-by-element tracking [5].

Action-angle variable truncated Power Series - Aps

Aps is an abbreviated name for the truncated power series in action-angle variable space. A class named Aps in Zlib C++ version is nearly completed. Some of the important member functions in this class are the nPB tracking and the extraction of the normalized resonance basis coefficients which was coded before in Fortran and have been used intensively for PEP-II lattice studies [7].

Lie Classes

Application of TPSA to nonlinear single-particle dynamics usually goes with the Lie algebraic analysis. Therefore, majority of the Zlib classes are to be for Lie algebras such as single Lie generators, Dragt-Finn factorizations [8], nonlinear normal forms [9], kick factorizations [10], integrable polynomial factorization [11], etc.

Parameterized maps - the Tps of Tps

In mapping analysis of a beam line lattice, in addition to the canonical phase-space variables, we often would like to have parameter variables which are constant but not specified with a value. Their values are either to be determined after the analysis or are dynamical (time dependent) to allow additional studies such as for synchrotron oscillation, power supply ripple, and ground motion at lower computational price. Treatment of such parameterized map in Fortran is tedious and usually uses semi-parameterization methods. Although some fully parameterized (coefficients of the power series in canonical space are treated as power series in parameter space) algorithms were written for treating both linear (but nonlinear in parameter space) and nonlinear cases [12], there were no implementation of such fully parameterized methods in the Zlib Fortran version. However, with the capability of the object orientation, it is easier to code such fully parameterized algorithms since one can consider each of the coefficients in the canonical space as an object of Tps in stead of a double. These fully parameterized mapping methods are currently under active development in Zlib C++ version.

Zlib Future Direction

While Zlib Fortran version will still be kept for optional use, there will be no more development. The future direction is to develop a more complete C++ version for Zlib. Currently, there are more than 30 classes in Zlib that are under active improvement and/or development.

Acknowledgement

Systematical programming use of differential algebras (truncated power series algebras) was introduced to the particle accelerator community by Berz [13]. On the physics side, the Lie algebraic application to accelerator beam dynamics was introduced by Dragt [14]. E. Forest and J. Irwin and others have also made contributions for more Lie algebraic applications. Michelotti made the most effort in advocating C++ programming for beam line studies and has used a link list method to develop a differential algebra package [15].

Highly acknowledgement is given to N. Malitsky for his participation in Zlib C++ version development.

References

[1] Y. Yan and C. Yan, "*Zlib — A Numerical Library for Differential Algebra,*" SSC Laboratory Report SSCL-300 (1990).

[2] Y.T. Yan, "*Zlib and Related Programs for Beam Dynamics Studies,*" in *Computational Accelerator Physics,* AIP Conf. Proc. No. 297, p.279 (1993), R. Ryne, eds.

[3] N. Malitsky, A. Reshetov, and Y. Yan, "*ZLIB++: an Object-Oriented Numerical Library for Differential Algebra,*" SSCL-659 (1994).

[4] N. Malitsky, in these proceedings.

[5] Y.T. Yan, "*Applications of Differential Algebras to Single-Particle Dynamics in Storage Rings,*" SSCL-500, in *The Physics of Particle Accelerator,* AIP Conf. Proc. No. 249, p. 378 (1992), M. Month and M. Dienes, eds.

[6] A. Dragt, private communication.

[7] Y.T. Yan, J. Irwin, and T. Chen, "*Resonance Basis Maps and nPB Tracking for Dynamics Aperture Studies,*" Particle Accelerators, Vol. 55, p. 263 (1996).

[8] A. Dragt and J. Finn, "*Lie Series and Invariant Functions for Analytic Symplectic Maps,*" J. Math. Phys., **17**, 2215 (1976).

[9] E. Forest, M. Berz, and J. Irwin, "*Normal Form Methods for Complicated Periodic Systems,*" Particle Accelerators, **24**, 91 (1989).

[10] J. Irwin, "*A Multi-Kick Factorization Algorithm for Nonlinear Maps,*" in *Accelerator Physics at the SSC,* AIP Conf. Proc. No. 326, edited by Y.T. Yan et al. (AIP, New York, 1995), p. 662; D. Abell and A.J. Dragt, to be published.

[11] J. Shi and Y.T. Yan, "*Explicitly Integrable Polynomial Hamiltonians and Evaluation of Lie Transformations,*" Phys. Rev. E, **48**, 3943 (1993).

[12] In Chapters 5 and 6 of Ref. 5.

[13] M. Berz, "*Differential Algebraic Description of Beam Dynamics to Very High Orders,*" Particle Accel. **24**, 109 (1989).

[14] A.J. Dragt, Annual Rev. Nucl. Part. Sci. **38**, 455 (1988).

[15] L. Michelotti, "*MXYZPPLK: a C++ version of differential algebra,*" Fermi National Accelerator Laboratory Report FN-535 (1990).

The Particle Beam Optics Interactive Computer Laboratory

George H. Gillespie, Barrey W. Hill, Nathan A. Brown,
R. Chris Babcock, Hendy Martono and David C. Carey*

G. H. Gillespie Associates, Inc., P.O. Box 2961, Del Mar, CA 92014, U.S.A.

*Fermi National Accelerator Laboratory, P. O. Box 500, Batavia, IL 60510, USA

Abstract. The Particle Beam Optics Interactive Computer Laboratory (PBO Lab) is an educational software concept to aid students and professionals in learning about charged particle beams and particle beam optical systems. The PBO Lab is being developed as a cross-platform application and includes four key elements. The first is a graphic user interface shell that provides for a highly interactive learning session. The second is a knowledge database containing information on electric and magnetic optics transport elements. The knowledge database provides interactive tutorials on the fundamental physics of charged particle optics and on the technology used in particle optics hardware. The third element is a graphical construction kit that provides tools for students to interactively and visually construct optical beamlines. The final element is a set of charged particle optics computational engines that compute trajectories, transport beam envelopes, fit parameters to optical constraints and carry out similar calculations for the student designed beamlines. The primary computational engine is provided by the third-order TRANSPORT code. Augmenting TRANSPORT is the multiple ray tracing program TURTLE and a first-order matrix program that includes a space charge model and support for calculating single particle trajectories in the presence of the beam space charge. This paper describes progress on the development of the PBO Lab.

INTRODUCTION

At the preceding (1993) conference in this series, the development of a unique graphic user interface (GUI) designed specifically for codes used in the accelerator community was described [1]. Known as the Shell for Particle Accelerator Related Codes (S.P.A.R.C.), the GUI has now been integrated with a number of application programs [2-5]. A primary motivation for the development of the S.P.A.R.C. GUI was to provide a tool which improves the productivity of scientists and engineers involved in the analysis or design of accelerators. One benefit of this capability is a significant reduction in the time required to train new researchers in the use of applications which operate within the S.P.A.R.C. environment. Implicit in this benefit is a potential role for the GUI as an educational tool. A new version of the GUI is now under development focused specifically on supporting the teaching of charged particle optics at the upper division undergraduate, graduate student, and post doctoral levels. This paper describes several of the goals of the development effort and the progress to date.

© 1997 American Institute of Physics

THE GRAPHIC USER INTERFACE (GUI) SHELL

The conceptual foundation for the PBO Lab GUI is derived from the S.P.A.R.C. software environment [1] which has been successfully integrated with several beamline and accelerator design codes [2-5]. S.P.A.R.C. was constructed using a modular approach and provides a good framework for rapidly prototyping new GUI concepts. Several new concepts were developed for a prototype of the PBO Lab that was used in the computer laboratory sessions of a particle optics course given by the U. S. Particle Accelerator School (USPAS) this past January. (Examples of some of these new features are described later in this paper.) However, all prior work using S.P.A.R.C. had been restricted to the Macintosh platform, and one of the key goals of the PBO Lab effort is to implement a cross platform version of the GUI. Figure 1 illustrates an initial test version of the PBO Lab GUI running on a Pentium using the WindowsNT operating system.

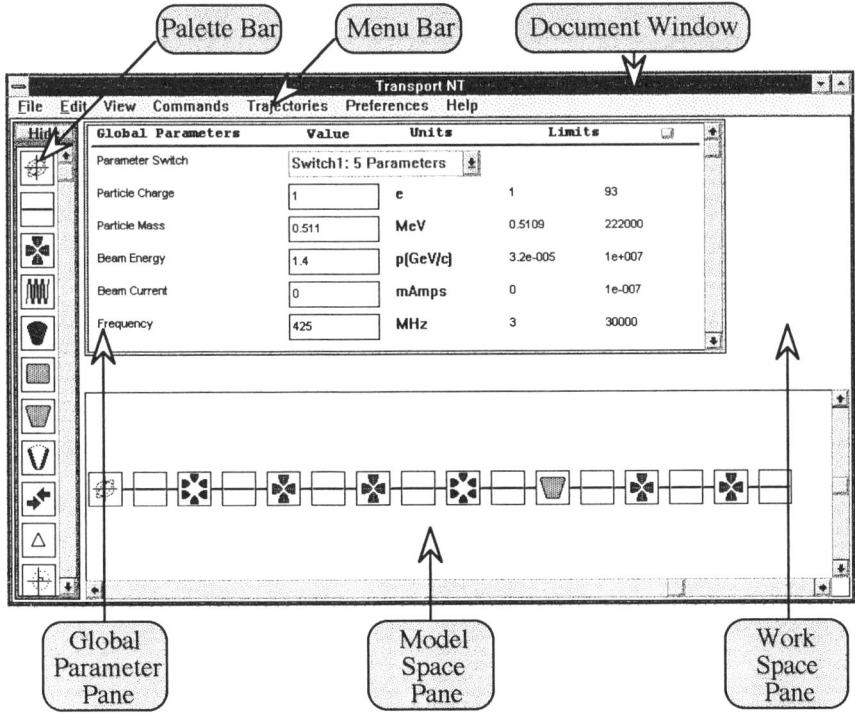

Figure 1. Example of the PBO Lab graphic user interface for TRANSPORT.

Parameter values are input using Data Tables in the Global Parameter Pane or in Piece Windows [1] for individual beamline components. A Piece Window is accessed by "double clicking" on the icon of the desired beamline component. The PBO Lab Piece Windows have several enhancements, described elsewhere [5], that distinguish them from those of previous S.P.A.R.C. applications.

PARTICLE OPTICS KNOWLEDGE DATABASE

Another goal of the PBO Lab effort is to integrate an information and knowledge database that focuses on (1) the fundamental principles of charged particle optics and (2) the technologies and hardware used in charged particle beamlines. Interactive, self-directed tutorials are used to present this material to the student. Much of the material developed for the PBO Lab tutorials was utilized as part of the USPAS optics course content, although the prototype software only included a limited number of examples. The information is generally presented in a "slide show" type of format, with hypertext links between different slides. In addition to text, graphics and hypertext links, the tutorials present useful formulae with numerical results for individual optics components that are based upon the user's input parameters for that component. Selected windows for the quadrupole tutorial in the prototype PBO Lab are shown in Figure 2.

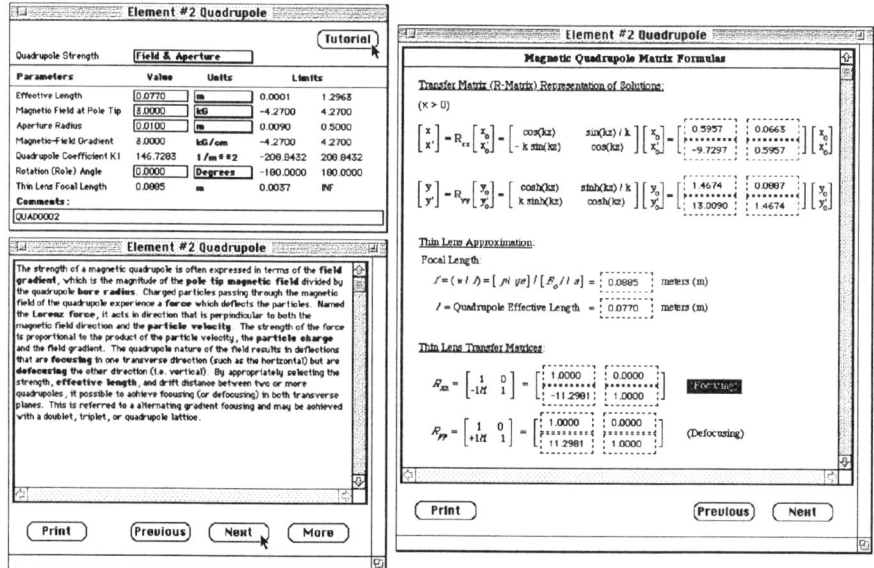

Figure 2. Example windows from the prototype PBO Lab interactive tutorials.

The PBO Lab quadrupole tutorial is opened by selecting the Tutorial button in the Piece Window (upper left in Figure 2) for any quadrupole element appearing on the Model Space or Work Space Panes. An initial tutorial window (not shown in Figure 2) is then displayed which provides the user with an outline of the quadrupole tutorial. The tutorial may be paged through (lower left in Figure 2), or bold (hyperlink) text may be selected to jump to other information. The right hand window in Figure 2 shows an interactive equation page, accessed using a hyperlink (**focusing** word in lower left window), that displays formulae and data for this particular quadrupole. Numerical data presented reflect the current values of the user's input parameters (upper left window). USPAS Students using the prototype PBO Lab often utilized the tutorials as a specialized particle optics calculator.

GRAPHICAL BEAMLINE CONSTRUCTION KIT

The construction of a beamline for a given problem follows the basic approach used in the S.P.A.R.C. environment [1]. A beamline is built by selecting and dragging (with the mouse) the icons from the Palette Bar for individual transport elements and dropping them onto the Model Space Pane. Graphical representations of components (called Pieces) are then displayed on this pane. Figure 1 illustrates one arc of a beamline composed in this way that contains drifts, quadrupoles, sextupoles, and one bending magnet. Each transport element (Piece) automatically snaps to one end of the beamline, the end closest to the "drop point" where the mouse button is released. Pieces may also be inserted into the middle of the beamline, by selecting a drop location that is near one of the connection lines between elements on the beamline. The connection line flashes, providing a positive feedback to the user, when the mouse cursor is on a valid insertion point.

Individual Pieces, or groups of Pieces ("segments"), on the beamline may be selected for use in other beamline construction tasks. Once a selection is made, it may then be Copied, Cut, Deleted, or dragged up to the Work Space Pane for future use. Pieces and segments may be inserted in, or dropped onto the ends of, the beamline on the Model Space. An advanced feature of the Work Space will allow beamline segments there to be defined as "Sublines" to be used, for example, to construct a beamline composed of repetitive elements (e.g. a lattice). Pieces that are placed on the Work Space as a group (segment or Subline) remain grouped and their configuration is fixed. A number of operations on Piece groups are being developed, such as Invert and Rotate operations. An "Alias" Piece may also be created for any Piece or Subline on the Work Space. As the name suggests, the Alias Piece has a pointer to the original Piece or Subline, but also has enhanced features that include the ability to specify parameter deviations (e.g. for error analysis). Alias and Subline Pieces may be manipulated as any other Piece, including the ability to be used in constructing other Aliases and Sublines. A powerful object model has been developed [6], which very efficiently describes either hierarchical or flat beamline representations, that forms the basis for this graphic functionality. Figure 3 illustrates an example of Subline use.

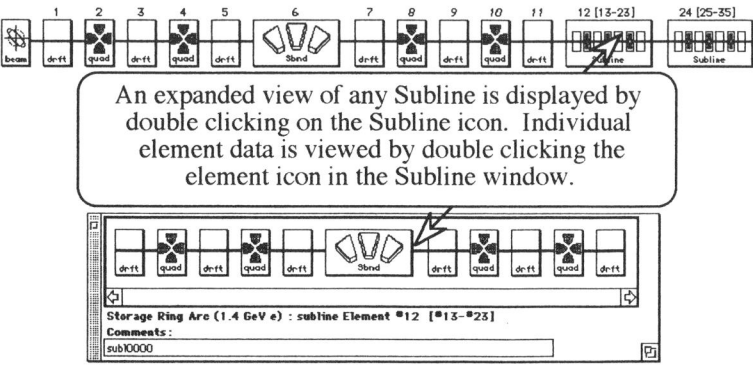

Figure 3. Sublines can be used to compactly display a group of optical elements.

PARTICLE OPTICS COMPUTATION ENGINES

The primary computation engine for the Particle Beam Optics Interactive Computer Laboratory is the TRANSPORT program. The version incorporated into the PBO Lab is a first-, second-, and third-order matrix code for designing magnetic optics systems [7]. Matrix models for more than a dozen transport elements are included in the program, and new elements may be readily incorporated through the use of a general matrix element. The prototype PBO Lab capitalized upon this capability, providing students at the USPAS with a generic thin lens optical element, and similar user-definable elements are anticipated for future versions. A large variety of beamline fitting and constraint problems may be addressed using TRANSPORT, including the ability to utilize user-defined algebraic relationships among parameters. This capability is fully supported by the PBO Lab GUI, and is one of the many features that help to provide an intuitive "front end" to TRANSPORT. Additional information on the TRANSPORT GUI is given in reference [5].

A second computation engine to be integrated with the PBO Lab is TURTLE [8]. TURTLE is a ray tracing program that utilizes the same description of beamlines as TRANSPORT and is frequently used in conjunction with TRANSPORT. For example, TRANSPORT can be used to provide general beamline design and problem solving support, such as fitting to constraints, and then TURTLE can be used to carry out multiparticle simulations of the resulting beamline performance. One goal of the present effort is to replace the existing character-based graphics output of TURTLE using a scatter-plot analysis tool developed for the PBO Lab.

The third major computation engine is a first-order matrix model developed specifically to run interactively using the PBO Lab. The model includes an envelope treatment of space charge, providing students with a tool to study the effects of space charge on optical beamline performance. Single particle trajectories may also be computed, both with and without the forces due to the beam envelope space charge. Figure 4 shows an example of the use of this engine.

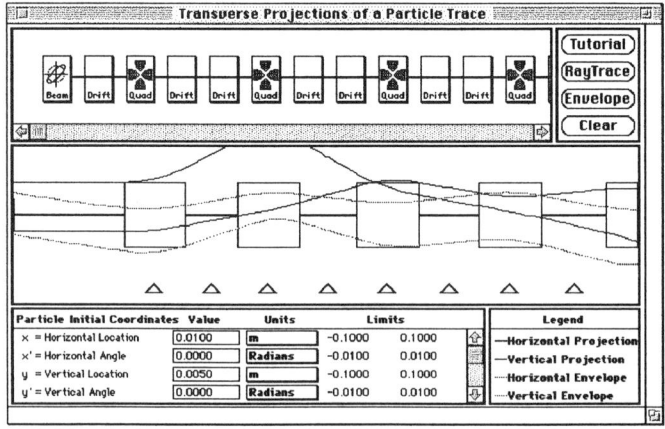

Figure 4. Using the PBO Lab to model beam envelope space charge effects.

SUMMARY

The Particle Beam Optics Interactive Computer Laboratory (PBO Lab) is a sophisticated graphic user interface being developed as a tool to aid in the teaching of charged particle optics and accelerator science. A prototype of the PBO Lab, linked with the TRANSPORT code, has been successfully utilized as the principal analysis tool in the computer laboratory sessions of a course at the U. S. Particle Accelerator School. A new software environment is currently under development that will extend the PBO Lab's capabilities to run under several operating systems including Windows and UNIX.

ACKNOWLEDGEMENTS

The authors are indebted to Karl Brown for useful comments on early versions of the PBO Lab and TRANSPORT GUI. The assistance of James Gillespie in the design of the software architecture for the PBO Lab is gratefully acknowledged. Portions of this work have been supported by the U. S. Department of Energy under SBIR grant number DE-FG03-94ER81767.

Macintosh is a trademark of Apple Computer, Pentium is a trademark of Intel, and Windows and WindowsNT are trademarks of Microsoft.

REFERENCES

1. G. H. Gillespie, "The Shell for Particle Accelerated Related Codes (SPARC) - A Unique Graphical User Interface," *AIP Conference Proceedings 297*, 576-583 (1993).
2. G. H. Gillespie and B. W. Hill, "A Graphical User Interface for TRACE 3-D Incorporating Some Expert System Type Features," *1992 Linear Accelerator Conference Proceedings 2*, AECL-10728, 787-789 (1992).
3. G. H. Gillespie and B. W. Hill, "An Interactive Graphical User Interface for the Linac Beam Dynamics Code PARMILA," *Proceedings of the 1994 International Linac Conference 2*, 517-519 (1994).
4. C. C. Paulson, A. M. M. Todd, M. A. Peacock, M. F. Reusch, D. Bruhwiler, S. L. Mendelsohn, D. Berwald, C. Piaszczyk, T. Meyers, G. H. Gillespie, B. W. Hill and R. A. Jameson, "Accelerator Systems Optimizing Code," *Proceedings of the 1995 Particle Accelerator Conference 2*, 1164-1166 (1995).
5. G. H. Gillespie, B. W. Hill, N. A. Brown, R. C. Babcock and D. C. Carey, "A Graphic User Interface for the Particle Optics Code TRANSPORT," to be published in the Proceedings of the 18th International Linac Conference held in Geneva, Switzerland, August 26-30, 1996, 3 pages.
6. B. W. Hill, H. Martono and J. S. Gillespie, "A Object Model for Beamline Descriptions," these proceedings, 6 pages.
7. D. C. Carey, K. L. Brown and F. Rothacker, "Third-Order TRANSPORT - A Computer Program for Designing Charged Particle Beam Transport Systems," Stanford Linear Accelerator Center Report No. SLAC-R-95-462, 295 pages (1995).
8. D. C. Carey, "TURTLE (Trace Unlimited Rays Through Lumped Elements) A Computer Program for Simulating Charged Particle Beam Transport Systems," Fermi National Accelerator Laboratory Report No. NAL-64, 45 pages (1978).

Automatic Differentiation and Lattice Function Matching

Jean-François Ostiguy, Leo Michelotti
and James A. Holt
Fermi National Accelerator Laboratory [*]
P.O. Box 500, Batavia, Illinois 60510

Abstract

Although popularized in accelerator physics for the calculation of the Taylor coefficients of phase space maps, automatic differentiation is a generic and powerful technique applicable to any problem where fast and accurate evaluation of (usually first order) derivatives is needed. Problems requiring the evaluation of sensitivities with respect to parameters of a model or the minimization of an error function permeate applications in accelerator physics.

The advent of languages with built-in support for operator overloading – such as C++ – results in greatly simplified syntax and highly maintainable software. To illustrate this point in practice, we apply automatic differentiation techniques to linear lattice function matching, an important problem encountered in the context of designing insertions or transfer lines between accelerators.

INTRODUCTION

As early as 1960, computer scientists and numerical analysts realized that derivatives could be computed efficiently using a technique now commonly referred to as "Automatic Differentiation." In essence, automatic differentiation applies the Liebnitz and chain rules to compute derivatives exactly (to machine precision). Note that "Automatic Differentiation" should not be confused with "Symbolic Differentiation" performed by programs such as MACSYMA, Maple or Mathematica.

---

[*] Operated by the Universities Research Association, Inc., under contract with the U.S. Department of Energy.

© 1997 American Institute of Physics

In a typical automatic differentiation application, a set of variables are selected as coordinates (independent variables) and the derivatives, up to a pre-specified order, are systematically and automatically computed every time an expression involving combinations of these variables is formed. Although it might appear at first glance that this is computationally more expensive than conventional finite differences, it can be shown that this is not the case, provided proper care is taken to eliminate redundant and unnecessary computations [1].

Perhaps because of limited access to computer resources and appropriate languages, until the early 1980's the optimization community took little notice of the developments in automatic differentiation. This state of affair is now rapidly changing.

Accelerator Physics

In the accelerator physics community, particle dynamicists were first to tap the potential of automatic differentiation. By design, storage rings and beamlines are very meticulously engineered to be as linear as possible from a beam optics point of view. Deviations from linearity are typically extremely small and therefore maps are often well represented by high order Taylor series expansions. In order to get reliable results, it is important to evaluate derivatives (i.e. the coefficients of the Taylor series expansion) as accurately as possible, an other area where automatic differentiation has a definite advantage over finite differencing. Perhaps because it has been historically identified with specialized nonlinear dynamics issues, there is a perception that automatic differentiation has little applicability to other sub-specialities of accelerator physics. This is definitely not the case. One interesting application is the solution of the the so-called matching problem, where one attempts to determine the settings of various optical elements to match the linear lattice functions to that of two circular machines at both ends of a transfer line or similarly, at on both sides of an insertion (e.g. a low beta insertion) in a circular machine. Other potential areas of applicability are magnet and accelerator cavity design where it is often needed to understand the sensitivity of field harmonics or resonant frequencies with respect to very small geometric errors due to imperfect machining or assembly. True optimization of various figures of merit for magnets and accelerating cavities is also within the realm of the possible, although somewhat more ambitious.

THE ROLE OF C++

Because of the large investment made by the numerical analysis and scientific communities in Fortran 77 code, the first attempts at using automatic differentiation were naturally executed in that language. It became quickly obvious that to avoid a substantial and error-prone rewriting effort, and to make newly developed code legible, maintainable and free of very obscure bugs one needed a way to automatically generate the required code. The pragmatic approach was to develop a preprocessor and generate standard Fortran 77 statements. This is still a viable approach, especially with legacy Fortran code. For new development, a language that supports complex abstract data types and operator overloading is more suitable and C++ supports all these features. One of the first usable and comprehensive implementations of automatic differentiation in C++ is the MXYZPTLK class library developed by Michelotti [2, 3]. The library uses the forward mode and while it does not claim to be the most efficient implementation, it has some advantages for people interested in interactive computation. MXYZPTLK is actively used at Fermilab. Using the library is a straightforward matter: to obtain derivatives, one simply declare the independent variables as `Coordinates` and all other dependent variables as `Jet` variables (`Coordinates` are implemented as type derived from `Jet`). `Jet` and `Coord` are simply abstract data type that carry derivative information along with the actual value of the variable. All fundamental and most commonly used operators and functions exist in overloaded versions and are compatible with scalar types (`int`, `double` and `complex`). Support is also available for `Jet` matrices.

THE BEAMLINE MATCHING PROBLEM

The matching problem presents itself in two distinct situations such as illustrated in Figures 1 and 2. In Figure 1, a beamline is inserted into an existing lattice. The problem consists in determining the strengths and positions of N optical elements (usually quadrupoles) such that the lattice functions at both ends of the insertion have values identical to those of the original lattice; in that case, the lattice functions outside the insertion region remain unchanged. Once usually also demands that the dispersion and its derivative remain unchanged to avoid uncontrolled dispersion outside the insertion region.

In Figure 2, a beam is transferred from machine "A" to machine "B" through a beamline. In the conventional uncoupled picture, the beam Twiss parameters at the exit of the transfer beamline must "match" the natural beam parameter of machine B to avoid emittance dilution. This means that for each degree of freedom, the transferred beam phase space ellipse must fall exactly on one of the invariant elliptical trajectories of machine B phase space. Any "mismatch" results in a tumbling ellipse which eventually fills out a larger area because of residual nonlinearities. Once again, one usually demands matching of the dispersion and its derivative.

Most established tracking codes, such as MAD [6], assume decoupled transverse dynamics. An error function based on conventional lattice functions, usually a weighted sum of quadratic error terms, is then minimized using variants of gradient or Newton based optimization algorithms. Derivatives are obtained by finite differences. Automatic differentiation is a generic technique that can be used in conjunction with any formulation of the matching problem. The main benefits are that the derivatives can be computed more efficiently and more accurately. Furthermore, the code is much easier to maintain.

We shall present here a formulation that does not make any assumption about coupling. This is not only of academic interest: a possible application would be the injection of a beam emerging from an uncoupled machine into a so-called "Mobius Ring" such as described by Talman [4]. In such a ring, special coupled optics would be used to mitigate beam-beam resonances. Although this has never been demonstrated in a real machine, it could lead to significantly higher luminosity in $\bar{p}$-p colliders such as the Tevatron, which is currently beam-beam limited.

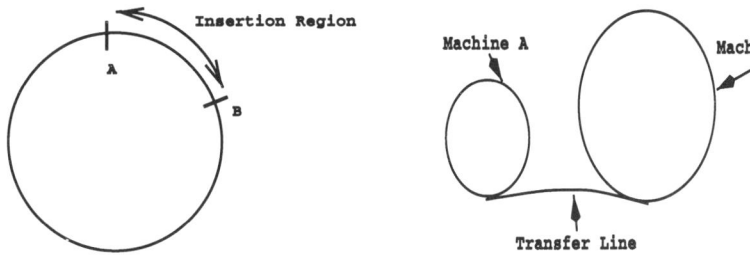

Figure 1: A beamline is inserted in an existing lattice between locations A and B.

Figure 2: The beam from machine A is transferred to machine B through a transfer line.

Although the familiar Twiss functions of uncoupled transverse space can be generalized to coupled transverse dynamics [5], these representations are not popular because, in contrast with the uncoupled situation, the relation with the dimensions and shape of the beam is not direct and simple. With this in mind, we adopt a different viewpoint. Consider a particle with initial transverse phase space coordinates

$$\mathbf{x}_1 = (x, x', y, y') \tag{1}$$

After propagating through a beamline, its phase coordinates become

$$\mathbf{x}_2(s) = M(s)\mathbf{x}_1 \tag{2}$$

where the matrix $M(s)$ depends, on the strengths and positions of various elements. Assuming the situation depicted in Figure 2, let M_A and M_B be the (linear) one-turn maps for machines A and B respectively and $\lambda_i^{(A)}$ and $u_i^{(A)}$ be the corresponding eigenvectors and eigenvalues. Then, for a beamline characterized by a transfer matrix M, a sufficient condition for the phase space of the beam emerging from machine A to be matched to the phase space of machine B is that M maps the (normalized) eigenvectors $u_i^{(A)}$ into the (normalized) eigenvectors $u_i^{(B)}$. Under this condition, the transformation M will obviously transform a stationary phase space distribution in machine A into a stationary phase space distribution in machine B. If $U^{(A)}$ and $U^{(B)}$ are matrices whose columns are the eigenvectors of M_A and M_B, then one must have

$$MU^{(A)} = U^{(B)} \tag{3}$$

Note that since the transfer matrices are real, the eigenvectors and eigenvalues always arise in complex conjugate pairs.

To solve a matching problem, one can now minimize the quadratic error function

$$F(q_1, q_2, ..., q_n) = [M(q_1, ...q_n)U^{(A)} - U^{(B)}]^\dagger W [M(q_1, ...q_n)U^{(A)} - U^{(B)}] \quad (4)$$

where W is a user-specified real positive definite weight matrix and $q_1, ...q_n$ are free parameters, usually the strengths and possibly the positions of a certain number of quadrupoles. In the context of automatic differentiation, $q_1, ...q_n$ are of type Coord and M is a matrix of Jet.

Including dispersion without treating it separately in an ad-hoc manner should be possible. The first thing that comes to mind is to include the longitudinal degree of freedom into the maps. To avoid dealing with RF and longitudinal matching, the phase space vector is augmented with the longitudinal momentum coordinate and matrices have dimension 5×5. All entries on last row are zero, except for the diagonal which is equal to 1. Remarkably, the additional eigenvalue is 1 and the four others remain unchanged. The eigenvector corresponding to the $\lambda = 1$ eigenvalue is $(\eta_x, \eta'_x, \eta_y, \eta'_y, 1)$. For the purpose of the matching algorithm, this eigenvector contains redundant information and is not needed. As long as the four remaining eigenvectors satisfy (3), the transfer -including dispersion- should be matched. Instead of dealing with augmented matrices, one might also consider treating the longitudinal momentum p as an additional parameter and minimize (4) with respect to variations of both p and q_i. In that case the matrices remain 4×4 but the eigenvectors $U^{(A)}$ and $U^{(B)}$ now also depend on p and also become Jet variables.

IMPLEMENTATION

A prototype C++ class using an algorithm based on the procedure outlined in the previous section has been developed and is currently being tested. Although dispersion has not been included yet, preliminary results obtained in matching a beam from a uncoupled to a coupled machine are encouraging.

Nonlinearities encountered in matching problems can be very severe. One often has to deal with cost functions that vary wildly and possess several local minima. These minima correspond to solutions that are not physically acceptable. Since different cost functions result in different types of nonlinearities, an interesting issue is to determine whether a cost function that attempts to minimize the error on eigenvectors mapping offers some advantages from that standpoint.

Another issue is that one usually wants to realize a match using a minimum number of variable elements. While the approach described here works, it is intuitively clear that in the uncoupled situation, the phase advance in a transfer line is inconsequential; therefore we may be over-constraining the problem unnecessarily. We are currently exploring other formulations based on matrix factorizations such as those developed by Irwin [7] that would allow us to formulate necessary and sufficient conditions for matching.

Finally, matching is very difficult and probably impossible to completely automate. Intervention by a knowledgeable user is essential to lead the optimization in a region of the parameter space where a physically acceptable solution exists. With currently available tools, this process can be very tedious. Our objective is to develop classes that would allow a lattice designer to stop the optimization when he desires, interactively change weight functions, constraints, optimization parameters and continue until a satisfactory solution is attained.

REFERENCES

[1] A. Griewank, *On Automatic Differentiation*, in Mathematical Programming: Recent Developments and Applications, M. Iri and K. Tanabe eds, Kluwer Academic Publishers, Dordrecht (1989), 83-108

[2] L. Michelotti, *MXYZPLTK: A pratical, User-Friendly C++ Implementation of Diffrential Algebra: User's Guide*, Fermi National Accelerator Laboratory, Batavia, IL, FN-535 (1990).

[3] L. Michelotti, *A C++ Hacker's Implementation of Automatic Differentiation*, Automatic Differentiation of Algorithms: Theory, Implementation, and Application, G. Corliss and A. Griewank Eds., SIAM Philadelphia, PA (1991)

[4] R. Talman, *The Mobius Ring*, Phys. Rev. Lett. **74** (1995), 1590-1593

[5] D. A. Edwards and L. C. Teng, *Parametrization of Linear Coupled Motion in Periodic Systems*, IEEE Transactions on Nuclear Science", **NS-20**, 3 (1973), 885

[6] Hans Grote, F. Christoph Iselin, *The Methodical Accelerator Design Program User's Reference Manual*, CERN, Geneva, Switzerland, SL 90-13 (AP)

[7] J. Irwin, Nuclear Instruments and Methods, **A298** (1990) 460-72

Computation and Analysis of Spin Dynamics [1]

Vladimir Balandin*[†], Martin Berz* and Nina Golubeva[†]

* Department of Physics and Astronomy and
National Superconducting Cyclotron Laboratory
Michigan State University, East Lansing, MI 48824
[†] Institute for Nuclear Research of RAS, 60th October Anniversary Pr., 7a
Moscow, Russia

Abstract. A method is described that allows the **computation and analysis of high-order spin maps for general non-autonomous optical systems.** It is shown how the equations of motion in curvilinear coordinates resulting from the Thomas-BMT equation can be solved within a differential algebraic framework in SU(2) and SO(3) representations.
The resulting maps are subjected to a spin-orbit normal form transformation, and the **nonlinear orbit dependencies of the invariant polarization axis as well as the orbit dependent spin tune can be obtained.** For the case of electron machines, the resulting invariant polarization can be used to determine the radiative equilibrium polarization via the Derbenev-Kondratenko approach. Both the computation of the spin map as well as the algorithm for the computation of the invariant axis have been implemented in the code **COSY INFINITY** [2] [3] [4].

THE ONE TURN MAP FOR SPIN-ORBIT MOTION

When viewing the motion "stroboscopically" at a fixed point in the ring (for example, at the point of physics experiment), it is convenient to view and analyze the motion in terms of a one-turn map instead of in terms of the differential equations of motion.

The equations of spin-orbit motion are **linear** in the spin, and hence the transformation of the spin variables can be described in terms of a **matrix**, the elements of which depend on the orbital quantities only. The orbital quantities themselves are unaffected by the spin motion, such that altogether the map has the form

---

[1] Supported by the US Department of Energy, Grant No. DE-FG02-95ER40931, and the Alfred P. Sloan Foundation

$$\begin{cases} \vec{x}_f = \mathcal{M}(\vec{x}_i) \\ \vec{s}_f = A(\vec{x}_i) \cdot \vec{s}_i \end{cases} \quad (1)$$

where $A(\vec{x}_i) \in \mathbf{SO}(3)$.

The **practical computation** of the spin-orbit map can be achieved in a variety of ways. Conceptually the **simplest way** is to interpret it as a motion in the **nine variables consisting of orbit and spin**. In this case, the DA method allows the computation of the spin-orbit map in the two conventional ways, namely via a propagation operator for the case of the z-independent fields like main fields, and via integration of the equations of motion with DA [5] [7]. However, in this simplest method, the number of independent variables increases from six to nine, which particularly in higher orders entails a rather substantial increase of computational and storage requirements. This limits the ability to perform analysis and computation of spin motion to high orders.

Due to the **special structure of the equations of motion**, it is possible to rephrase the dynamics such that it is still described in terms of **only the six orbital variables**. For this purpose, we derive the equation of motion for the individual elements of the matrix $A(\vec{x})$. To this end, we write $\vec{s}_f = A(\vec{x}) \cdot \vec{s}_i$ and insert this into the map (1). Integrating the equations of motion for the matrix $A(\vec{x})$ along with the orbital equations now allows the computation of the spin motion based on only six initial variables. In practice, the **orthogonal symmetry also may be used** to save computational expense.

INVARIANT FUNCTIONS AND THE STABLE DIRECTION OF POLARIZATION

A function $V(\vec{x}, \vec{s})$ is called an **invariant function** of the map (1) if

$$V(\vec{x}, \vec{s}) = V(\mathcal{M}(\vec{x}), A(\vec{x}) \cdot \vec{s}) \quad (2)$$

Taking into account that in the map (1) the orbital part is independent of the spin motion and the linearity of the map (1) with respect to spin variables, it is enough to consider functions of the form

$$V(\vec{x}, \vec{s}) = b(\vec{x}) + \vec{g}(\vec{x}) \cdot \vec{s}. \quad (3)$$

If we substitute the expression (3) in the condition of invariance (2), we will see that $b(\vec{x})$ is the usual invariant function of the orbital part of the map. It means that one can find $V(\vec{z}, \vec{s})$ as

$$V(\vec{x}, \vec{s}) = \vec{g}(\vec{x}) \cdot \vec{s}. \quad (4)$$

Equation (2) for the function (4) becomes

$$A(\vec{x}) \cdot \vec{g}(\vec{x}) = \vec{g}(\mathcal{M}(\vec{x})). \quad (5)$$

We will call the function $V(\vec{x}, \vec{s})$ in (4) **nondegenerate** if $|\vec{g}(\vec{0})| \neq 0$. In this case, due to the property of the function $|\vec{g}(\vec{x})|$ to be an invariant of the orbital part of the map (1), it is possible to assume that $|\vec{g}(\vec{x})| = 1$. If the matrix $A(\vec{0}) \neq I$, then there is not more than one independent nondegenerate invariant function (up to multiplication with the invariant function of the orbital map).

The **existence of an invariant function** V in the form (4) gives us a relatively **simple geometrical picture** of the behavior of the spin vector during iterations of the map (1): the projection of the spin vector $\vec{s}$ on the vector $\vec{g}(\vec{x})$ is preserved.

If V is a nondegenerate invariant function and the orbital motion has a stable fixed point $\vec{x} = \vec{0}$, then the vector $\vec{g}(\vec{0})$ defines the direction along which the polarization of a particle is conserved. More precisely, for any $\epsilon > 0$ there exists $\delta > 0$ such that if $|\vec{x}_0| + |\vec{g}(\vec{0}) - \vec{s}_0| < \delta$ then for all n

$$|\vec{x}_n| + |\vec{g}(\vec{0}) - \vec{s}_n| < \epsilon,$$

where n is the number of the map iterations.

The existence of nondegenerate invariant functions is a very important property which allows us, for example, to inject the polarized beam in the storage ring and then guarantee polarization conservation for a large number of turns.

This definition (introduced in [6]) of the vector $\vec{g}(\vec{x})$ and the stable direction of polarization do not depend on the selection of the coordinate system and on the Hamiltonian form of the orbit motion as in original paper of Derbenev and Kondratenko [1]. In the case when the action-angle variables I, φ for the orbital motion and the Derbenev-Kondratenko vector $\vec{n}(I, \varphi)$ exist, the introduced vector $\vec{g}(\vec{x})$ gives the one-turn boundary conditions for $\vec{n}$.

COORDINATE TRANSFORMATIONS AS A METHOD OF FINDING THE INVARIANT FUNCTION

A more general way to find the invariant function is to guess it for some simple situations, and then try to reduce the initial problem to the investigated one by a coordinate substitution which in many cases is the result of a normal form transformation (**normal form method**).

Example: If the matrix $A(\vec{x})$ in (1) has the form

$$A(\vec{x}) = \begin{pmatrix} +\cos(\lambda(\vec{x})) & +\sin(\lambda(\vec{x})) & 0 \\ -\sin(\lambda(\vec{x})) & +\cos(\lambda(\vec{x})) & 0 \\ 0 & 0 & 1 \end{pmatrix} \quad (6)$$

then $V = s_3$ is the invariant function and $\vec{g}(\vec{x}) = (0, 0, 1)$. The projection of the spin vector on the vertical direction is preserved during the iterations of the map (1) with the matrix A in the form (6), and the projection on the transverse plane is rotated with frequency $\lambda(\vec{x})$, the so-called **spin tune**. This example is simple but very important; as we will see below, even in the general situation the matrix $A(\vec{x})$ in (1) may be reduced to the form (6) by means of coordinate transformations with very high accuracy.

NORMAL FORM ALGORITHM FOR THE SPIN-ORBIT MAP

Introduce new variables $\vec{y}, \vec{\xi}$ by the equations

$$\vec{y} = \mathcal{K}(\vec{x}), \qquad \vec{\xi} = C(\vec{x}) \cdot \vec{s} \tag{7}$$

where $C(\vec{x}) \in SO(3)$. In the new variables the map (1) becomes

$$\begin{cases} \vec{y}_f = \mathcal{K}(\mathcal{M}(\mathcal{K}^{-1}(\vec{y}_i))) = \mathcal{N}(\vec{y}_i), \\ \vec{\xi}_f = C(\mathcal{M}(\mathcal{K}^{-1}(\vec{y}_i))) \cdot A(\mathcal{K}^{-1}(\vec{y}_i)) \cdot C^{-1}(\mathcal{K}^{-1}(\vec{y}_i)) \cdot \vec{\xi}_i = \bar{A}(\vec{y}_i) \cdot \vec{\xi}_i. \end{cases} \tag{8}$$

Our aim now is to show that if there are no resonances up to order m, then the coordinate transformation (7) may be chosen in such a way that the matrix $\bar{A}$ in (8) will have the form (6) up to order m. So in the new variables, we will have the approximate invariant function $V = \xi_3$. We will find the coordinate transformation (7) as a composition of two successive transformations: the first one changes the orbital variables and second one changes the spin part of the map.

DA Normal Form Algorithm for the Orbital Map

The first process of successive coordinate substitutions to obtain the invariant functions requires to perform the transformation of the orbital map. The starting step consists of the **fixed-point transformation** and the **linear diagonalization**. All further steps are purely nonlinear and no longer affect the linear part.

We begin the mth step by splitting the momentary map $\mathcal{M}$ into its linear and nonlinear parts $\mathcal{R}$ and $\mathcal{S}_m$. Then we perform a transformation using a map $\mathcal{K}_m = \mathcal{E} + \mathcal{T}_m$ where $\mathcal{T}_m$ vanishes to order $m - 1$. To study the effect of the transformation, we now infer up to order m:

$$\mathcal{K} \circ \mathcal{M} \circ \mathcal{K}^{-1} =_m \mathcal{R} + \mathcal{S}_m + (\mathcal{T}_m \circ \mathcal{R} - \mathcal{R} \circ \mathcal{T}_m). \tag{9}$$

A close inspection of the equation (9) reveals that $\mathcal{S}_m$ can be simplified by choosing the commutator $\{\mathcal{T}_m, \mathcal{R}\} = \mathcal{T}_m \circ \mathcal{R} - \mathcal{R} \circ \mathcal{T}_m$ appropriately. The detailed description of this procedure can be found in [7,8].

The Connection between the SO(3) and SU(2) Groups and the SU(2) Representation for the Spin Part of the Map

To reduce the calculations we note that while the orthogonal 3×3 matrix $A(\vec{x}) \in \mathbf{SO}(3)$, which defines the spin part of the map (1), consists of 9 elements, it can be described completely using a smaller number of parameters. We will realize it using the connection between the **SO**(3) and **SU**(2) groups.

For a matrix U in **SU**(2), we have $\det U = 1$ and $U \cdot U^* = I$. Thus any such $U \in \mathbf{SU}(2)$ has the form

$$U = \begin{pmatrix} a & b \\ -b^* & a^* \end{pmatrix}, \quad a \cdot a^* + b \cdot b^* = 1 \tag{10}$$

Corresponding to the vector $\vec{s}$, we define the matrix

$$L = \begin{pmatrix} s_3 & s_1 + is_2 \\ s_1 - is_2 & -s_3 \end{pmatrix}$$

and represent the map (1) in the form

$$\begin{cases} \vec{x}_f = \mathcal{M}(\vec{x}_i) \\ L_f = U(\vec{x}_i) \cdot L_i \cdot U^*(\vec{x}_i) \end{cases} \tag{11}$$

where $U(\vec{x}) \in \mathbf{SU}(2)$ is given by (10). A connection between the matrix U in (11) and the matrix A in (1) may be easily established using the following formula

$$A = \begin{pmatrix} Re(a^2 - b^2) & -Im(a^2 + b^2) & -2 \cdot Re(ab) \\ Im(a^2 - b^2) & Re(a^2 + b^2) & -2 \cdot Im(ab) \\ 2 \cdot Re(ab^*) & -2 \cdot Im(ab^*) & aa^* - bb^* \end{pmatrix} \tag{12}$$

Normal Form Algorithm for the SU(2) Representation of the Spin Part of the Map

We start from the spin-orbit map in the form

$$\begin{cases} \vec{x}_f = \mathcal{N}(\vec{x}_i) \\ L_f = U(\vec{x}_i) \cdot L_i \cdot U^*(\vec{x}_i) \end{cases} \tag{13}$$

where $\mathcal{N}(\vec{x})$ is the normal form of the orbital part of map up to order m.

Consider the coordinate transformation from the old matrix L to the new matrix $\bar{L}$ by the equation

$$L = C(\vec{x}) \cdot \bar{L} \cdot C^*(\vec{x}), \quad C(\vec{x}) \in \mathbf{SU}(2). \tag{14}$$

In the new variables the map (13) becomes

$$\begin{cases} \vec{x}_f = \mathcal{N}(\vec{x}_i) \\ \bar{L}_f = \bar{U}(\vec{x}_i) \cdot \bar{L}_i \cdot \bar{U}^*(\vec{x}_i) \end{cases} \tag{15}$$

where $\bar{U}(\vec{x}) = C^*(\mathcal{N}(\vec{x})) \cdot U(\vec{x}) \cdot C(\vec{x})$. If there are no resonances between orbital and spin tunes up to order m, we can find a matrix $C(\vec{x})$ so that the matrix $\bar{U}(\vec{x})$ will be diagonal up to order m and will have the form

$$\bar{U}(\vec{x}) = \mathbf{diag}(\exp(i\ae(I)), \exp(-i\ae(I))) + O(|\vec{x}|^{m+1}), \tag{16}$$

where I are the invariants for orbital motion. In this case

$$\bar{A}(I) = \begin{pmatrix} +\cos(\lambda(I)) & +\sin(\lambda(I)) & 0 \\ -\sin(\lambda(I)) & +\cos(\lambda(I)) & 0 \\ 0 & 0 & 1 \end{pmatrix} + O(|\vec{x}|^{m+1}), \tag{17}$$

and we will have **the approximate invariant function and spin tune**

$$V = s_3, \quad \lambda(I) = 2 \cdot \ae(I).$$

REFERENCES

1. Derbenev, Ya.S., and Kondratenko, A.M., Sov. Phys. JETP. **35** (1972) 230, **37** (1973) 968.
2. Berz., M., COSY INFINITY Version 6 reference manual. Technical Report MSUCL-869, National Superconducting Cyclotron Laboratory, Michigan State University, East Lansing, MI 48824, 1993.
3. Berz., M., COSY INFINITY Version 6. In *Berz, M., Martin, S., and Ziegler, K., (Eds.), Proc. Nonlinear Effects in Accelerators*, page 125. IOP Publishing, 1992.
4. Berz., M., New features in COSY INFINITY. In *Third Computational Accelerator Physics Conference*, page 267. AIP Conference Proceedings 297, 1993.
5. Berz., M., Differential algebraic description of beam dynamics to very high orders. *Particle Accelerators*, 24:109, 1989.
6. Balandin, V., and Golubeva, N., in *Proc. of the 1993 Part. Accel. Conf., Washington*, 1993.
7. Berz. M., *High-Order Computation and Normal Form Analysis of Repetitive Systems*, in: M. Month (Ed), *Physics of Particle Accelerators*, volume AIP 249, page 456. American Institute of Physics, 1991.
8. Berz. M., Differential algebraic formulation of normal form theory. In *M. Berz, S. Martin and K. Ziegler (Eds.), Proc. Nonlinear Effects in Accelerators*, page 77. IOP Publishing, 1992.

Hamiltonian Methods for the Study of Polarized Proton Beam Dynamics in Accelerators and Storage Rings

Vladimir Balandin*[†] and Nina Golubeva[†]

*Department of Physics and Astronomy and NSCL,
Michigan State University, East Lansing, MI 48824
[†] Institute for Nuclear Research of RAS,
60th October Anniversary Pr., 7a, Moscow, Russia

Abstract. The equations of classical spin-orbit motion can be extended to a **Hamiltonian system** in 9-dimensional phase space by introducing a **coupled spin-orbit Poisson bracket** (3) and **Hamiltonian function** (5). After this extension it becomes possible to apply the **methods of the theory of Hamiltonian systems** to the study of polarized particles beam dynamics in circular accelerators and storage rings. Some of those methods have been implemented in the **computer code FORGET-ME-NOT** [1], [2].

SPIN-ORBIT MOTION EQUATIONS

The quasi-classical description of the motion of particles with spin in accelerators and storage rings includes the equations of orbit motion which we will take in the Hamiltonian form

$$\frac{d\vec{q}}{dt} = \frac{\partial H_{orbt}}{\partial \vec{p}}, \qquad \frac{d\vec{p}}{dt} = -\frac{\partial H_{orbt}}{\partial \vec{q}} \qquad (1)$$

and the Tomas-BMT equation [3], [4] for the classical spin vector $\vec{s}$

$$\frac{d\vec{s}}{dt} = \left[\vec{W} \times \vec{s}\right]. \qquad (2)$$

Here

$$H_{orbt} = c\sqrt{\vec{\pi}^2 + m_0^2 c^2} + e\Phi,$$

$$\vec{W} = -\frac{e}{m_0 \gamma c}\left((1+\gamma G)\vec{B} - \frac{G\left(\vec{\pi}\cdot\vec{B}\right)\vec{\pi}}{m_0^2 c^2 (1+\gamma)} - \frac{1}{m_0 c}\left(G + \frac{1}{1+\gamma}\right)\left[\vec{\pi}\times\vec{E}\right]\right),$$

and t is the time, $\vec{q}$ and $\vec{p}$ are canonical position and momentum variables, and $\vec{s} = (s_1, s_2, s_3)$ is the classical spin vector of length $\hbar/2$ in a fixed Cartesian coordinate system, e and m_0 are the charge and the rest mass of the particle, c is the velocity of light, $G = (g-2)/2$ which quantifies the anomalous spin g factor, γ is the Lorentz factor, $\vec{\pi}$ is kinetic momentum vector, $\vec{E}$ and $\vec{B}$ are the electric and magnetic fields, and $\vec{A}$ and Φ are the vector and scalar potentials.

Lateron we will refer to the system (1)-(2) as the **triangular system** (the equations of spin motion contain the orbital variables but the evolution of orbital variables does not depend on the spin degree of freedom).

HAMILTONIAN EXTENSION OF THE EQUATIONS OF CLASSICAL SPIN-ORBIT MOTION

Introduce the Poisson bracket

$$\{f(\vec{z}), g(\vec{z})\} = f_{\vec{q}} \cdot g_{\vec{p}} - f_{\vec{p}} \cdot g_{\vec{q}} + [f_{\vec{s}} \times g_{\vec{s}}] \cdot \vec{s} \qquad (3)$$

in 9-dimensional phase space $\vec{z} = (\vec{x}, \vec{s})$ of 6 orbital $\vec{x} = (\vec{q}, \vec{p})$ and 3 spin $\vec{s}$ variables and consider a Hamiltonian system of ordinary differential equations

$$\frac{d\vec{z}}{dt} = \{\vec{z}, H\} \qquad (4)$$

with the Hamiltonian function

$$H = H_{orbt}(\vec{x}, t) + \vec{W}(\vec{x}, t) \cdot \vec{s}. \qquad (5)$$

In variables $\vec{q}$, $\vec{p}$ and $\vec{s}$ the system (4) can be written as

$$\frac{d\vec{q}}{dt} = \frac{\partial H_{orbt}}{\partial \vec{p}} + \frac{\partial (\vec{W} \cdot \vec{s})}{\partial \vec{p}}, \quad \frac{d\vec{p}}{dt} = -\frac{\partial H_{orbt}}{\partial \vec{q}} - \frac{\partial (\vec{W} \cdot \vec{s})}{\partial \vec{q}} \qquad (6)$$

$$\frac{d\vec{s}}{dt} = [\vec{W} \times \vec{s}] \qquad (7)$$

and we will understand the equations (6)-(7) as the **Hamiltonian extension of the equations of classical spin-orbit motion** (1)-(2).

CONNECTION BETWEEN TRIANGULAR SYSTEM AND ITS HAMILTONIAN EXTENSION

Not ascribing any physical sense to the spin dependent members in the right sides of equations (6), we wish to point out some ways for establishing the connections between properties and solutions of system (6)-(7) and initial triangular system (1)-(2) (**truncation procedures** [1]).

1. If $\vec{z}(t, t_0, \vec{z}_0) = (\vec{x}(t, t_0, \vec{x}_0, \vec{s}_0), \vec{s}(t, t_0, \vec{x}_0, \vec{s}_0))$ is the solution of (6)-(7) which passes through the point $\vec{z}_0 = (\vec{x}_0, \vec{s}_0)$ when $t = t_0$, then

$$\vec{z}_*(t, t_0, \vec{z}_0) = (\vec{x}_*(t, t_0, \vec{x}_0), \vec{s}_*(t, t_0, \vec{x}_0, \vec{s}_0)),$$

where $\vec{x}_*(t, t_0, \vec{x}_0) = \vec{x}(t, t_0, \vec{x}_0, \vec{0})$ and $\vec{s}_*(t, t_0, \vec{x}_0, \vec{s}_0)) = \left.\frac{\partial \vec{s}}{\partial \vec{s}_0}\right|_{\vec{s}_0 = \vec{0}} \cdot \vec{s}_0,$

gives us the solution of (1)-(2).

2. If the system (6)-(7) admits an invariant function $V(\vec{z}, t)$ which can be represented in the form

$$V(\vec{z}, t) = V_m(\vec{z}, t) + V_{>m}(\vec{z}, t),$$

where V_m is a homogeneous polynomial of degree m in variables $\vec{s}$, and

$$\lim_{|\vec{s}| \to 0} \frac{V_{>m}}{|\vec{s}|^m} = 0,$$

then $V_m(\vec{z}, t)$ is a first integral of the system (1)-(2).

3. If $\vec{x}(t, t_0, \vec{x}_0) \stackrel{\text{def}}{=} \vec{\phi}(t, t_0, \vec{x}_0)$ is the solution of (1), then the system (1)-(2) can be written as a family of Hamiltonian systems of the type (4) depending on parameters $\vec{x}_0$, t_0 with Hamiltonian function

$$H = \vec{W}\left(\vec{\phi}(t, t_0, \vec{x}_0), t\right) \cdot \vec{s}$$

HAMILTONIAN METHODS FOR EXTENDED SYSTEM

Degenerate Poisson Brackets for Global Variables, or Local Darboux Coordinates ?

The spin-orbit Poisson bracket (3) is **degenerate**. It has the nontrivial **Casimir function** $|\vec{s}|^2$ and on the level surface $|\vec{s}|^2 =$ const> 0 its rank is constant and is equal to 8. This means that we can decrease the dimensions of the system (4) by introducing the atlas of local **Darboux coordinates** in which the structural matrix of the spin-orbit Poisson bracket will have the form

---

[1] Note that some authors consider the spin depended members in the right sides of equations (6) as the possibility to treat the quasi-classical effect of the spin on the orbit motion, and in this case no truncations of results will be needed.

$$\begin{pmatrix} O_{44} & I_4 & O_{41} \\ -I_4 & O_{44} & O_{41} \\ O_{14} & O_{14} & O_{11} \end{pmatrix} \quad (8)$$

where O_{kl} is a $k \times l$ zero matrix and I_n is a $n \times n$ identical matrix. Thus in local coordinates we obtain the **classical Hamiltonian system with 4 degrees of freedom depending on one parameter** $|\vec{s}|^2$. It is clear that the Darboux coordinates are not unique and may be introduced in different ways. We consider only one typical example.

Let $\vec{i}, \vec{j}, \vec{k}$ be an arbitrary orthogonal system of unit vectors in three dimensional space R^3 satisfying the condition $\vec{i} \cdot [\vec{j} \times \vec{k}] = 1$. We introduce three new spin variables I, J, ψ by the equations:

$$\vec{s} \cdot \vec{i} = J, \quad \vec{s} \cdot \vec{j} = \sqrt{I - J^2} \cos(\psi), \quad \vec{s} \cdot \vec{k} = \sqrt{I - J^2} \sin(\psi). \quad (9)$$

Here J is the projection of the spin vector on the $\vec{i}$-axis, ψ is the polar angle in the transverse plane and $I = |\vec{s}|^2$. In the new variables the spin part of equations (6)-(7) becomes

$$\dot{\psi} = H_J, \qquad \dot{J} = -H_\psi, \qquad \dot{I} = 0, \quad (10)$$

where the Hamiltonian (5) takes on the form

$$H = H_{orbt} + \left(\vec{W} \cdot \vec{i}\right) J + \sqrt{I - J^2} \left(\left(\vec{W} \cdot \vec{j}\right) \cos(\psi) + \left(\vec{W} \cdot \vec{k}\right) \sin(\psi)\right).$$

Unfortunately, when $\left(\vec{W} \cdot \vec{j}\right)^2 + \left(\vec{W} \cdot \vec{k}\right)^2 \neq 0$, this coordinate system cannot be extended onto the whole sphere $I =$const> 0, since it has a singularity for $I - J^2 = 0$. This means we need to have a whole atlas of local coordinates systems (at least two local coordinate systems defined by different vectors $\vec{i}_1$, $\vec{j}_1, \vec{k}_1$ and $\vec{i}_2, \vec{j}_2, \vec{k}_2$ in (9)) for a complete description of spin motion on the sphere in the electric and magnetic fields depending on time and position of the particle. We will have the same difficulties with any other Darboux coordinates because they are defined by the topological properties of the sphere. So the **way pointed by the Darboux theorem does not look like the most natural or straightforward approach to the problem of investigation of polarized beam dynamics** and we prefer to study the equations of spin-orbit motion using initial global variables $\vec{x}, \vec{s}$ and the Poisson bracket (3) (degenerate).

Canonical Transformations and Introduction of Machine Coordinates for Circular Accelerators

In the theory of circular accelerators it is useful to describe the spin-orbit motion in terms of a suitable curvilinear coordinate system. Corresponding variables transformation can be chosen to be **canonical** in respect with Poisson brackets (3), and it allows us to make coordinate transformations working not with equations of motion, but directly with the Hamiltonian function (5).

If we linearize the resulting Hamiltonian equations in respect to spin variables and then neglect the effect of spin on the orbit motion (**triangular truncations procedure**) we get the transformed form of triangular system too.

Methods of Numerical Integration and Normal Forms for Spin-Orbit Maps and Hamiltonians

Today **symplectic tracking** methods for orbital motion (methods which conserve the classical Poisson bracket) are common tools in accelerator physics and we believe that such methods, which **conserve the Poisson bracket (3) exactly or with high accuracy**, are the most suitable ones for numerical simulation of the equations (6)-(7).

Those methods, of course, are mainly a point of theoretical interest, but in application to the triangular system (practical interest) the Hamiltonian approach allows us to reduce the initial problem of numerical integration of the system (1)-(2) to that of symplectic integration of the equations of orbital motion (1) only (see [5], [6]).

The detailed description of **normal forms** algorithms for spin-orbit maps and Hamiltonians can be found in [2], [5].

COMPUTER IMPLEMENTATION

The computer code **FORGET-ME-NOT** [1] has been written for the study of polarized beam dynamics and includes the following options:

1. Calculation of **strengths of imperfection spin resonances** and **first order intrinsic spin resonances** with betatron oscillations with the help of an **averaging method**.
2. Calculation of **one-turn Taylor maps for orbit and spin motion** up to arbitrarily high order with respect to the amplitudes of the betatron and synchrotron oscillations and determination of:
    2.1. **Invariant functions of the orbit motion.**
    2.2. **Equilibrium polarization direction.**
    2.3. **Dependence of orbit and spin tunes on invariants of orbit motion** (spread of orbit and spin tunes).
3. **Numerical tracking of particles with spin** in accelerators and storage rings:
    3.1. **Symplectic with respect to the 6-D orbit motion.**
    3.2. **Orthogonal with respect to the 3-D spin motion.**

All options use the same physical model. The use of various approaches allows the computed results to understand from various points of view. FORGET-ME-NOT has been applied to the investigation of scheme for preserving polarization in the TRIUMF KAON Booster [6], [7] and to the investigation of spin motion at high energy in the HERA proton ring [8].

Note that the study of dynamics of particles with spin is available also in **COSY INFINITY** [9], including efficient methods to treat fringe fields, a variety of methods to directly use measured field data, and several optimizers. The details about algorithms for analysis of spin dynamics, build in COSY INFINITY, can be found in [10], [11], and a short review of possibilities of another computer programs in this field in [11], [12].

REFERENCES

1. Balandin, V.V., and Golubeva, N.I., Proc. of the XV Int. Conf. on High Energy Particle Accelerators, Int. J. Mod. Phys. A, 2B,998 (1992).
2. Balandin, V.V., and Golubeva, N.I., Particle Accelerator Conference Washington USA (1993).
3. Thomas, L.H., *Phil. Mag.,* **3**:1, 1927.
4. Bargmann, V., Michel, L., and Telegdi, V.L., *Precession of the polarization of particle moving in a homogeneous electromagnetic field*, Physical Review Letters, 2(10):435-436, May 1959.
5. Balandin, V.V., and Golubeva, N.I., *Hamiltonian Methods for the Study of Polarized Beam Dynamics in Accelerators and Storage Rings*, DESY Report in preparation.
6. Balandin, V.V., Golubeva, N.I., *Investigation of Spin Motion in the Booster Lattice*, TRIUMF Report TRI-DN-93-K236, (1993).
7. Balandin, V.V., Golubeva, N.I., *Fast Betatron Tune Jumps and Partial Siberian Snakes for Preserving the Polarization in the Booster Lattice*, TRIUMF Report TRI-DN-93-K240, (1993).
8. Balandin, V.V., Barber, D.P., Golubeva, N.I., *Studies of the Behaviour of Proton Spin Motion in HERA-p at High Energies*, DESY M 96-04, (1996).
9. Berz, M., COSY INFINITY Version 6 reference manual. Technical Report MSUCL-869, National Superconducting Cyclotron Laboratory, Michigan State University, East Lansing, MI 48824, 1993.
10. Balandin, V., Berz, M., Golubeva, N., *Computation and Analysis of Spin Dynamics*, this proceeding.
11. Balandin, V., Berz, M., Golubeva, N., *Description and Analysis of High-Order Spin Dynamics in General Particle Optical Systems*, in preparation.
12. Barber, D.P., Heinemann, K., and Ripken, G., *A canonical 8-dimensional formalism for classical spin-orbit motion in storage rings. (II) Normal forms and n-axis*, Zeitschrift fur Physik, C(64):143-167, 1994.

SIMULATIONS
FOR CONTROL SYSTEMS

Evaluation of a Server-Client Architecture for Accelerator Modeling and Simulation*

B. A. Bowling, W. Akers, H. Shoaee, W. Watson, J. van Zeijts,
S. Witherspoon

Thomas Jefferson National Accelerator Facility, Newport News, VA 23606 USA

Abstract. Traditional approaches to computational modeling and simulation often utilize a batch method for code execution using file-formatted input/output. This method of code implementation was generally chosen for several factors, including CPU throughput and availability, complexity of the required modeling problem, and presentation of computation results. With the advent of faster computer hardware and the advances in networking and software techniques, other program architectures for accelerator modeling have recently been employed. Jefferson Laboratory has implemented a client/server solution for accelerator beam transport modeling utilizing a query-based I/O. The goal of this code is to provide modeling information for control system applications and to serve as a computation engine for general modeling tasks, such as machine studies. This paper performs a comparison between the batch execution and server/client architectures, focusing on design and implementation issues, performance, and general utility towards accelerator modeling demands.

INTRODUCTION

Traditional approaches to computational modeling and simulation often utilize a batch method for code execution using file-formatted input/output. This method of code implementation was generally chosen for several factors, including CPU throughput and availability, complexity of the required modeling problem, and presentation of computation results. With the advent of faster computer hardware and the advances in networking and software techniques, other program architectures for accelerator modeling have recently been employed. Jefferson Laboratory has implemented a client/server solution for accelerator beam transport modeling utilizing a query-based I/O. The goal of this code is to provide modeling information for control system applications and to serve as a computation engine for general modeling tasks, such as machine studies. This paper performs a comparison between the batch execution and server/client architectures, focusing on design and implementation issues, performance, and general utility towards accelerator modeling demands.

---

*work supported by US DOE contract# DE-AC05-84ER40150

© 1997 American Institute of Physics

II. TRADITIONAL ARCHITECTURES

In the last thirty years of simulation and modeling work, a few primary implementation architectures were employed for many codes. The simplest architecture form, viewed from implementation and execution standpoints, are the codes in which the configuration and lattice information are integrated and compiled internally with the algorithm. This implementation form is easy to execute, since the program contains all information required to perform the calculations. However, this approach is highly application-specific, with any changes in configuration requiring direct editing of the source code.

A natural extension to the above is the segregation of the configuration from the actual algorithm. This method is generally implemented using data files which contain configuration, lattice, and operation commands, and is often referred to as decks. Most of the simulations and models which are in use today employ this architecture, and there have been standards developed within the modeling community which describe the format of the data decks, for example the MAD and ZCEDEX formats[1]. The separation of the configuration/operation parameters from the actual executed code allows for the development of generic modeling and simulation applications which can be used at many accelerator sites.

Results of calculations are usually produced in a tabulated output format, often maintained as disk files. This form is convenient in that results for a particular input configuration can be maintained for future review and/or analysis. However, if it is required that computed results be available for other programs, such as machine control processes, then interface codes which operate on computed model/simulation files must be developed.

III. CLIENT/SERVER ARCHITECTURE

In the client/server approach, the application is distributed between at least two processes, the server and the client. The server is the unified source of information or algorithmic calculations, and the client is the process or processes which perform requests to the server in order to obtain the information. In the case of modeling or simulation, the server contains the calculation engine and configuration/lattice information. The client, which is part of an application process, performs queries towards a server, requesting information or calculation results. Both the server and the client require a communication functionality and protocol to be established.

Client/server architectures naturally lead to extensions which enhance performance and functionality. At the simplest level, a single server can service a single client. Adding the capability of multi-user access to the server allows for the simultaneous servicing of multiple clients from one server. Multiple servers can also be implemented, each handling its own set of clients. The use of local area and wide area networks allows distribution of clients and servers on different hardware platforms. Callback mechanisms can be implemented on the server which can provide automatic update mechanisms to attached clients.

A server for modeling and simulation purposes is basically a batch model architecture with the addition of a high-level event/callback processing manager, communications interface for client transaction, and command message decoding and encoding. Multi-user servers must keep track of requests from multiple clients and prevent inadvertent interactions between clients, as well as preserving the state and integrity of the model or simulation. It is sometimes advantageous to structure the model or simulation code such that static information, like lattice construction and static calculations, be generated at server start-up. Hence, client requests to the server require the minimum of computational processing, thus improving server response.

IV. EXAMPLE CLIENT/SERVER MODEL

An illustration of a client/server architecture is the ARTEMIS (Accelerator Real TimE Modeling Information Server) modeling environment, currently in use at Jefferson Lab for accelerator controls and analysis. ARTEMIS provides various model information (twiss parameters, transfer matrices, etc.) and supporting computations (e.g. quad strength calculation for matching) for all model-driven facilities, including automated beam steering, beam position and energy feedback, and beam diagnostic and optimization procedures [Fig. 1]. Implementation of ARTEMIS as a server allowed for the centralization of the model calculations, which provided a uniform and consistent data source for these and other applications, while eliminating the need for redundant calculations by different application software.

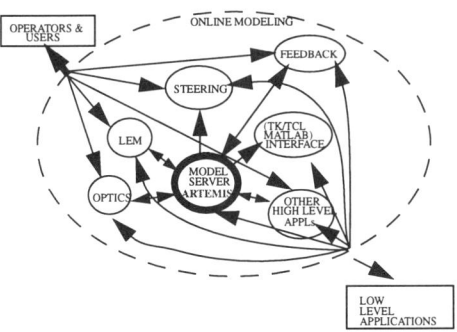

Figure 1: Modeling Environment Overview

ARTEMIS is based on standard second-order transfer matrix calculations, and provides the following functionality:

- Generation of first and second-order transfer functions.
- Provisions for inclusion of higher order models as needs dictate.
- Correct treatment of the acceleration process in the linac cavities, including effects of adiabatic damping and cavity focusing effects.
- Providing lattice and input settings to clients, which has the functionality of an on-line database. This includes hardware information (e.g. list of quadrupoles, BPMs, etc. in a given region, with the ability of providing wildcard input) and element operational settings.
- Model update mechanisms which include user-initiated, periodic, and input event triggered.
- Model updates triggered by an external event.
- User initiated model calculations (on-demand modeling).

ARTEMIS provides the capability for one to instantiate four distinct machine model servers:

- Golden model: includes a set of machine parameters and settings which have been verified and deemed reasonable by an authorized expert.
- Design model: this is a machine model based on the paper design and initial setpoints of the accelerator.
- Current real-time model: this is a machine model representing the current accelerator setpoints; the update mechanism may be event-driven or user initiated.
- Simulation model: used to determine the outcome of *what-if* scenarios as applied to the accelerator lattice.

The implementation of the server section of ARTEMIS is illustrated in figure 2. It consists of several subassemblies: a lattice database, which performs the input-output lattice management for the server, a client communications section, the model engine controller process (GENX), and CAUListener which provides the interface to machine settings. The static lattice information is maintained in a common shared-memory region, which includes all of the element setpoints and pre-computed local transfer matrices. Individual requests from clients invoke the server to apply the appropriate algorithm against the information contained in the common classes.

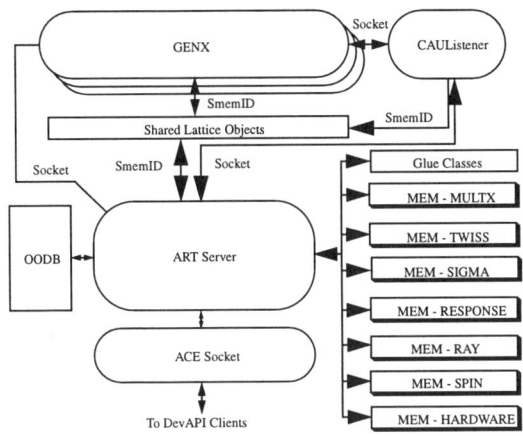

Figure 2: ARTEMIS Server Architecture

The communications and server/event management for ARTEMIS is provided by cdev (control device or common device), which is a C++ library toolkit initially developed at Jefferson Lab. Cdev provides a standard application programming interface (API) to one or more underlying packages, typically control system interfaces. It consists of two layers: the uppermost layer is used directly by applications, and provides an abstraction of the underlying package, and the second layer (service layer) provides the interface to one or more underlying packages, and is implemented as optional loadable libraries.

A server engine and client interface has been integrated into the cdev interface. Based on the ACE C++ wrapper classes for communication of cdev data packets(employing TCP/IP), the interface greatly simplifies the construction of servers and clients. The cdev server interface includes input/output messaging queues, built-in event and monitor mechanisms, and multi-user processing. The client interface contains the server connect/disconnect functions and callback mechanisms which are available to application programs.

The format of a cdev transaction consists of a device name, message, and attribute. The device name represents any logical or virtual device, for instance a name of a magnet or monitor. The message string indicates which operation is desired, and the attribute allows for specification of indentifiers. Cdev also allows the message string to specify application operational parameters, such as which model server to use. The cdev string "model "get element" "device=Quadrupole"" is interpreted as a command to the default model server to retrieve the logical element names for every quadrupole in the server's lattice. Another example is: "IPM1L01 "get betax" "model=DESIGN"", which returns the machine horizontal beta function for the element names IPM1L01, retrieved from the model server DESIGN.

V. EVALUATION

It is evident that implementing a client/server model or simulation requires more code development than the traditional batch architecture. New factors, such as communications and model coherency, must be addressed for the client/server. Additionally, the server must always be available to the end user, thus requiring robust code. Corrupted message commands, run-away calculations, arithmetic errors, etc. must be identified and trapped by the server. In our experience, this has been one of the important factors to deal with, and identifying such problems can be difficult at times. The server also has to operate orthogonally between clients, so the design phase of the server must address this. The ARTEMIS server handles this by performing the computation/access for every message received.

The advantages of the client/server model are apparent on the client end. Once the communications and messaging interface is established, the client application can immediately take advantage of the server. Therefore, the client application is not burdened with the implementation details of the model or simulation. This fact allows for rapid application development.

The client/server architecture also allows existing toolkit applications to use server results. An client application interface to TCL/TK and MATLAB has been developed for use with ARTEMIS, and is extensively used for control and analysis. Applications which require modeling information can be realistically implemented and under test in a matter of hours.

An interface which allows server access using the World-Wide Web has also been implemented. Cdev provides a gateway process which is easily interfaced to CGI applications, which allow for the creation of HTML documents from user-initiated requests. The web-based applications allows the user to retrieve parameters such as transfer matrices and Twiss parameters from any of the executing model servers. The pages are easy to use, and have proven beneficial in many situations.

VI. REFERENCES

[1] G. Morpurgo et al, "*Super-ZCEDEX User's Guide*", CERN Internal Document, 1986.

Accelerator Physics Computing in a Control System Environment

J. A. Holt, A. Braun, L. Michelotti, M. Martens

Fermi National Accelerator Laboratory[†]
P. O. Box 500, Batavia, IL 60510, USA

October 3, 1996

Abstract

Connecting accelerator physics models to a control system to interactively simulate an operating accelerator presents software architectural and algorithm development challenges to produce results in a timely and effective manner. Using the C++ class libraries BEAMLINE and MXYZPTLK (differential algebra) on-line models of the Tevatron and several of the beam transfer lines are under development and are being embedded into the Fermilab control system. These models interact with the real machines and the control system in two modes. The first mode is by running as a pseudo hardware frontend where all control system application's input and output are redirected to the model. In the second mode, the model can read all parameters of the real machine but the user interacts with the model via a separate application which enables the user to do special calculations not neccesarily associated with current machine operation.

Introduction

Every laboratory in the design of new accelerators has expended a great deal of effort in the development of new acclerator physics software tools. These tools are rarely used in the comissioning of the new accelerator or to model the presently existing accelerators. Most online models presently in use are wrapped versions of batch-oriented accelerator codes. A comprehensive modeling environment in C++ which is connected to the control system is under

---

[†] Operated by the Universities Research Association,Inc, under contract with the U.S. Department of Energy

development at Fermilab. This paper briefly describes the class libraries used in the model system as well as the model system achitecture.

Class Libraries

There are seven layers of class libraries used. MXYZTPLK [1] which is an automatic differentiation library, BEAMLINE which models the accelerator, Machine library which encapsulates and manipulates a whole accelerator, server classes which are the model, a TCLGUI library which contains GUI classes using Tcl/Tk, a TCP/IP socket class library for communication, and a database library which encapsulates the Sybase client library routines.

The fundamental MXYZPTLK class is called a Jet which is essentially a truncated power series object containing the reference point to be expanded about. These Jet objects can calculate derivatives to any order up to machine precision. Things like beam transport and the Taylor expansion of control parameters in terms of other variables have found to be very useful.

The BEAMLINE class library is used to model the accelerator elements in the tunnel. Every accelerator element inherits from one base class. One can also have beamlines of beamlines. There are Particle and JetParticle classes which propagate through each element for tracking or for the creation of maps. Very important for an interactive application are methods for editing beamlines and beamline elements (including missalignments and splitting).

The Machine classes are used to manipulate a beamline in a global way. The beamline pointer inside the class is made public for ease of access of the beamline methods. Examples of global manipulation of a beamline, would be for example, turning off and on of separators in the Tevatron machine class or the building and maipulation of elements wired together in circuits. High level calculations such as Luminosity would logically be performed here as well.

Each accelerator model has one server class. This class initializes the machine class, handles all of the TCP/IP communication via the socket class library, parses the input from the socket and calls the proper Machine, beamline and database class methods. After a request is processed, The server I/O blocks on the input socket, waiting for the next request.

Tcl/Tk is our preferred choice for all graphics because of its ease of development and use. Inside the control room however, the standard control system graphics is used when possible. A Tcl/C++ interface class is used to allow C++ to access Tcl variables and allow the Tcl interpreter to call C++ methods. Three types of graphical beam monitors have been developed. The first type is a single particle phase space monitor, which displays as a point the particle trajectory in the monitor. The user can also click inside the plot to change the particle trajectory. The second type of monitor is a particle bunch monitor where the trajectory of each particle in the bunch is plotted as a pixel on a graph. There is no user input on this monitor. The last type of monitor is a histogram which models a multi-wire monitor by plotting the x and y of each particle in separate histograms. This monitor is very useful in one pass beamlines such as extraction beamlines. A single particle monitor shown in Figure 1.

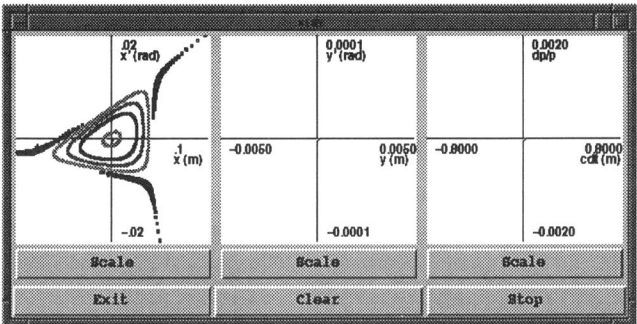

Figure 1: *A single particle monitor showing third order resonant extraction*

A set of database classses has been developed to encapsulate the many details needed to communicate with a Sybase relational database which is used for model data storage and retrieval.

System Architecture

The controls group has implemented the facility to redirect control system (console) applications from the real machine to a software "virtual" machine. This permits the testing and debugger of console applications without wasting valuable beam time. In addition to this benefit new accelerator operators can be trained and be allowed to make mistakes on this "virtual" machine without impacting accelerator operations.

In the Fermilab ACNET control system, a console application communicates to the hardware in the field through a process called the Data Pool Manager (DPM). The DPM translates a console application reading, setting, control or status request into a structure which contains information about the characteristics of the device. This information is then sent to a hardware front end for processing. The data can be intercepted at this point, the location field in the structure is changed and redirected to process called an open access server. This server passes the request to an appropriate client (called an open access client). This client in turn translates the request information into a format that the model understands which is sent via a TCP/IP socket.

The data which comes back from the model is translated back into the format the control system recognizes and is stored in a data pool for reading by the console application. When an open access client is started up, this data pool can be populated with data from the real machine. Figure 2 shows a block diagram of the system. The only way the user knows that the console application is being redirected is that a yellow diagonal line is drawn in the application window.

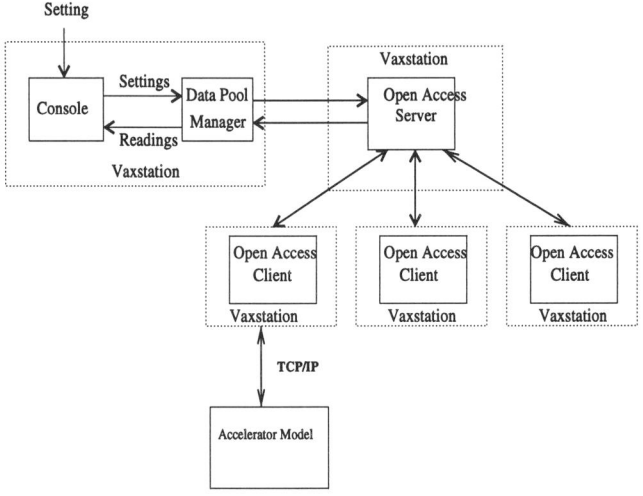

Figure 2: *Redirection of a control system application to a model*

The other style of model in the control system is run as a standalone model client application. The user runs the appropriate model console application which, when it starts up, sends a message via a TCL/IP socket to a predetermined port on a model compute server. There is a master server process which is listening on this port, parses the messsage and starts up the appropriate accelerator model server. This process is assigned a unique port number and a database table number. The port number is sent back to the client application which from then on uses only this port to communicate to the model. Since the client application is in the control system enviroment, real data can be aquired using the standard control system data aquisition routines. Data such as magent currents, measured beam sigmas or BPM data can be sent to the model for calculation or comparison.

When the model finishes a calculation, the data is copied into a relational database instead of being directly sent back to the client application. A set of temporary tables are assiged for each instance of a model and the client application makes a SQL query to these database table instead. This was found to save time and made selective queries of data for presentation to the user much easier. Relational databases are optimized for these types of tasks. The model just does a non-selective bulk copy into the database without concern of which part the user wants to see. For example if the user wants the Twiss parameters just at the BPM's, the model calculates the parameters everywhere, copies them into the database and the client application makes a SQL query for the Twiss parameters just at the BPM's. At the present time, Twiss parameters, closed orbit position, element strength and missalignment, and the R-matrix are stored in database tables. Data for the design lattices are stored in separate tables and copied to the temporary tables upon model startup to improve response time for a design lattice request. Figure 3 shows a block diagram between the various processes.

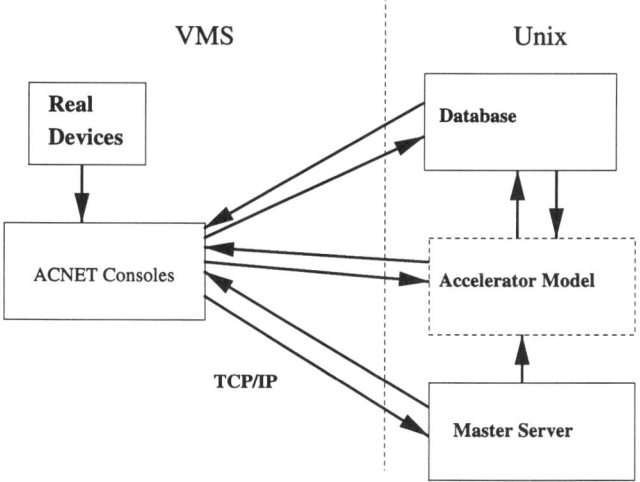

Figure 3: *Online model block diagram*

Present Status

An open access client is being tested for the 8 GeV transfer line from the booster to the Main Ring. The beamline BPM page was redirected successfully with no noticeable time delay in data acquisition. Except for the diagonal line drawn through the console application, one can't tell that the application has been redirected. For a large ring however, time response from the model will be a problem. A ramping accelerator in particular will be difficult. Development has started in this area.

The standalone model has been used extensively in the last Tevatron collider run and compared to real accelerator data. For example, the difference between model and machine helical orbits was 300μm rms. The calculated luminosity was within 15% of the measured luminosity. Confirmation of the effect of a rolled low-beta quadrupole on luminosity was duplicated with the model.

The standalone model is currently being used to simulate the Tevatron fixed target run. The graphics monitors in particular have shown that the model is close to the measured extracted beam widths and beam shapes in the switchyard beamlines.

References

[1] L. Michelotti, *MXYZPTLK and Beamline: C++ Objects for Beam Physics*. In *Advanced Beam Dynamics Workshop on Effects of Errors in Accelerators, their Diagnosis and Correction*. (held in Corpus Christi, Texas, October 3-8, 1991) Published by American Institute of Physics, as Conference Proceedings No. 255. 1992.

Abductive Model Refinement for Accelerator Control

Carl Stern, William Klein, George Luger, Mike Kroupa

Vista Control Systems, Inc.,
134 B Eastgate Drive, Los Alamos, New Mexico 87544
email stern@cs.unm.edu

Abstract. Many aspects of accelerator control require a complex procedure that includes planning, control, and re-evaluation of the process model. As control actions are performed new information is obtained from the system which allows the model to be adjusted. In many cases, observed errors in the model suggest certain control actions for gathering new information used for further refining the model. The process of comparing predicted with observed behavior to produce testable hypotheses for adjusting the predictive model is called *abductive model refinement*. This paper describes our ideas for applying abductive model refinement to beamline tuning tasks, including minimum steering through a set of quadrupole lenses and developing a waist at a specified location in a beamline.

TRADITIONAL AND KNOWLEDGE BASED CONTROL

The development of a theory and methodology for model refinement comprises one element of a larger research project: the development of an hierarchically structured architecture for distributed adaptive control. Our approach, described elsewhere (1), integrates knowledge based methods, including the explicit representation of control knowledge and control models, to support an adaptive capability. The view of control upon which this architecture is based is significantly different from the traditional view.

Traditional control theory views process and controller as strongly separated components of a control system. The process takes inputs from the environment and the controller and produces outputs which are then operated on by the controller to produce new inputs. Missing from this model of control is a view of the process and controller together as comprising a mutually dependent system with potentially time-variant behavior. Even in adaptive systems, where the controller adjusts its response function to minimize system error, the control system lacks the ability to recognize that a particular control method can no longer function successfully and to modify its internal representations and control algorithms accordingly. This is an acceptable model for control in simple stable systems where predictable low order functions can be minimized by traditional control methods (e.g., PID, FAM, etc.). *Figure 1a* illustrates the traditional control paradigm.

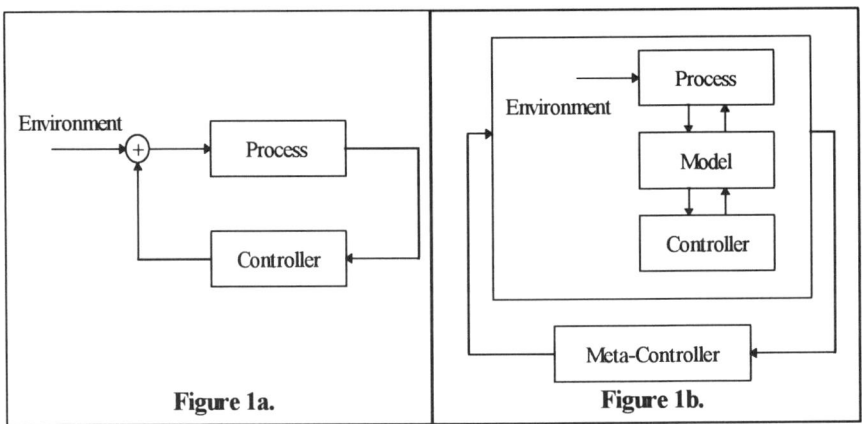

Figure 1a. **Figure 1b.**

The whole system view of control includes structures for modifying internal control processes based on information about both the process and the controller. An important aspect of this paradigm is the use meta-level control in conjunction with an explicit process model. The model is built to reflect both the state of the physical system and the relationships between the physical process, the control hardware, and the control software. The model provides a shared representational framework between process and controller and facilitates intelligent decision making and reasoning. *Figure 1b* illustrates the whole system view of control.

In the whole system view, it is also useful to employ confidence metrics encoding the reliability or effectiveness of control methods relative to observable states of the system. Knowledge based control systems can then select control algorithms by matching knowledge about the system's current state against a set of available control algorithms indexed by their estimated effectiveness under the current set of conditions.

From our viewpoint, model refinement is not a persistent process continually tracking the system's current state. Rather, it is triggered by the inability of the system to satisfy a goal using the current model. This may occur because of the failure of attempted control actions or due to lack of confidence in the effectiveness of potential actions. In many cases, it may be sufficient to update model data by making new system measurements. In other cases, however, failure to achieve a goal is directly correlated to a mismatch between the model and the process. Abductive model refinement is then used to diagnose and correct the deviation.

COGNITIVE FOUNDATIONS

Human experts in many domains exhibit the ability to effectively improve their model of a given environment through exploratory action. This involves the use of an initial rough model as a basis for planned testing and interaction followed by a process of evaluation. The attempt to explain discrepancies between expectations and observations generates a new understanding of the environment that results in a "refined model." In many cases, this process must be repeated a number of times before a satisfactory model of the environment is found. This cycle of exploration and model adaptation is called *abductive model refinement* because it depends on abductive reasoning, i.e. reasoning that attempts to *explain* the source of the differential between expectations and experience.

In our study of expert human performance in the area of particle accelerator control we have encountered precisely this pattern of problem solving. For example, we have found that accelerator physicists commissioning a new or reconfigured beam line will try to understand the behavior of the system by conducting a carefully designed sequence of experiments. Typically this involves an attempt to produce some standard set of beam conditions at specific locations. They employ an initial model of the accelerator beam line in conjunction with a software modeling code such as TRANSPORT to compute some configuration of magnet field strengths that is expected to produce the desired beam condition. This prediction then serves as the basis of an experiment to verify the expected beam condition based on the computed magnet settings. If the experimental finding fails to fall within an acceptable range of accuracy, this is in fact a very useful result. The physicist then analyzes the results in order to generate hypotheses regarding the source of the discrepancy between prediction and observation. This in general leads to a cycle of experimentation and explanatory hypotheses that eventually results in an improved beam line model.

REPRESENTATIONAL AND ALGORITHMIC ISSUES

We are currently developing a set of representational schemes and algorithms necessary to support a computational implementation of abductive model refinement in the context of accelerator control. Active study and analysis of the problem solving of expert accelerator physicists [1] has played a key role in shaping our current approach. We are also collaborating with these same physicists in the development of mathematical methods for diagnostic estimation. What follows is a brief account of our current insights and achievements.

Implementing a Model Refinement Algorithm

Abductive model refinement includes the following sequence of interrelated steps (although not necessarily in the given order):

1. Recognition of a discrepancy between model-based prediction and observation;
2. Generation of hypotheses for explaining this discrepancy;

---

[1] Most notably Andrew Jason at LANL and Xijie Wang at BNL.

3. Planning and execution of sequences of actions to test explanatory hypotheses;
4. Modification of a model based on a verified hypothesis.

Each of these steps presupposes a complex set of capabilities. Developing a computational implementation of the model refinement process has required a detailed understanding of patterns of reasoning involved in the second, third, and fourth steps in particular. This has entailed a number of related research goals:

- the application or modification of artificial intelligence planning methods to model the design of useful experiments;
- the application of abductive reasoning to generate plausible hypotheses that explain observed discrepancies;
- a form of model-based reasoning in which conjectured model revisions are tested for consistency with known data and causal relationships.

The high degree of complexity of the model refinement task necessitates the use of certain simplifying assumptions. We follow the traditional model based reasoning approach in assuming that certain dimensions of the causal connection structure of the physical system are fixed and known (2). These define a fixed causal framework for reasoning without which too many diagnostic possibilities and combinations would have to be considered.

Following the traditional representation used in model based reasoning, we consider a model of a physical system to be a set of elements together with a causal connection structure between elements. Each element has a set of inputs and outputs where each output is defined as a transfer function over the inputs together with the element's internal state. The elements together with their causal connections constitute a network as seen in *Figure 2*.

An issue of great concern for model refinement in complex systems such as particle accelerators arises from the dependencies between beliefs. An inference regarding some property of an element, such as a remnant field or misalignment in a magnet, often depends upon a good deal of assumed knowledge about other elements. If these assumptions are incorrect or of limited accuracy, the inference itself is tainted. Since incremental model refinement involves the construction of chains of inferences, it is necessary to keep track of the evidence for inferred beliefs as well as the degree of certainty and accuracy that such evidence warrants. For this reason we cache inferences, recording the dependencies between assumptions and conclusions. We use a network representation similar to that used by a Justification Based Truth Maintenance Systems (JTMS) architecture (3) to organize cached inferences. This allows us to immediately determine the set of beliefs to which a revised belief is linked and reevaluate the status of those dependent beliefs accordingly.

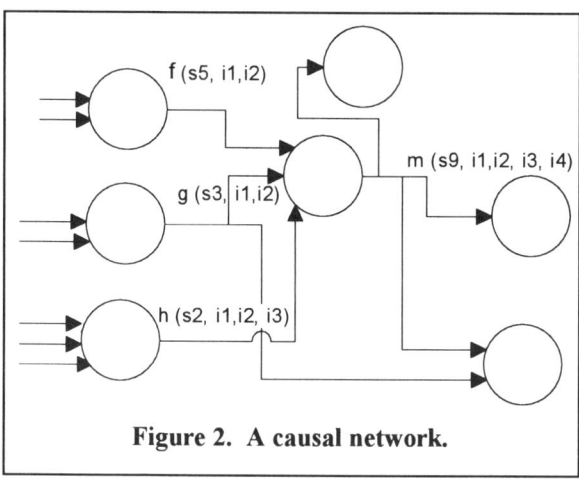

Figure 2. A causal network.

Two key issues in model revision present significant challenges. The first, the problem of cyclic dependencies in beliefs, poses a problem from the algorithmic point of view. Algorithms that update the numerical probabilities or confidence levels associated with beliefs in a belief network behave erratically in the presence of cycles, e.g., when belief A supports belief B, B supports C, and C supports A. When A supports C and C also indirectly supports A this can lead to an incorrect calculation of the evidential weights supporting both A and C. Unfortunately, belief cycles tend to permeate reasoning about models.

A second more fundamental issue is related to the problem of multiple source errors. A recognized discrepancy between prediction and observation can derive from a combination of errors in a model, or even worse, a sequence of causal interactions that magnifies minor inaccuracies into significant errors. Finding explanations for multiple source errors is not only difficult for humans; it also poses a serious computational challenge for most abduction algorithms.

We do not believe that there is a general solution to the second problem, but rather heuristics that are effective in finding explanations in many cases. Our current approach is to focus on knowledge engineering with expert accelerator physicists in order to discover such heuristics.

One heuristic that we are currently exploring is to refine the beamline model by identifying "islands" in the beamline that can be measured and studied in relative isolation, i.e., in a way that is minimally dependent on assumptions about the rest of the beamline. We have a identified a few experimental methods that support such independent calibration techniques. Once islands of high accuracy and confidence are constructed, the generally strategy is to cautiously extend them, verifying them by generating predictions through the use of the usual beam propagation and fitting algorithms and then testing those predictions against actual beam measurements.

Consider a simple example of a method that implements the strategy described above, i.e., to begin from and gradually extend islands in the model that have been calibrated with a high degree of accuracy. In this example we begin by measuring the beam itself using a method that assumes nothing about the rest of the model. We start with a location in the beamline that contains two screens with a drift space of length L between them. This is illustrated in the right half of *Figure 3*. Working in one dimension, we then produce a minimum spot size at the second screen, measuring both Xs and Xs' under this condition.

Using the matrix for a drift space and setting $\delta\sigma 11'/\delta\sigma 12 = 0$ corresponding to a minimum spot size at the second screen, we can determine a number of useful parameters: $\sigma 11$ at the first screen as a linear function of $\sigma 12$, beam emittance (ϵ), as well as the position of the beam waist. Varying the current to quads α and β (left side of *Figure 3*) and applying curve fitting to measured beam sizes on the first screen allows us to calculate actual field strengths as a function of current as well as remnant fields in α and β.

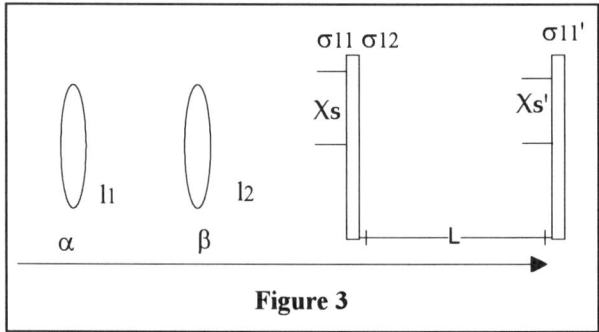

Figure 3

CONCLUSION

We are currently planning to test our ideas in abductive model refinement through a series of tests at Brookhaven National Laboratory's Accelerator Test Facility. There is reason to believe that the most difficult of these tests, production of a specified waist condition in an undulator cavity of a Free Electron Laser, can only be accomplished by refining the current beamline model. Future work involves the extension and generalization of our current system into a general architecture for accelerator control.

REFERENCES

1. Klein, W.B., Westervelt, R. T. , & Luger, G.F., *Journal of Intelligent & Fuzzy Systems*, in press, (1996).
2. Davis, R., *Artificial Intelligence*, 24, 99-94 (1983).
3. Doyle, J., *Artificial Intelligence*, 12, 231-272 (1979).

A Simple Tool for Beamline Commissioning and Transport Optimization

Luciano Catani

Istituto Nazionale di Fisica Nucleare - Sezione di ROMA II
Via della Ricerca Scientifica, 1 ROMA I-00133 ITALY

Abstract. The paper presents experimental results, simulations and some considerations about the development of a software tool for on-line beam transport optimization at LISA 25 MeV linear superconducting accelerator of INFN-LNF. The tool uses an optimization algorithm borrowed from Evolutionary Strategies techniques that allows the definition of an optimization procedure for which the knowledge of beamline's properties is unnecessary. Some of its limitations are also discussed. Preliminary experiments with such a system demonstrated quite some advantages such us the ease of development and implementation, the independence from beamline errors and elements drifts and a high degree of flexibility that makes this tool useful for the beamline set up at commissioning and for later optimization.

INTRODUCTION

One of the first operational problem to face when commissioning a linear accelerator is the transport of the first beam along the beamline. This task is usually performed by operators that, looking at signals from diagnostic devices, set magnetic elements to transport the beam along the beamline sections step-by-step. During commissioning the procedure is usually repeated every time the machine is modified and the transport conditions as well as the optics setup are changed. Later optimizations must be repeated whenever the beam experiences different regimes of transport (e.g. different current, bunch length, rep. rate etc.) and to compensate drifts of control devices and parasitic effect due to interaction of the beam with environment.

Automatic procedures to optimize the transport usually rely on the knowledge of the beamline characteristics, but those must be updated if the machine is modified or its working regime is changed, and do not usually allow for drifts and parasitic effects. A different approach to commissioning and optimization could be the use of optimization techniques known as Evolutionary Strategies [ES] or the more famous Genetic Algorithms to which ES can be somehow related. In our task, Evolutionary Strategies techniques can be used to define an optimization procedure that works on a set of optics parameter (such us correcting coils) in order, for instance, to minimize displacement from main orbit. The implementation of this procedure is independent from both the knowledge of relationships between the parameters themselves and the transport condition chosen to as quality factor to be optimized.

This could be the great advantage of this approach: realize an optimization procedure that works for a particular kind of system, such us an accelerator beamline, and can be applied to different configurations, operation modes or even accelerators, without the need to reconfigure. One could even work on a different set of optics elements (e.g. quadrupoles instead of correctors) to optimize a different parameter (beam current on a current monitor or beam spot position on a viewscreen instead of beam offset) it being possible to perform the optimization of non homogenous transport parameters at the same time.

The work presented here shows the preliminary experimental application of a simple ES algorithm [1] to a set of correcting coils of the 1 MeV 180° arc of the LISA injector [3]. The results evidenced some possible improvements of the method that have been then studied by simulation.

The latter were used to analyze the features of the algorithm with the aim to find the best tuning of its set-up for this particular application.

APPLYING EVOLUTION STRATEGY TO BEAM TRANSPORT OPTIMIZATION

The basic idea of the algorithm used for the transport optimization comes from the so called Evolution Strategy [2] that is an attempt to translate into a mathematical form the nature's optimization method. The interest of the method arises from the observation that the evolutionary strategy itself could be considered as the optimized optimization method.

Following the ES, the optimization of a set of N variables x_i (point P in a N-dimensions space) starts from an initial population of M possible configurations P_j, j=1,..M. These will be the "parents" of a new generation of D new points P'_k, k=1,..D produced applying the ES operators to them. Using the Genetic Algorithm terminology those operators will be defined selection, reproduction, crossover, mutation and others. In general they select the parental points, mix their values x_i, add a random mutation and evaluate for every P'_k its fitness or quality factor, the latter being the only information given to the algorithm to perform optimization.

Between the D new points, the M best will be chosen as the new parents for the next generation. The key of the ES algorithm is that the generation of the descendants is performed considering not only the values of the parents variables but also the mutation amplitudes that generated them. In this way one selects both the best values and the most effective mutation amplitude toward the optimized value.

Genetic Algorithms have a slightly different but more rigorous definition. The variables are coded in a string of elements of a finite alphabet to whom the stochastic GA operators are applied.

The descent to the deepest point of a surface in a N-dimensional space is the pictorial view of the process. If the surface is smooth and represented by a monotone function of all x_i (as in the minimization of N independent real numbers) then the procedure will easily find the optimized parameter's configuration.

However, in the optimization of a set of beam correctors minimizing the RMS displacements Δ of some downstream Beam Position Monitors [BPM], the quality factor defined is obviously not a monotone function of the correctors values x_i. If we choose for example two correctors and plot Δ as function of their settings we identify a curve that defines the relation between the correctors for the minimum Δ value. If another preceding corrector of the beamline is varied, this minimum-Δ curve will drift changing the actual quality factor of the first two correctors. Even if the change of the third corrector was toward its optimum value the resulting Δ could be worse if the first two now have their optimum values lying on a different curve. Clearly the larger is the number of the correctors, the harder will be the optimization.

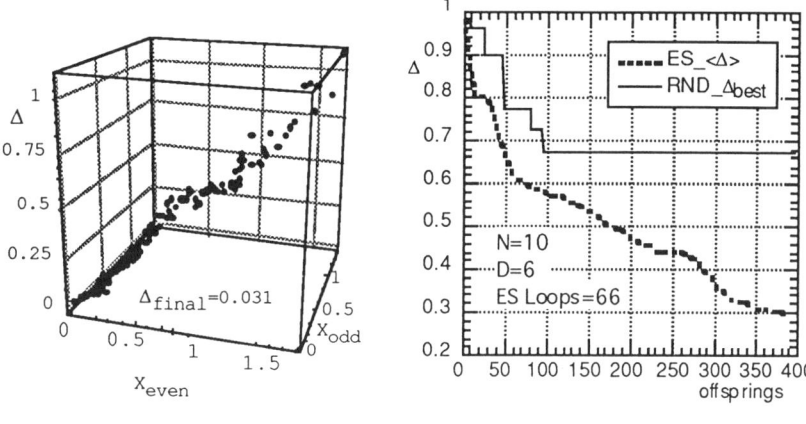

Figure 1a   Figure 1b

In spite of that, the procedure has proven to be able to find the way to achieve a better correctors configuration. Fig.1a shows the 3D graph of the optimization path (the points give the average of parents values for all generation loops) obtained from the simulation for a system of 8 correctors trough 200 generation loop. In that graph $X_{even} = \sqrt{\sum_i \tilde{x}_{2i}}, i = 1,.. \frac{N}{2}$, $X_{odd} = \sqrt{\sum_i \tilde{x}_{2i+1}}, i = 1,.. \left[\frac{N}{2}\right] - 1$ where $\tilde{x}_i = x_i - x_{i_f}$ and x_{i_f} is the final value of i-th corrector.

From the graph we can see that the procedure defines a search path to the optimized configuration. The result of the application to the same problem of a simple random search would have been, in the same 3D graph, a cloud of scattered point. The effectiveness of ES respect to random search RND is again evident in Fig.1b. The random search is represented by the best Δ produced after a number of generated offsprings. For the ES curve, the 396 offsprings are produced in 66 loops of 6 descendants. For every loop the parents average is shown.

Parasitic effects like electrostatic charging of the beam tube or power supply drift can be considered as weak, slow varying extra correctors. Their effect will also be compensated by the procedure.

EXPERIMENTAL RESULTS

In the experimental set up N was the number of controlled coils (N=6) and Δ the RMS displacement of the beam to be minimized on the N downstream BPMs. M=4 parents (at the beginning all identical and equal to original corrector's values) and D=12 descendants were used.

The results of one of the tests are presented in Fig.2 showing the Δ vs. loop-number curve and the initial and final values for the chosen BPMs.

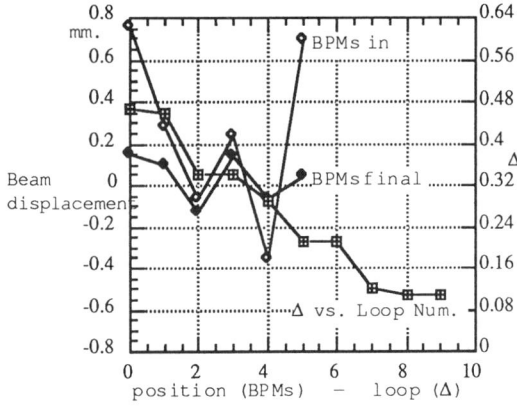

Figure 2

In this particular case the optimization started from a bad transport condition and after a few loops a better corrector's configuration was found. The Δ curve stopped decreasing when the values of the BPM signals were of the same order of the readout noise.

The noise of instrumentation, as well as the modifications of the beamline optic set up while optimizing, have been evidenced as a source of disturbances for the process. Nevertheless they do not prevent the evolution toward an optimized configuration, although they can in some case influence its convergence speed.

For the applications of this method to the on-line optimization it is also required, for practical reasons, that the evolution to the optimized configuration has a reasonable speed. The evaluation of quality factors requires every new configuration to be applied to the beamline and the beam response to be measured; this is an unavoidable limitation of this method.

Approximately 6 minutes were needed to run trough the 9 generations shown in Fig.2 because with the injector pulsed at 1Hz repetition rate, 3 sec. delay was allowed between the setting of new values of correctors and the readout of BPMs.

For a given algorithm the highest evolution speed should follow from the best compromise between the number D of descendants (i.e. the evaluations to be performed) at every generation loop, and the fact that a limited descendants (and parents) population will result in a slower evolution towards the optimized configuration.

SIMULATIONS AND ALGORITHM OPTIMIZATION

The Evolutionary Strategies techniques have two main kinds of "tuners" that allow to better adjust the algorithm for the particular application. The first are the operators that we apply at every loop to a population (of corrector's configurations) to generate and select new offsprings. The second is the definition of the quality factor that best represents the system and gives proper feedback to the optimization algorithm.

For instance, in the original version of the procedure every new offspring was generated selecting all of its x_i from one of the actual parents at random; then a random mutation was applied. We could argue that this works well when the x_i are independent but when the value of a corrector strongly depends on the preceding elements this mix of variables wouldn't be appropriate and one should use something like a clonation instead. The selection of the parents to be duplicated could be guided by their quality factor and the latter could even be used to control the mutation amplitude, in order to have a fine search when the procedure is approaching the optimum. Defining the quality factor it could be advantageous to use instead of a pure RMS a weighted root-mean-square displacement at the BPMs. Appropriate weight functions will assign to the procedure a priority in the optimization, starting from the elements at the beginning of the beamline as an operator would do.

To analyze these modifications four types of optimization algorithms have been defined and tested. Simulations have been performed applying those to a test beamline and the results of many runs have been statistically analyzed.

For all runs the number of loops n_L allowed depended on the number of offsprings to be produced and tested at every generation loop in such a way to have $D \cdot n_L = h$ (in those runs h=240 was chosen). Doing so the evaluation of the different configurations has been performed assuming that they would have an equal execution time in the on-line application.

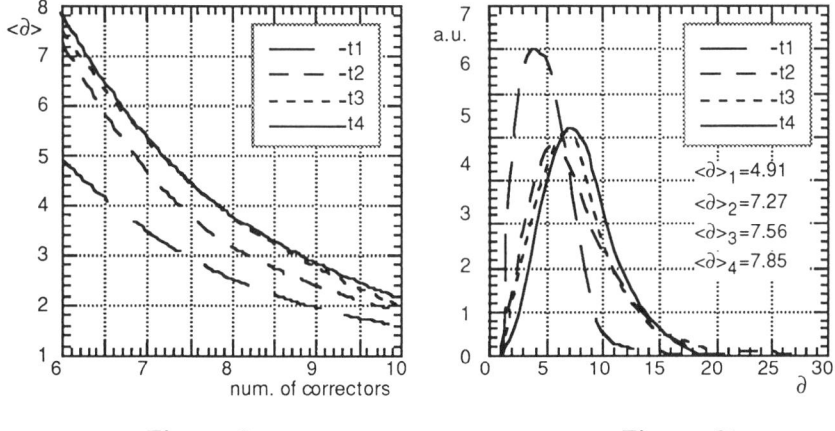

Figure 3a   Figure 3b

The graph in Fig.3a shows $\partial = \langle \Delta_{initial}/\Delta_{final} \rangle$ (i.e. the average over all runs of the ratio between the initial and final RMS beam displacement) as function of the number of correctors for the different types of algorithm. Poorer performance of type t1 are due to the random selection of corrector values from different parental configuration as in the original ES algorithm. In Fig.3b distribution curves of ∂ values are shown for all examined types of optimizing algorithm.

Fig.4 shows distribution curves of ∂ values for type t4 and six correctors but different weighting functions used in the calculation of the quality factor. The weight function used is $\exp(-i/N \cdot \tau)$ where N and i represent respectively the number of correctors and their order as seen from the beam.

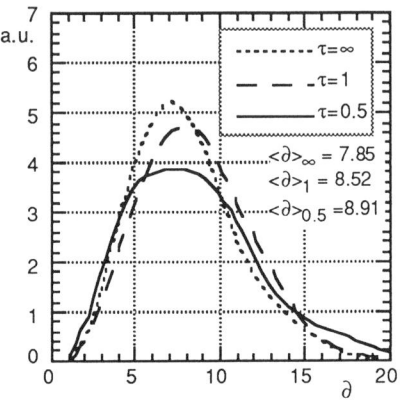

Figure 4

The parameter τ is the variable of the weight function. The lower is the value of τ, the higher will be the weight of first correctors in the quality factor Δ. The graph shows that an appropriate value for τ gives to the algorithm better on-line performances; in other words it finds more quickly better correctors configurations.

A weak dependence on the number of descendants D has been evidenced especially for low number of correctors showing in those cases that, for a given D·Loops=cost, it is better to have larger number of generations.

Random fluctuation of readout signals (noise) proved to be less dangerous than expected. RMS value of noise mainly determines the lower Δ value achievable.

Improvements to the performance of the basic ES algorithm have been obtained; however almost all configurations gave acceptable result showing the particular robustness of the method.

IMPLEMENTATION OF THE TRANSPORT OPTIMIZATION TOOL

The LISA multiprocessor Control System [4] is based on a three level hierarchical structure with the intermediate level being the Supervisor and concentrator for the communication between the consoles (first level) and the I/O to the accelerator elements (third level). The consoles and the VME processors on second and third levels are linked to each other through a memory-mapping (VME vertical bus) from upper to lower levels that allows fast communication and data exchange between them. The consoles are Macintosh computers running LabVIEW as environment for the graphic interface tools.

Because the Transport Optimization panel must be able to work on optic elements controlled by different CPUs of the Control System, the required position for this procedure in the hierarchical structure is at the top, the console level, where all the elements are accessible in the same way.

The program created for the experimental test allows the optimization of either vertical or horizontal correctors, but not simultaneously, and use BPMs readout to evaluate the quality factor for the optimization.

As the elements to be optimized are selected, their actual values are read from control system and will become the k initial parents. The use of optic elements of different kind has been foreseen. In that case the setting value of different elements had to be normalized or properly coded in order to create a suitable population of numbers for the ES algorithm. Fig. 5 shows the user interface of the transport optimization tool

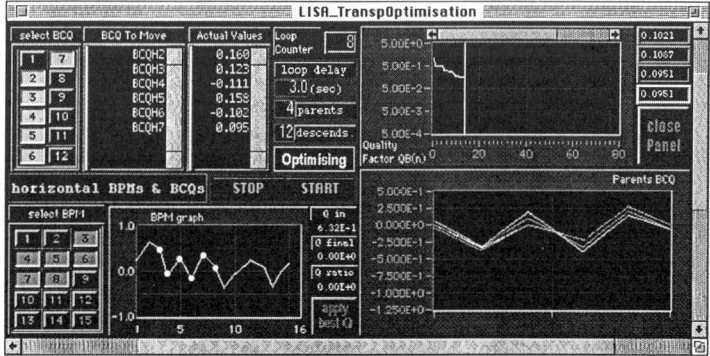

Figure 5

CONCLUSIONS

During the activities with LISA we observed that operations with an experimental accelerator can be made easier by a tool that help operators to find the correct set up of the beamline and later optimizes it.

Even if stochastic technique is not commonly used for on-line optimization, especially for particle accelerator, the results obtained here with Evolution Strategy algorithm could be of some interest for other accelerator physicists. The simplicity of the basic algorithm makes feasible a test of the technique to verify its advantage.

The long execution time of the optimization procedure is the main inconvenient of this solution. A faster control systems together with high repetition rate accelerators will reduce the evaluation time of corrector configurations, and thus the overall execution time.

Obviously because the settings produced during the process are not predictable, a low power operation mode must be available so we can drive the beam to any trajectory and keep safe the machine from any damage.

Further studies of the technique are foreseen to investigate other possible applications and to improve its feature, mainly the speed and the efficiency.

REFERENCES

1. I. Rechenberg, "Evolution Strategy: Nature's way of Optimization", From: "Optimization: Methods and Applications, Possibilities and Limitation" — Lecture Notes in Engineering, Vol.47: Ed. H.W.Bergmann. Springer, Berlin 1989
2. I. Rechenberg, "The Evolution Strategiy. A Mathematical Model of Darwinian Evolution" — Springer Series in Synergetics, 22, Springer, Berlin-Heidelberg
3. M.Castellano et al. "Commissioning and Operating Experience with the LISA Superconducting Linac" — LNF 96/028, Submitted to Nucl. Instr. Meth. in Phys. Res.A
4. L. Catani et al., "Commissioning and Operation of the Control System of the LISA Superconducting Linac" — Nucl. Instr. Meth. in Phys. Res.A 352 (1994) 71-74

Comparison Between Various Beam Steering Algorithms for the CEBAF Lattice[*]

M. Chowdhary, Y-C Chao, S. Witherspoon
Thomas Jefferson National Accelerator Facility, Newport News, VA 23606 USA

Abstract. In this paper we describe a comparative study performed to evaluate various beam steering algorithms for CEBAF lattice. The first approach that was evaluated used a Singular Value Decomposition (SVD) based algorithm to determine the corrector magnet setting for various regions of the CEBAF lattice. The second studied algorithm is known as PROSAC (Projective RMS Orbit Subtraction And Correction). This algorithm was developed at TJNAF to support the commissioning activity. The third set of algorithms tested are known as COCU (CERN Orbit Correction Utility) which is a production steering package used at CERN. A program simulating a variety of errors such as misalignment, BPM offset, etc. was used to generate test inputs for these three sets of algorithms. Conclusions of this study are presented in this paper.

INTRODUCTION

The CEBAF accelerator consists of a 45 MeV injector, two side-by-side superconducting linacs, and 9 recirculation arcs that recirculate the beam through the linacs up to 5 times for 4 GeV total energy. Beams of different energies are separated at the first spreader and are transported through isochronous arcs to the recombiner at entrance of second Linac. At the exit of second Linac, the beams of different energies are separated again to be sent to either Experimental Halls or through the recirculation arcs. An orbit correction system is required at CEBAF to increase the machine aperture and to steer the beam through any portion of the accelerator for a desired beam delivery objective. A variety of beam steering algorithms of varying characteristics and complexity are available. In this study we have compared three such algorithms.

SVD Based Algorithm

Consider there are M beam position monitors and N corrector magnets available to the beam steering algorithm. Changes in corrector strength $\Delta\theta$ (vector of length N) will reduce the closed orbit error Δx (vector of length M). These two vectors are linearly related through a response matrix R_{ij} as indicated by

$$\Delta x_i = \sum_{j=1}^{N} R_{ij} \cdot \Delta\theta_j \qquad (1)$$

---

[*]work supported by US DOE contract# DE-AC05-84ER40150

The response matrix can be written in terms of betatron amplitude, phase advance and the tune of the machine as

$$R_{ij} = \sqrt{\beta_i \beta_j} \sin(\psi_i - \psi_j) \quad (2)$$

where β_i, β_j are the betatron amplitudes at i th BPM and j th corrector. The response matrix can be experimentally determined by changing the strength of j th corrector by unit excitation and measuring the resulting beam motion at all BPMs while rest of correctors are set to 0. Orbit correction using SVD is inverse of this process of experimental determination of response matrix. Singular Value Decomposition of R_{ij} can be written as

$$R = U \cdot W \cdot V^T \quad (3)$$

where U is M x M unitary matrix, W is M x N diagonal matrix that contains all the singular values, and V is an N x N unitary matrix. The inverse of matrix R_{ij} can be obtained from

$$R^{-1} = V \cdot W^{-1} \cdot U^T \quad (4)$$

where W^{-1} can be constructed by inverting the singular values and then taking a transpose of the matrix. If any of the singular values are zero, then this singularity can be removed by setting the inverse of that singular value to be zero rather than a large number in the inverse matrix. Including all non-zero singular values to determine the inverse will result in most accurate correction of the closed orbit error. However, if the R_{ij} matrix is nearly singular then, this correction might require unreasonably large corrector settings on a few correctors. This condition can be avoided by eliminating the smallest singular values from the inverse calculation until the corrector settings requirement enters a more reasonable range. The first step in the orbit correction process is obtaining the difference orbit Δx, which is the difference between the orbit measured by BPMs and the desired reference orbit. Next, the corrector settings are computed using

$$\Delta \theta = R^{-1} \cdot \Delta x \quad (5)$$

If the corrector settings turn out to be over the saturation limit of power supplies, then the R^{-1} is recalculated by eliminating the lowest singular value from inverse calculation. Once satisfactory corrector settings are obtained, they could be applied to reduce the closed orbit error. See reference (1) for an implementation of this algorithm at Advanced Photon Source at Argonne National Lab.

PROSAC algorithm

This algorithm (2) utilizes the projection of j th corrector on the closed orbit as a parameter for selecting the *best* corrector and iteratively determining its settings to progressively reduce the closed orbit error. That corrector magnet, in N dimensional space of all available magnets, which has the largest projection on the vector for closed orbit error is considered *best*. The projection V_j of j th corrector effect on orbit is given by

$$V_j = C_j \cdot \Delta x \qquad (6)$$

where Δx is the closed orbit error and C_j is the unit response on M- BPMs by the j th corrector. Now the setting for this corrector is calculated using

$$\Delta \theta_j = W_j \cdot \Delta x \qquad (7)$$

where W_j is the j th column in the response matrix R_{ij}.

This algorithm has a variety of options available for implementing orbit correction scheme. The first option in this algorithm allows for either using the projection V_j or the normalized projection $V_j / |C_j|$ for selecting the *best* corrector. The second set of options allows for three different starting conditions for iterations. First condition is to start with all correctors set as-is and then incremental corrections are applied to reduce the closed orbit error. Second condition requires that all correctors are set to zero and then incremental corrections are applied to reach the desired reference orbit. Third condition starts with performing a simple least squares fit for the current settings of BPMs and correctors and then either one of the above mentioned conditions in the first option could be applied to reach the reference orbit.

The iterative process of reducing the closed orbit error continues with selecting the best corrector and applying the correction until the closed RMS orbit error is reduced to 20% or a user defined fraction of the initial value.

COCU

COCU (Closed Orbit Correction Utilities) is a comprehensive collection of orbit correction algorithms unified under a standard user interface. It has been a major orbit correction tool used at CERN and several other accelerators. The repertoire of algorithms include MICADO, a minimum corrector number routine, MINIMO, an algorithm looking for absolute best corrector combinations, SIMPLEX, a minimization program capable of inequality corrector constraints, and a number of other algorithms. It also performs harmonic analysis in the case of closed orbit in circular machines and conditioning of the input beamline layout to avoid near-degenerate configurations. A detailed description of COCU can be found in reference (3) and references therein. We have linked the majority of the core COCU program with a graphical user interface to facilitate the data transfer between simulation and orbit correction algorithms.

SIMULATION PROGRAM FOR TEST INPUTS

To compare the performance of the various orbit correction algorithms, simulation program was developed with the CEBAF accelerator as a test bed. Approximately 100 simulation files were generated mimicking all conceivable errors in the machine. These include isolated and distributed optics errors, isolated and distributed misalignment errors in all coordinates, isolated and distributed monitor errors, injection errors in all coordinates, initial corrector kicks and errors, and earth field effects. The use of simulation data helps provide a measure of how each algorithm has performed everywhere against the uncorrected orbit, including areas inaccessible to orbit monitors in the real machine. It also allows creation of special cases where the near-degeneracy of the orbit correction system and corrector magnitude limits are put to test.

RESULTS AND CONCLUSIONS

We have tested over 100 simulation files against the various orbit correction algorithms mentioned above. In most cases all the algorithms produced similar results in terms of the final orbit and overall corrector strengths. We will briefly describe cases where performance of these algorithms differ:

(a). With its inclination to find the minimal set of correctors to control the orbit, in a few cases the MICADO line of algorithms tend to concentrate too much strength into a small number of correctors, as shown in Figure 1.

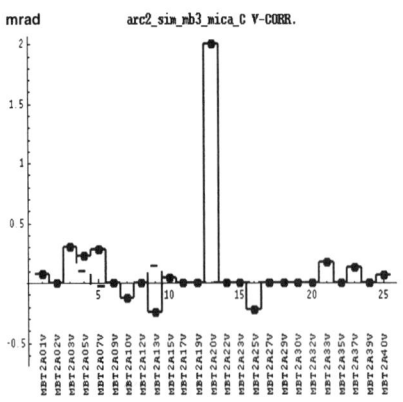

Figure 1. Corrector strengths MICADO

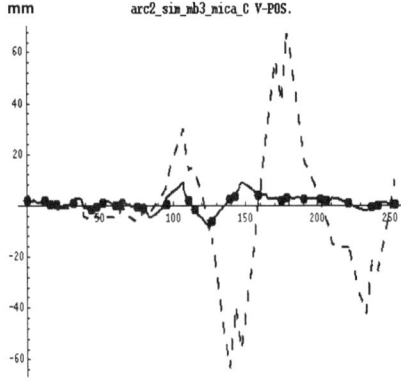

Figure 2. Orbit correction by MICADO

Figure 2 shows the orbit before (dotted line) and after (solid line) correction. Corresponding cases using an SVD based algorithm is shown in Figures 3 and 4, which are similar to results from PROSAC.

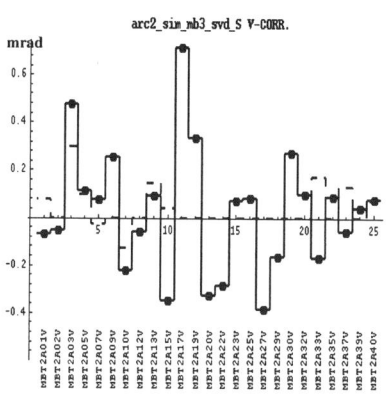

Figure 3. Corrector strengths SVD

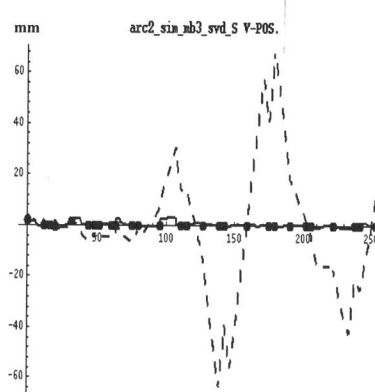

Figure 4. Orbit correction by SVD

(b). All algorithms appear to produce large "fighting" correctors due to near-degeneracy of the monitor-corrector response matrix, although PROSAC appears to be the least vulnerable. Figures 5 and 6 show the correction result of an SVD based algorithm where no singular value has been excluded, meaning virtually no constraint on corrector strengths. It can be seen that some correctors conspire to create large local orbit bumps without detection by BPMs (BPM readings are indicated by solid circles). If one proceeds to eliminate singular values such that corrector limits of 1 mrad is imposed, no correction can be successfully accomplished. The MICADO line of algorithms produced results more close to those from SVD algorithms.

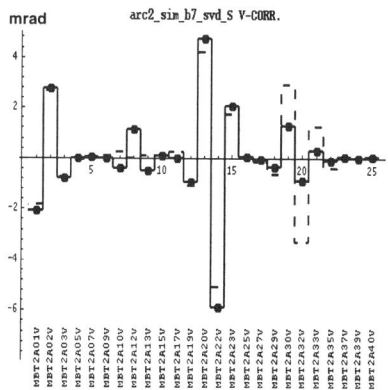

Figure 5. Corrector strengths SVD

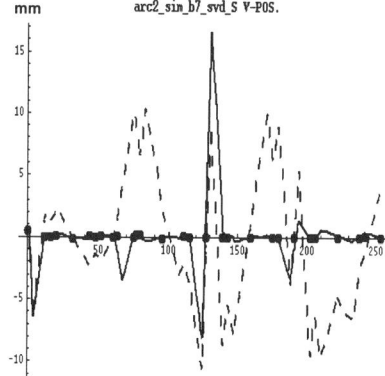

Figure 6. Orbit correction by SVD

In this particular case the algorithm PROSAC succeed in reducing the orbit significantly within the corrector limits, without inducing orbit bumps. We are aware of similar problems with PROSAC, but the occurrence is less frequent.

(c). We also compared the behavior of the various algorithms in a region with insufficient monitors such that the information derived from these monitors is inadequate for orbit reconstruction on the order of 5-10 mm. This can be equivalent to evaluating the error handling ability of these algorithms when some monitors are not working. Our conclusion is that all algorithms tested can easily produce undetectable after-correction orbit errors on the same order as the uncertainty in orbit reconstruction by any method. In such cases algorithmic ingenuity apparently can not compensate for fundamental lack of information. Figure 7 shows such a case by SIMPLEX, a COCU algorithm, where a missing BPM towards the end of the line caused undetectable orbit excursion of about 7 mm. Such problems can be rectified only by more BPMs or drastically changed optics.

In summary, we have tested various orbit correction algorithms against simulated orbits. The relative pros and cons are discussed above which may help accelerator controllers in choosing the optimal method to use. Keeping a wide variety of algorithmic options and careful conditioning of the monitor-corrector system to avoid near-degeneracies appear to be the best policy when it comes to orbit correction.

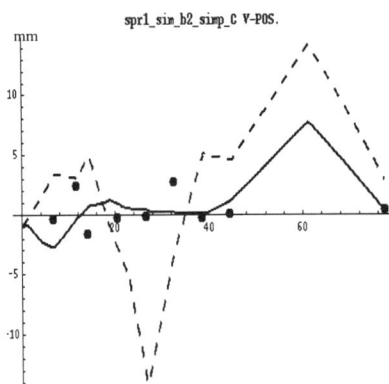

Figure 7. Uncorrectable orbit due to BPM deficiency

REFERENCES

1. Chung, Y., Decker, G., Evans, K. Jr., "Closed Orbit Correction Using Singular Value Decomposition of the response matrix," *Proceedings of 1993 Particle Accelerator Conference*, 1993.
2. Chao, Y-C, "Orbit Correction Scheme," *TJNAF Technical Note TN 96-043*, 1996
3. Herr, W., "Algorithms and procedures used in the orbit correction package COCU," *CERN Report SL/95-07*, 1995.

NEW COMPUTER TECHNIQUES
AND ENVIRONMENTS

The Classic Project

F. Christoph Iselin

CERN, SL Div., AP Group, CH-1211 Geneva 23, Switzerland

Abstract. Exchange of data and algorithms among accelerator physics programs is difficult because of unnecessary differences in input formats and internal data structures. To alleviate these problems a C++ class library called CLASSIC (Class Library for Accelerator System Simulation and Control) is being developed with the goal to provide standard building blocks for computer programs used in accelerator design. It includes modules for building accelerator lattice structures in computer memory using a standard input language, a graphical user interface, or a programmed algorithm. It also provides simulation algorithms. These can easily be replaced by modules which communicate with the control system of the accelerator. Exchange of both data and algorithm between different programs using the CLASSIC library should present no difficulty.

INTRODUCTION

During accelerator design it is often necessary to switch back and forth between various computer programs. The widely different input formats used to describe the accelerator lattice make the exchange of input data between programs difficult, and the risk is important that data used for different programs are inconsistent.

The data structure describing the accelerator lattice in computer memory also differs widely from one program to another, making the exchange of algorithms between programs impossible. A common data structure for general use in optics programs would be of inestimable help.

Many different methods have been used to represent transfer maps and to analyse optical behaviour of the machine. A uniform representation of maps would allow the transmission of maps between programs, so one program might be optimised for map generation and another for map analysis.

Optics programs often run in control computers where they are used to study the effects of varying machine parameters without disturbing operation.

This mode of operation requires a flexible interface connecting the optics program with the accelerator control system.

The CLASSIC library attempts to address all above problems. Its development started at a workshop held in August 1995 at SLAC, where a collaboration was set up. This collaboration includes in alphabetical order: CERN, DESY, FNAL, SLAC, and TJNAF.

Earlier designs of C++ class libraries for accelerator physics include the packages BEAMLINE and MXYZPTLK, published in 1990 (1), and a development version of MAD, started in 1994 (2).

STRUCTURE OF THE "CLASSIC" LIBRARY

Input Language

The input language problem was already addressed by a workshop held at SLAC in 1984, where a format-free "standard input language" was defined with mnemonic type codes for all accelerator elements. This language (3) has been implemented in several programs. The CLASSIC library will provide a parser, modelled after the parser for the C++ version of MAD (2). The language is somewhat extended with respect to the standard, and its syntax has changed to be closer to the C or C++ syntax.

Beam Line Elements

In the "CLASSIC" library beam line elements are represented by C++ objects. All elements are derived from a common superclass "Component", they can thus be used polymorphically. Each element class has an abstract base class which defines its interface, and a concrete representation which implements it. This approach effectively decouples the element implementation from the algorithms.

Beam Lines

A beam line is built as a composite structure. From the programmers point of view it is a template class containing a vector of "beam line objects". Each beam line object contains a pointer to a single element or to another beam line. Beam lines can be defined recursively. The template approach gives the programmer complete freedom to attach any data to a position in a beam line, e. g. misalignments or precomputed maps.

The CLASSIC library provides three different methods for building the accelerator lattice structure.

By calling library functions: A "Factory" object (4) can be instructed to make a copy of an already known element. The element attributes are set to particular values after constructing the element. A "Beam Line Builder" object constructs beam lines from already defined elements and/or beam lines.

By a Graphical User Interface: CLASSIC will provide a simple interface permitting to build accelerator structures.

By a Standard Input Language Parser: See the previous section for details.

All element attributes can easily be accessed by their name. A matching algorithm can thus easily vary a set of variables, run an algorithm, and observe a set of output variables.

Imperfections

Misalignments can be attached to a beam line object as described in Section "Beam Lines". To attach field errors to an element, it is embedded in a "wrapper" object. The wrapper class is derived from the abstract interface class of the element, it contains a pointer to the actual element and a list of "field modifiers". A field modifier can define an additive systematic or random error, a scale factor, or any other change to be applied to the field of the element. This mechanism allows different instances of identical elements to share their representation, but to have different modifiers. The algorithms see the wrappers as if they were the actual elements. When requested for its field, a wrapper forwards the request to its element, applies all modifiers to it, and returns the result.

Transfer Maps

Transfer maps are represented in CLASSIC as truncated power series maps. Great emphasis is being put on providing fast methods to manipulate such maps.

Algorithms

Algorithms are represented by C++ objects which walk through a beam line in the fashion of the "Visitor" pattern (5). A beam line "accepts" an algorithm and dispatches it in turn to each of its "beam line objects". If a beam line object has a misalignment, it takes care of it by calling the proper methods of the visitor before and after traversing the element. Otherwise it sends the visitor to its element pointer.

Single elements then request the visitor to perform the algorithm on their particular type, thus implementing the C++ double dispatch required to make the computation depend on both the element type and the algorithm. The visitor in turns asks the elements for their length and other data.

Sub-lines further dispatch the visitor to their components. This process is depicted in Fig. 1.

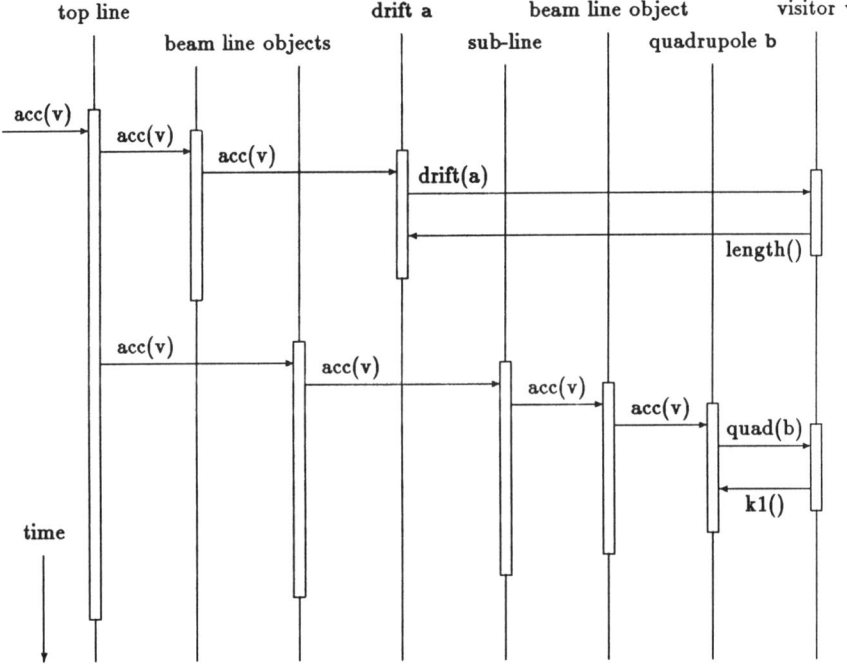

FIGURE 1. Illustration of Visitor Pattern. This figure illustrates the principle of message passing between objects, the actual messages passed will be different. The vertical lines represent the object life times, and the boxes represent the activation time of each object.

A tentative list of algorithms includes the following:

Surveyor: Computes positions and orientations of elements referred to a global geometry.

Lattice Functions : Computes lattice functions at all element positions.

Tracker: Calculates six dimensional phase space vectors at all element positions in a beam line, or tracks particles through this beam line.

Linear Mapper: Calculate the linear map from the entrance of a beam line to all element positions. This algorithm can also be used to compute the transfer matrix for a beam line.

Mapper: Calculate the truncated power series maps of a given order from beam line entrance to all element positions, or to the exit of the beam line.

Setup Segment Map: Builds the transfer maps for given segments of the beam line, and builds another beam line consisting of "macro-elements" represented by those transfer maps.

Set Mover: Walks through the beam line and assigns desired displacements to selected elements.

Set Element Strength: Modify the strengths of selected elements.

Alignment Error: Assigns random alignment errors to selected elements.

Field Error: Assigns field errors to selected elements.

The Visitor pattern allows to expand this list easily.

Control System Interface

The interface to the accelerator control system goes through CDEV (6). This system has been chosen because it has been widely used in several laboratories around the world and has proven to be very flexible. Its precise function remains to be defined. It will permit to read machine parameters from the control system and to feed them into the computer model. The computer model is then used to simulate the accelerator state, and some results may be fed back to the control system. On both sides of the interface parameters can be referred to by their names.

STATUS OF THE CLASSIC PROJECT

The abstract interfaces and the representations for beam lines and beam line elements have been implemented and are being tested in the framework of a new MAD version. The mechanisms for handling of machine imperfections are completely defined, but need to be implemented. Several algorithms and the CDEV interface are in the design phase.

REFERENCES

(1) L. Michelotti, *MXYZPTLK: A practical user-friendly C++ implementation of differential algebra: user's guide*. FN 535, FNAL, 1990.

(2) H. Grote and F. C. Iselin, *The MAD Program, User's Reference Manual.* CERN/SL/90-13 (AP) (Rev. 4). Also available at the URL http://hpariel.cern.ch/fci/mad/mad.html.

(3) D. C. Carey and F. C. Iselin, *A Standard Input Language for Particle Beam and Accelerator Computer Programs*. 1984 Summer Study on the Design and Utilisation of the Super-conducting Super Collider, Snowmass, Colorado.

(4) E. Gamma, R. Helm, R. Johnson, and J. Vlissides, *Design Patterns*, New York, Addison-Wesley, 1994, pp. 107-116.

(5) E. Gamma, R. Helm, R. Johnson, and J. Vlissides, *Design Patterns*, New York, Addison-Wesley, 1994, pp. 331-344.

(6) Jie Chen, Walt Akers, Graham Heyes, Danjin Wu, and Chip Watson. *An Object-Oriented Class Library for Developing Device Control Application*, in *Proceedings of the International Conference on Accelerator and Large Experimental Physics Control Systems*, Chicago, 1995, to be published on CD-ROM.

A Consistent Interface between PIC-Simulations

U. Becker*, M. Dohlus **, T. Weiland *

* Technische Hochschule Darmstadt, Fachbereich 18, FG TEMF,
Schloßgartenstr. 8, 64289 Darmstadt, Germany

** Deutsches Elektronen Synchroton, DESY,
Notkestraße 85, D-22603 Hamburg, Germany

Abstract. Many applications for PIC-simulations are limited by cpu-time, computer- memory or unphysical noise which is caused at large numbers of timesteps by the movement of particles. Therefore the aim is to divide the calculation area into different smaller parts to reduce the calculation time of each part. This subdivision requires a consistent, physical interface, which handles not only the particle properties but also the electromagnetic fields in the interface plane. Here we present an interface, implemented in a two dimensional (rz-geometry) and a three dimensional (xyz-geometry) PIC-code. The data stored during a two dimensional PIC-simulation can be reused for either another two dimensional or a three dimensional PIC-simulation. The following conditions, which have to be fulfilled at the interface plane, are checked during the simulation: the particle-movement through the plane has to be one-directional and the electromagnetic field of present resonances has to be negligible in the interface plane. We demonstrate the functionality of the interface for different cut-planes, applied to PIC-simulations of high-power tubes.

INTRODUCTION

The Particle-In-Cell (PIC) method, handling the interaction of charged particles with electromagnetic fields, is widely distributed for the simulation of different kinds of high-power tubes. This method, applied in time-domain, couples the relativistic equations of motion selfconsistently with the time-integration loop for the fields. This is done using the well known leap-frog scheme [1]. There are several disadvantages which complicate or even prevent the application of such PIC-simulations to any kind of tube:

- A large structure needs a long simulation time. For a high number of integration timesteps unphysical highfrequent noise arises due to the movement of discrete macroparticles in the dispersive spatial grid. The noise destroys the particle- and field properties.

- A high power tube with mainly rotational symmetric properties but 3-dimensional effects at the end has to be simulated completely in xyz-geometry, if those 3d-effects shall be studied.

- An optimization of the output cavity wastes a lot of cpu-time, because the lower part of the structure has to be simulated each time.

- The first part of a tube may require a high spatial resolution (due to a high acceleration or a fine resolution of the cathode surface). The fine mesh implies a short timestep, which has to be used during the whole simulation. Thus the simulation of upper regions of the structure, where a much larger timestep would be sufficient, is very inefficient.

An improvement on these points could be achieved by subdividing the structure into smaller computation areas, which can be optimized separately. The most critical point of such subdivisions is the handling of the boundary-conditions, especially of those where particles are traversing. The boundary plane has to simulate the real environment without disturbing the properties of the previous run.

Therefore a selfconsistent interface between two different PIC-simulations has been developed which maintains the particle-properties as well as the electromagnetic fields.

METHOD

The most realistic way to define a boundary condition including moving charges and their space charge fields is to store the particle and field information in a previous run at the particular z-position. This is done for a specified time range, in case of periodic particle bunches for one rf-period in steady state. Figure 1 shows a principle sketch of the interface-handling. In a first PIC-simulation of a rotational symmetric structure, the interface data is written. The stored data can be reused in either a 2d-rotational symmetric or a 3d-cartesian PIC-run, where the fields and particle currents are turned on smoothly to not excite unwanted frequencies by the switching-on time behaviour. The second run can use a different timestep. Therefore the monitored time functions are mapped to the new timestep. The unphysical noise, which has been arised during the first run, is reduced by lowpass-filtering the stored electric and magnetic time signals.

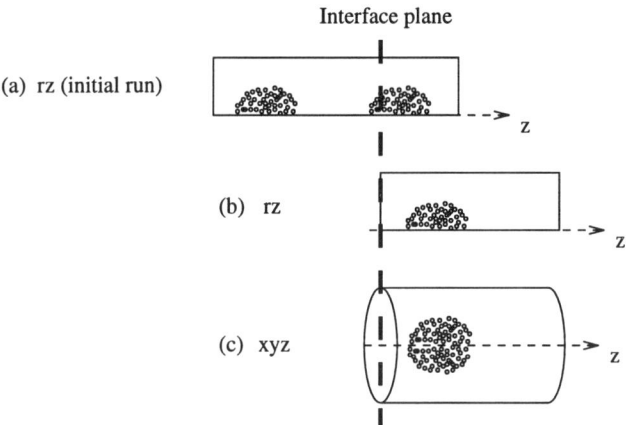

FIGURE 1: The initial run (a) writes the interface data which can be read by a 2d-rotational symmetric (b) or a 3d-cartesian PIC-simulation (c).

Applying this procedure we assume, that there is no backward-coupling from the second to the first run. Further cavity modes and other resonances have to vanish at the interface-plane.

In the following two paragraphs a detailed description of the particle and field part is given.

Particle Part

For a specified time range in the initial run, each particle passing the interface plane is monitored. The stored properties position, momentum, time and charge are used to reinject the particle in the second run. Whereas the reinjection is straightforward for the rotational symmetric case, the 3d-cartesian case requires a transformation of the ring-shaped 2d-charges to 3d-point charges. Figure 2 shows on the left hand side the partitioning of the ring-charge, which depends on its radius and the xy-grid resolution. Herewith one reaches a nearly constant density of particles in the xy-plane, which yields in a smooth current excitation.

Field Part

The boundary condition for the leap frog time integration scheme requires one tangential boundary field over the entire time range (either electric $\vec{E}_t$ or magnetic $\vec{B}_t$ field). Here the electric field is chosen due to the fact, that radial and azimuthal electric field components are allocated at the same z-coordinate. The two dimensional calculation stores those components for one

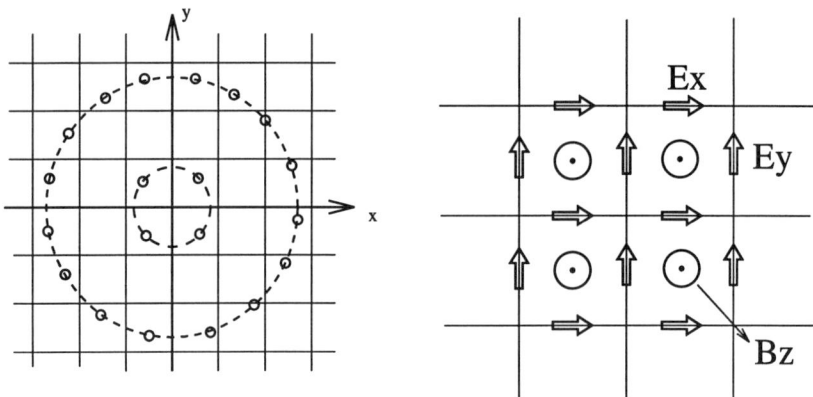

FIGURE 2: Each ring-charge is replaced by a number of point-charges, depending on its radius and the xy-grid resolution. On the right hand side the allocation of electric and magnetic field components in the interface plane is shown.

rf-period. In order to reduce the unphysical noise, these signals are lowpass filtered. By applying the filter twice, forward and backward, a time delay of the filtered signals can be avoided, which is essentiell for conserving synchronism of particles and fields.

Calculating the boundary field for the new run, one has to take care of two other points: a) The tangential field has to be interpolated to the new grid, which can cause unphysical longitudinal components of the static magnetic field rising linearly with time. Therefore the interpolation is done separately for the static and the dynamic parts. The time-independent part is interpolated using a scalar potential, whereas the dynamic part is interpolated in the usual way. b) The longitudinal component of the static magnetic field has to be adjusted by some switching periods with non-vanishing divergence of the elctric boundary field. This procedure is described in [3].

EXAMPLES

The first example shows the application of the interface on a pure dc-beam right behind a Pierce-like electron gun. The second example compares the particle properties of a periodic bunch train for different kinds of the interface.

DC-Current

The time-domain simulation of an electron gun for a S-band klystron, as shown in figure 3, requires a short time step due to fine resolution of the

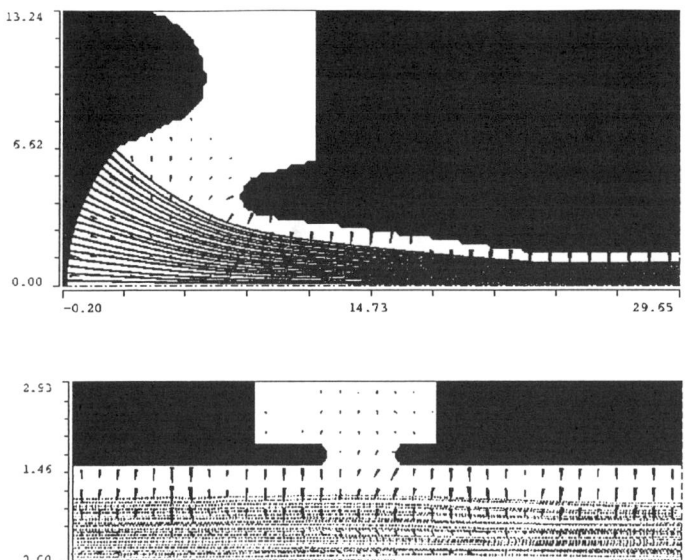

FIGURE 3: The upper plot shows the gun-simulation. The interface data, stored at the position z=25cm, serves as boundary condition for the following part, the input cavity of a klystron.

emitting cathode surface. The interface allows a higher timestep for the following parts of the tube. Space charge effects are dominating the simulation of the 700A dc-current, so that a consistent handling of the boundary fields is necessary.

Train of Gaussian Bunches

This example examines a periodic train of gaussian shaped bunches inside a closed beam pipe with radius 1.5 cm. Each bunch carries a total charge of 20 nC. The interface plane is located at z=0m. Figure 4 shows that the particles neither gain nor loose energy through the interface. In this example the macroparticles represent different number of electrons. The center of each bunch carries the highest amount of charge according to the gaussian distribution in time. Further one can show, that also the transversal movement of the particles is conserved.

CONCLUSION

An interface has been developed, which is able to subdivide the simulation area of a PIC-calculation into smaller parts. The stored properties of particles

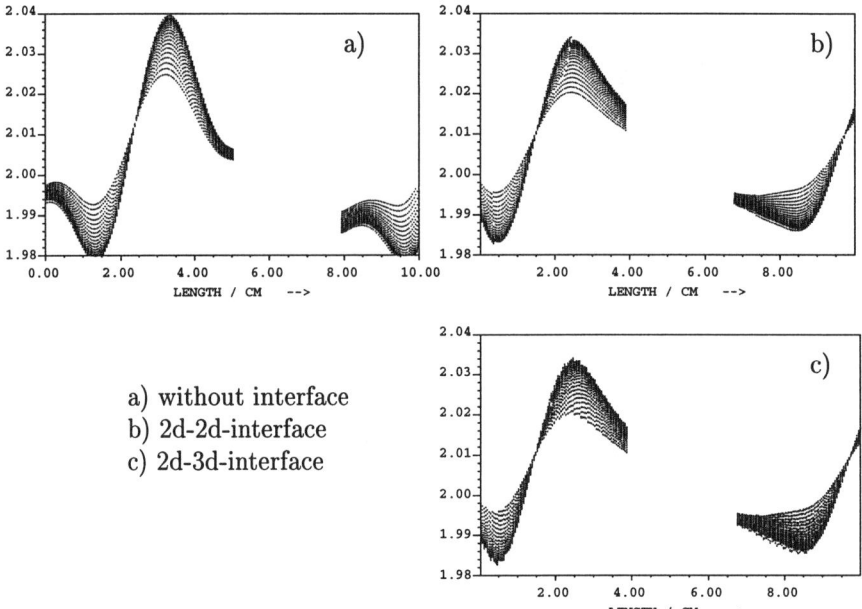

FIGURE 4: The particle energies (relative mass gamma) of gaussian shaped bunches inside a closed beam pipe are compared. The interface plane is located at z=0cm.

and space-charge fields are transformed in a consistent way into a boundary condition of a following PIC-simulation. Backwards travelling particles and overlapping resonances in the interface plane are neglected. The examples show the conservation of particle-energy as well as the tranversal and longitudinal momenta. For an optimization of large structures a lot of cpu-time and memory can be saved.

REFERENCES

1. C.K.Birdsall, A.B.Langdon, *Plasma Physics via Computer Simulation*, Adam Hilger Book Company, New York 1991.
2. O.Buneman, *Relativistic Plasmas*, page 205, Benjamin, New York, 1968.
3. U.Becker, M.Dohlus, T.Weiland, *Three Dimensional Klystron Simulation*, Particle Accelerators, 1995, Vol.51, pp.135-154
4. The MAFIA collaboration, *User's Guide MAFIA Version 3.x*, CST GmbH, Lauteschlägerstr.38, 64289 Darmstadt, Germany.

Unified Accelerator Libraries *

Nikolay Malitsky and Richard Talman

Laboratory of Nuclear Studies, Cornell University, Ithaca, NY 14853

Abstract

A "Universal Accelerator Libraries" (UAL) environment is described. Its purpose is to facilitate program modularity and inter-program and inter-process communication among heterogeneous programs. The goal ultimately is to facilitate model-based control of accelerators.

The performance of an accelerator depends to a large extent on the quality of the theoretical algorithms built into its control system. Evolution of new computing technologies and accumulating accelerator experience supply the necessary conditions for the next phases of integration—simulation facilities and (later) artificial-intelligence based control systems. However, several intermediate steps must be taken first, the foremost being the design and implementation of a framework for distributed accelerator programs.

There is a complementarity between the inherent heterogeneity of accelerator problems and personalities and the homogeneity that is essential for sharing codes. The Unified Accelerator Libraries (UAL) aim to create an environment in which diverse accelerator algorithms can be used together, connected to arbitrary data sources and integrated with other computer software. Each accelerator program is considered as a separate self-contained class, that may have its own internal organization and methods. Connection with the UAL is by common data objects (element attributes, beam parameters, *etc.*) On the one hand, this provides flexibility for each physicist to develop his or her own ideas and algorithms. On the other hand, one

---

*This manuscript has been authored under contract number DE-FG02-95ER40920 with the U.S. Department of Energy.

can consider the UAL to be a tool for comparing different approaches and selecting the most effective ones.

The *description* of accelerator structures is a key part of accelerator programs. There are many elements of different physical types, each having many attributes, all organized in a more or less hierarchical fashion. This has resulted in the creation of diverse lattice description languages and formats, each requiring a parser to perform the conversion from human readable ASCII files to the internal data structures needed to harness powerful computation algorithms (often written in FORTRAN or C in the past.) A disadvantage of such a "proprietary", "embedded" parser is that it makes the code "closed", impeding its re-use and extension. One result is that small, *ad hoc*, special-purpose, codes proliferate, often in the form of hard-to-support, "unofficial" modifications of existing programs. The development time of such codes/versions is typically short and, regrettably, so usually is their useful lifetime. In this way implementation of proprietary algorithms without an architectural focus impedes their integration with other software, leaving similar problems for subsequent developers.

The promise (some might say triumph) of object-oriented technology is to overcome this problem by enabling the replacement of the proprietary input format by one of the object-oriented programming languages. This can exploit the similarity of object-oriented methodologies to mathematical (algebraic) abstractions to make the application programming interface (API) simpler and more "physical" than present, code-specific, grammars. In principle (and in practice) the accelerator can be described using only overloaded assignment and arithmetic operators. "Addition" enables one to construct physical elements from some simple "bricks" (element attributes), and these elements can be concatenated into beam lines by "multiplication". Adoption of a standard programming language with its well-defined, highly disciplined grammar and lexical elements brings many benefits: support of major (conditional, iterative, *etc.*) statements and expressions, extensions in many different areas, interfacing to popular freeware and commercial applications (GUI, databases, communication, *etc.*), stability and portability, development and maintenance support by thousands of users and vendors, and natural connection to new tendencies and technologies in computer science.

Because of their common nature, the choice among different object-oriented programming languages is not essential; it depends on the particular task and environment. We feel the software should be built using two languages, with the core of algorithm's (developer's world) written in compiled languages while their invocation (user's world) is controlled by an interpreted language. In summary, since the latter constitutes the "input language" for accelerator programs, that language should be a standard, object-oriented, extensible, interpreted language. At this time, of the three candidates having similar functionality, Python, Scheme, and PERL, we have chosen PERL because of its popularity and better (for us) syntax.

For accelerator modeling the API is conventionally divided into two main parts, Accelerator Description and Actions. Certainly, the Accelerator Description must not be determined by the particular interpreted language. It should be based on the same object model as the one recognized by the core algorithms and, as such, it should be considered to be a "wrapper" around this common object-oriented accelerator description. From the software developer's perspective, the accelerator is characterized by two contradictory peculiarities. On the one hand, the accelerator is built from ideal standard elements with well-defined design principles. On the other hand, the resulting model is to be used as a tool for predicting and improving performance in the presence of non-ideal, perhaps not previously classified, devices. (This complementarity has contributed in the previously mentioned proliferation of special purpose codes.) We support an appreciable range of such possibilities by dividing all accelerator elements into three conventional categories, MAD, COSY (describable by Taylor series map), and WILD (a catch-all for special-purpose elements such as internal targets, electron cooling inserts, etc.)

The second main API feature, Actions, have traditionally consisted of an even more heterogeneous collections of commands and directives; again this is due to the variety of physical missions. The power of a modern computer language (in our case PERL) then makes it possible to merge these diverse descriptions seamlessly into a model is which beam dynamics in different sectors of the same lattice is described by different computation algorithms. A typical application might describe most of the lattice by TEAPOT conventional algorithms and one or more sectors by truncated Taylor series maps or WILD elements. As well as supporting this modularity, PERL automatically supplies the conditions for studying "dynamic" accelerator performance in which some of the optical parameters are varied, either adiabatically or impulsively. This permits the modeling, and eventually control, of effects like energy ramping, transition crossing, tune modulation, resonant extraction, and so on.

At this time the UAL joins three object-oriented accelerator programs as shown (Fig. 1.A): Platform for Accelerator Codes (PAC), Thin Element Program for Optics and Tracking (TEAPOT), and Numerical Library for Differential Algebra (ZLIB). The PAC is a collection of common data objects that can be shared, exchanged, or converted by other codes and processes. The central place in this architecture is taken by an accelerator object model, the Standard Machine Format (SMF), that is an extension of the Standard Input Format (SIF)[1]. In comparison with the design[2] SMF has been enhanced to speed up the access to element attributes (Fig. 2). Its source code and a functioning example (the Cornell Electron Storage Ring, CESR) are available now from the UAL site (http://w4.lns.cornell.edu/~tpot). Two other parts, the TEAPOT tracking engine and the ZLIB library, have been used recently and successfully at Brookhaven National Laboratory, for evaluating performance with a helical magnet insert[3] [4] and in investigations of tune modulation[5].

The next stage of UAL development, integration of TEAPOT matching and correction algorithms, requires the specification of new PAC classes (Adjusters, Buses, and Detectors) as well as the choice of an underlying communication facility. Because the direct integration of high level application algorithms with a particular communication layer reduces portability and runs the risk of obsolescence, we plan to employ the Common Object Request Broker Architecture (CORBA), an industry standard for distributed objects. CORBA provides an abstract specification that can be used on top of virtually any communication layer; it isolates the application software from the communication system, hiding heterogeneous features such as low level protocols, server-host hardware, operating systems, *etc.* Also this allows developers to connect applications written in different computer languages (C++, Java, *etc.*) The similarity of our architecture to CORBA-based frameworks such as Data Interchange and Synergistic Collateral Study (DISCUS)[6] makes straightforward the connection of UAL with distributed control systems (Fig. 1.B).

References

[1] D.C. Carey and F.C. Iselin, *Standard Input Language for Particle Beam and Accelerator Computer Programs*, Snowmass, Colorado, 1984

[2] N. Malitsky, R. Talman, *et.al.*, *A Proposed Flat Yet Hierarchical Accelerator Lattice Object Model*, BNL RHIC/AP/74, 1995

[3] N. Malitsky, *Application of a differential algebra approach to a RHIC helical dipole*, BNL RHIC/AP/51, 1994

[4] W. Fischer, *Preliminary Tracking Results with Helical Magnets in RHIC*, BNL RHIC/AP/110, 1996

[5] W. Fischer and T.Satogata, *A Simulation Study on Tune Modulation Effects in RHIC*, BNL RHIC/AP/109, 1996

[6] T.J. Mowbray and R. Zahavi, *The Essential CORBA*, John Wiley & Sons, Inc., 1995

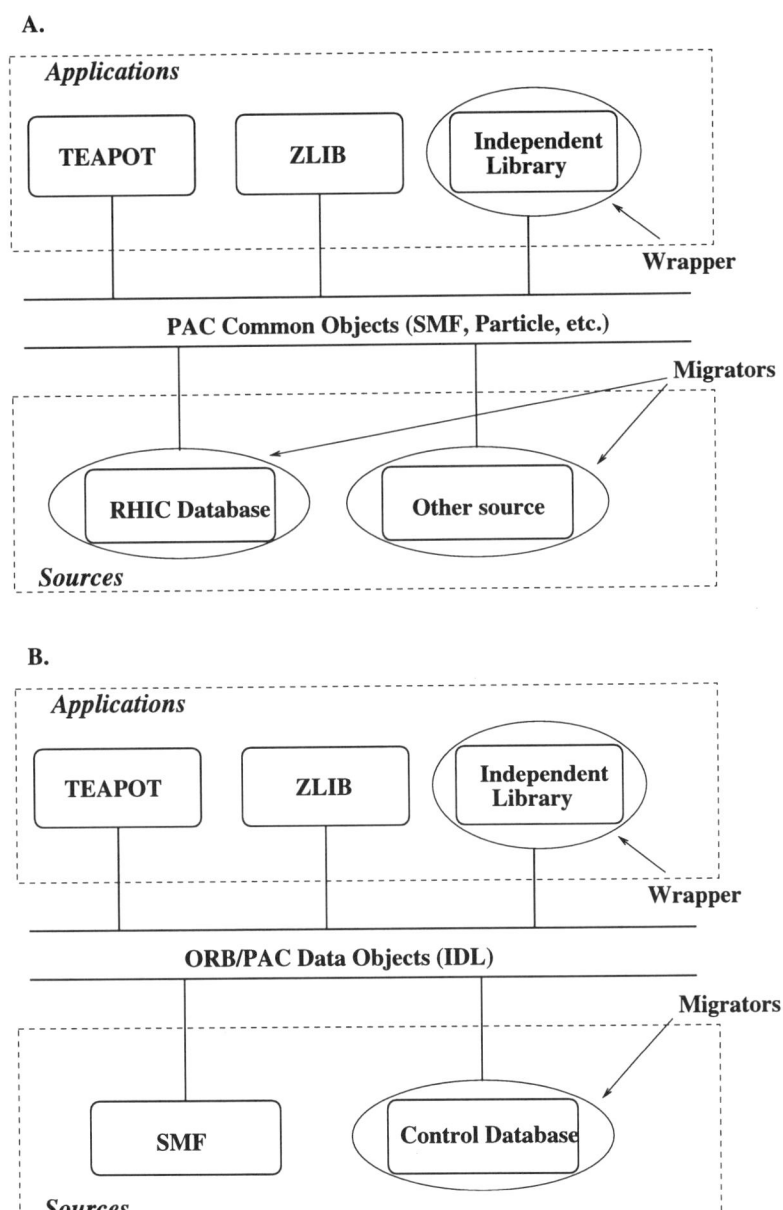

Figure 1: Unified Accelerator Libraries. (A)The present architecture. (B)CORBA-based framework.

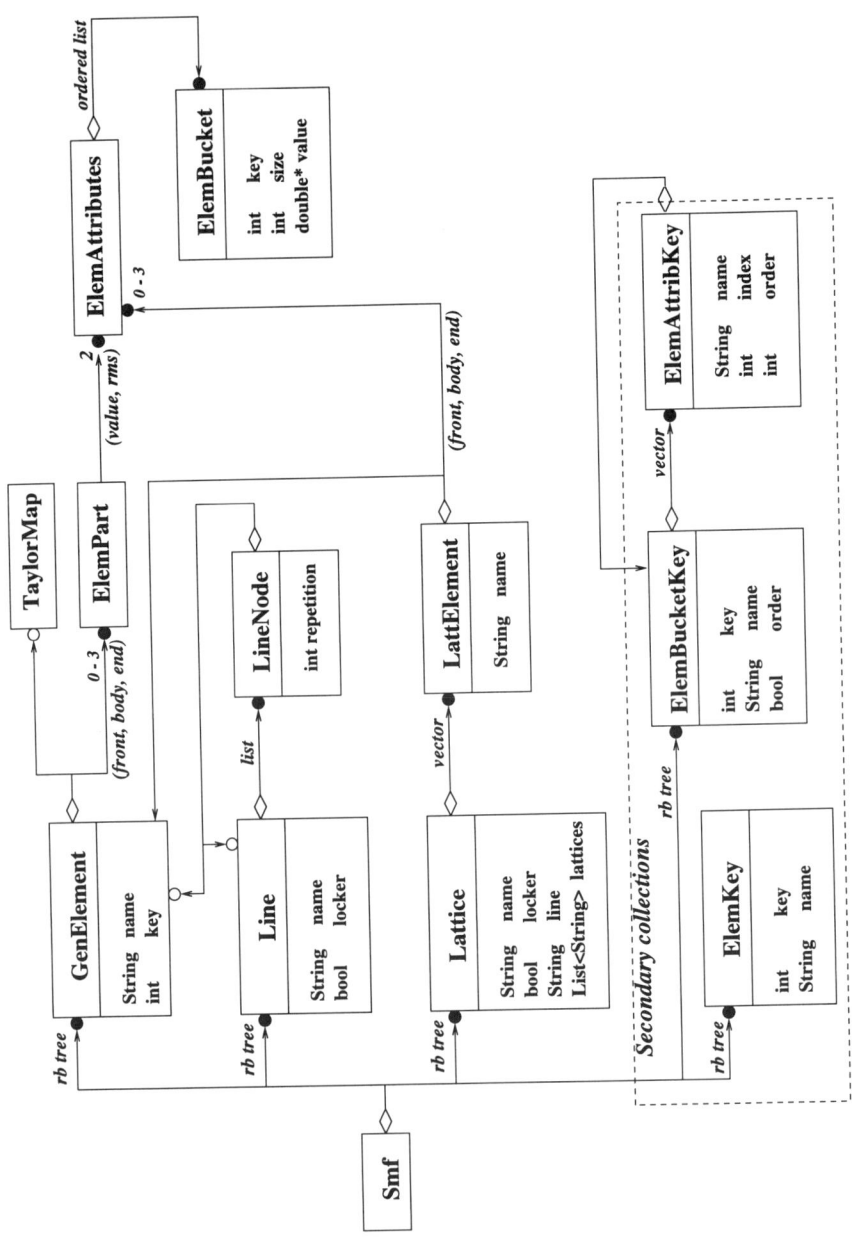

Figure 2: Standard Machine Format, an accelerator lattice object model. (*Vector, list, red-black tree* are containers of the Standard Template Library.)

An Object Oriented C++ Class Library for Solving Electromagnetic Time Domain Problems

Hiroshi Abe

C&C Research Laboratories, NEC Europe Ltd., Sankt Augustin, Germany

Abstract. The authors has developed a C++ class library to solve three dimensional electromagnetic transient problems. All the data and method for solving field are defined as classes and users can build their specific application using these classes of the library. In the first part of this paper, the algorithms and class structure of the library are described. Three application examples are shown in the rest part. Showing these examples the ease of merging with the advanced computer environment and the enhancability of the library to a specific problem are shown.

INTRODUCTION

Object Oriented Programming Style is getting popular in scientific computing field. Although the computation speed of generated module by C++ is still slower than that by FORTRAN, the ease of describing mathematical expression into program using C++'s operator overloading is attractive for physicists (1). Encapsulation of data and method using the class provide a safety in maintaining a large system. C++'s good reusability reduces the efforts to generate new classes that are derived from the existing classes.

The author adopted C++ as a main program language and have developed a C++ class library to solve three dimensional electromagnetic transient problems. We focus on high frequency domain problems, such as microwave, optical devices so far.

In this paper, the author describes the library's algorithms and class structures. Several application examples using the library are shown.

LIBRARY FOR ELECTRO-MAGNETICS

Library for Electro-Magnetics, LEM, is a C++ class library to solve three dimensional electromagnetic transient problems. LEM is not an application but a library to provide the users the necessary set of functionality to build their specific applications. In this chapter, the algorithms and class structures of LEM are described.

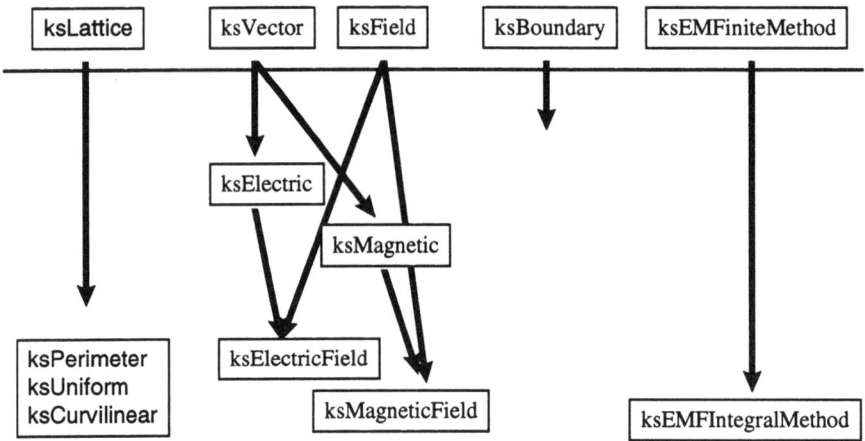

FIGURE 1. Typical class hierarchy of LEM.

Algorithms

LEM adopts Finite Volume Method on Yee Lattice for spatial discretization (2) and Leap-Frog scheme for time marching. The lattice is structured lattice, currently only support the orthogonal linear lattice. There are six different classes to generate different lattice objects which have different spacing of grid-points. To support arbitrary shaped devices with the orthogonal linear lattice, Mask method (3) is adopted for modeling.

An absorbing boundary condition (ABC) is supported to simulate the boundary that is terminated by waveguide (4). Using this ABC, the users can get S parameters with small calculation in high accuracy (~0.1%).

Class Structure

In the LEM, several base classes are defined. The most base classes provide the data and method which can be shared within the derived classes (Fig. 1). This gives the users common interfaces within a class family.

The ksVector class, a mathematical class, defines all the operations for vector, such as inner product, outer product, etc. The ksField class, a template class, provides data structure and operations for managing physical field. The ksElectrics and ksMagnetics are derived from ksVector and they inherit all the necessary functions from ksVector. They also may add their specific data and methods. Using ksField class with these ksElectrics and ksMagnetics classes, ksElectricField and ksMagneticField classes can be defined. The combination of ksVector and ksField classes derives a vector field class and the derived class can be applicable to another physical vector field, like velocity field.

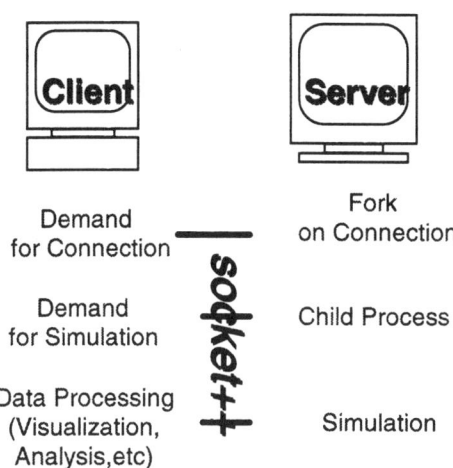

FIGURE 2. The server client real time simulation example.

Almost all classes in the LEM belong to their hierarchy. The enhancability of a class family depends upon the abstraction of the most base class.

APPLICATION EXAMPLES

Three examples using LEM are shown in this chapter. The first example shows an applicability of LEM to server-client real time simulation. In a heterogeneous computer environment, the tasks for simulation should be distributed onto their suitable machines. The solver task should be on high speed machine, the rendering for visualization should be on graphic workstations, for instance.

The next example shows the applicability to parallelization using MPICH, a standard message passing interface.

The last example shows the applicability to a real application, CD-ROM pickup simulation.

Server Client Real Time Application

In the heterogeneous computer environment, the data communication between computers is the important technique. This server-client model could be the base of the applications in this environment.

In this model, the solver program of the simulation is on the server computer, the visualization and analysis are done on client computer. To eliminate the unnecessary effort, we use IRIS explorer from NAG for visualization and the graphical user interface, and we use socket++ library from Swaminathan for the data communication. The procedures in this application are shown in Fig. 2. The

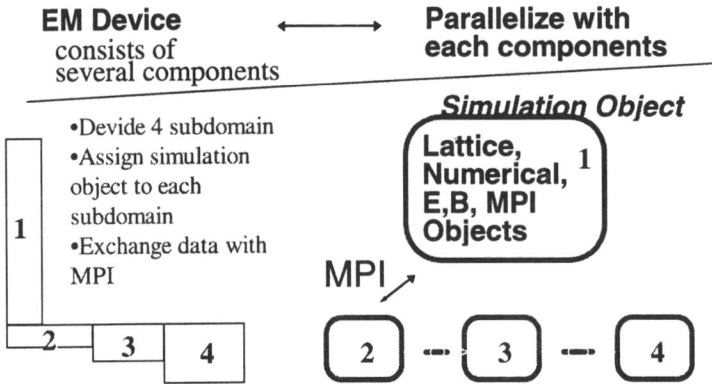

FIGURE 3 Parallelization using MPI class.

solver runs on the server computer as a daemon waiting for the connection request from a client. When a request comes, the solver daemon forks itself to generate its child process and establishes the connection. The solver daemon does simulation and returns data by the request from the client. With the received data from the solver daemon, the client does the visualization and the data analysis. Although this example is very simple one, the program model can be applied and extended to real applications.

Parallelized Application with MPICH

This example is a case of parallelization using MPICH (5). We adopt domain decomposition to parallelize. We newly define a simulation object which consists of a complete set of objects for simulation such as ksLattice, ksElectricField and so on. Each sub-domain is assigned to the simulation object and communicates via ksMPI class which is the top of MPICH. MPICH is an implementation of MPI-1 standard (6). The ksMPI class deals with all the functions for communication between the simulation objects. Besides, the ksMPI class interpolates the received data to generate the data for the object, so each simulation objects can have its own lattice independently upon the other objects. The communication number is once per one time step because LEM adopts explicit scheme for time marching. All the data to communicate is sent using same named method in ksMPI class by C++'s function overloading.

Enhancement Example (CD-ROM pickup simulation)

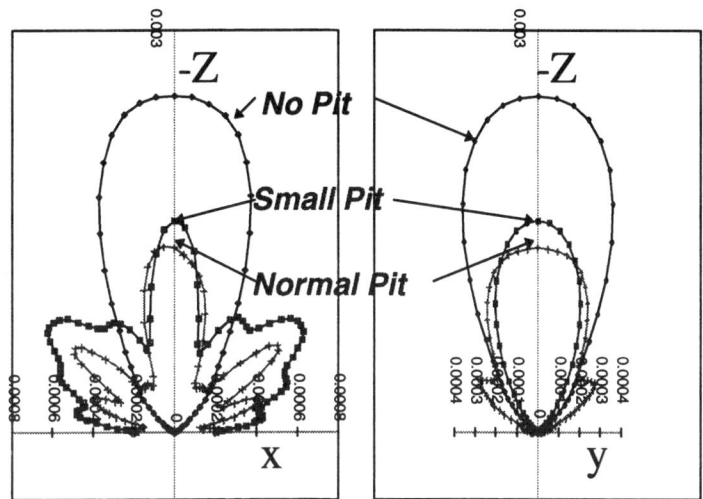

FIGURE 4 CD-ROM pickup simulation results. The plot lines indicate the strength of the y component of the reflected laser in each directions. CD-ROM's surface is on x-y plain. The pit is located at the origin. The grids is (61,61,31). The time step width is 1.6E-16 sec. The applied laser has only y-component with Gaussian spatial distribution centered at the origin.

As an example of a real application using LEM, CD-ROM pickup simulation example is described. This example might not be suitable one in accelerator physics, however, we would like to show it as an example which LEM can apply to different applications with small enhancements. The objective of this simulation is to observe the pit size dependency against laser with certain wavelength. In order to simulate this problem using LEM, ksOpenBoundary to simulate free space (7), ksBeam to apply laser and ksFar to simulate far zone analysis are developed (8). The other parts of the simulation are included within LEM.

The laser apply in z direction. As is shown in Fig.4, almost all of the applied laser is reflected in opposite direction with no pit case. With normal sized pit, the reflected laser in -z direction is reduced. If the difference of the reflected lasers between without pit and with pit is large enough to be detected by the laser detector, the pick up process works properly. As for the small pit case, the reflected laser becomes larger than the normal pit case. This tendency makes the detect-process difficult. This is actually encountered in real experience with designing the next generation of CD-ROM. The quantitative discussion needs the more accurate simulation, however, we success to simulate the phenomena using LEM with small enhancement.

CONCLUSION

The author has developed a C++ class library for solving three dimensional electromagnetic transient problems. The library, LEM, can be got the benefit of heterogeneous and parallel computer environment with small efforts by the benefits of object oriented programming style. The LEM is shown to be applicable to real applications with small enhancement. The enhancement is easily done since the C++'s good reusability of existing codes.

The LEM version 1.1.1 is available via World Wide Web at:

http://www.ccrl-nece.technopark.gmd.de/~habe/LEM/LEM.html.

ACKNOWLEDGMENTS

The CD-ROM pick-up simulation are done by a collaboration work with Computer System Research Laboratory of NEC Japan. The author is grateful to Dr. Doi, Mr. Oba of PCC Development Laboratories and other colleagues in NEC Corporation.

REFERENCES

1. Haney, S. W., *Computers in Physics* **8**, 690-694 (1994).
2. Weiland, T., *Particle Accelerators* **15**, 245-292 (1984).
3. Tsuboi, K., Miyakoshi K., and Kuwahara, K., "Incompressible flow simulations with solid models in rectangular grid system," in *Third International Conference in Numerical Grid Generation in Computational Fluid Dynamics and Related Fields, Barcelona, Spain*, June 1991.
4. Abe, H., and Hayami, K., "A Three Dimensional Maxwell Equation Solver for Arbitrary Devices with Inhomogeneities" in *Proceeding of Fourteenth Symposium on Calculations in Electrical and Electronics Engineering, Japan Society for Simulation Technology*, 1992, pp. 117-122 (in Japanese).
5. Gropp, W., Lusk, E., and Skjellum, A., "MPICH - A Portable Implementation of MPI", at *http://www.mcs.anl.gov/home/lusk/mpich/* 1996.
6. Message Passing Interface Forum, "MPI: A Message-Passing Interface Standard", 1995.
7. Mur, G., *IEEE Trans. EMC* **23**, 377 (1981).
8. Kunz, K. S., and Luebbers, R. J., *The Finite Difference Time Domain Method for Electromagnetics*, CRC Press, 1993, ch. 7, pp. 105-118.

Multilevel Codes RFQ.3L For RFQ Designing

Boris I. Bondarev, Alexander P. Durkin, Stanislav V. Vinogradov

Moscow Radiotechnical Institute
Warshawskoe shosse, 132, 113519, Moscow, Russia

The RFQ linac is employed widely in the most of accelerator centers over the world as an accelerator for the low-energy range. The RFQ quadruples offers the good compromise on particle acceleration and focusing for intense beam with moderately low injecting energy (up to 100 KeV) and good output beam quality. Many well-known codes are used now for RFQ designing and beam simulation. They include build-in algorithms for working out main regularities of intervane voltage, modulation factor, synchronous phase and average bore radius variations along the linac. The good choice of parameter arrays must give a possibility for beam high transmission concurrently with minimal length of a linac.

The new criteria is arose when high-current high-energy linac for transmutation problem is considered. The main requirement for such linacs is an action without any beam losses. In such cases the above algorithms can lead to undesired versions of a linac. The new concept of parameter choice based on scientific visualization for space charge-dominated beam acceleration and focusing.

The our multilevel codes the authors propose a new RFQ codes package which gives users a possibility to go successfully the all way from input data (operating frequency, input energy, beam current and possible level of intervane voltage) up to accelerating/focusing channel designing and CSD-beam simulation. There are two main requirements: maximum of scientific visualization on each calculation step and a possibility to cut off undesired linac versions long before the time-consumed real calculations will start.

The new package contains codes with three level of used mathematical model complexity. The first level codes make only a preliminary choice of main parameter arrays on the base of physical process simplified model. Codes of this level are richly supplied with visual information. The information helps to find the shortest way to the better linac version.

The second level codes are used for channel data correct calculations with the real shape of vanes. The information from the first level codes is used as input data. The third level codes are based on the information from the second level codes and on complex PIC-models. They are needed for beam accurate simulation in the chosen channel version.

In principle, at the first level the channel data is chosen so that beam transmission is 100 % and there is a some reserve for beam size increasing due to nonlinear effects taking into account at the third level. If during the third level calculations the impossibility of full beam transmission will be found then it is needed to restart the procedure: return to the first level codes and make adjustment in order to increase the difference between beam and bore radii. In practice, the recurrence is needed only in rare cases.

The main characteristics of mathematical models for each of three levels as well as main equations are presented below.

THE FIRST LEVEL CODES GENERAL DESCRIPTION

The functions for $U(n)$, $m(n)$, $j_s(n)$ and $R_0(n)$ where n is a period number must be selected for the given beam input parameters (energy, charge, rest mass, current, emittance) in such a manner that to maximize the output beam current with required parameters.

The tool contains:

- calculation of allowable ranges of interwane voltage U_u and bore radius R, determined by field strength limitation and requirement of transverse motion stability;
- approximate (fast) and accurate (slow) calculations of period lengths for given variation curves $U(n)$, $m(n)$, $j_s(n)$ and $R_0(n)$;
- the stability diagrams calculation;
- beam simulation in the best RFQ-accelerator version.

At all stage of calculations there are assumed that vanes have "ideal" shape and the electric RF-field has one harmonic in z-direction

$$E_z = -\frac{1}{2} U k \sigma I_o(kr) \cos(kz) \cdot \cos(\omega t),$$

$$E_x = -\frac{1}{2} U \left(\frac{x}{R^2} + \sigma k \frac{x}{r} I_1(kr) \sin(kz) \right) \cdot \cos(\omega t),$$

$$E_y = -\frac{1}{2} U \left(-\frac{y}{R^2} + \sigma k \frac{y}{r} I_1(kr) \sin(kz) \right) \cdot \cos(\omega t).$$

Here $k=2p/L$, L - modulation period, s - interaction factor depending on vane modulation m and minimum bore radius a, $r^2=x^2+y^2$,

$$\sigma = \frac{m^2 - 1}{m_0 I_0(ka) + I_0(mka)},\ I_0(r),\ I_1(r)\ -\ \text{Bessel function}.$$

The fast channel calculation is made using the relations

$$\begin{cases} L_i = \beta_{ii}\lambda\left[\dfrac{2\pi}{\Delta\varphi} + \dfrac{UT(i)}{W_0\beta_{ii}^2}\cdot\left[\dfrac{\sin(\varphi_s(i))}{2\pi} - \cos(\varphi_s(i))\right]\right] \\ \beta_{fi} = \beta_{ii}\left[1 + \dfrac{2UT(i)\cos(\varphi_s(i))}{W_0\beta_{ii}^2}\right] \end{cases}$$

where b_{ii} and b_{fi} - relative velocities at the input and output of modulation period with order number i, λ - RF wave length, $\Delta\varphi = \varphi_s(i+1) - \varphi_s(i)$, $T(i) = \pi\sigma(i)/4$ - the accelerating factor, $W_0 = m_0c^2$ - ion rest energy.

The above relations give a quite good approximation for accelerator length, period number and synchronous energy increments.

Visual information indicated after fast calculation makes it possible to estimate total length, modulation period number and n-, z-, W-dependence of main parameters.

The channel accurate calculation uses an iteration process for following equation solution

$$\frac{d^2z}{d\varphi^2} = \frac{eU\sigma\lambda^2}{4\pi\gamma^2 L}\cos\left(\frac{2\pi z}{L}\right)\cdot\cos(\varphi)$$

with boundary condition $z(j_i) = 0$, $z(j_f) = L$, where $\Delta\varphi = \varphi_f - \varphi_i = \varphi_s(i+1) - \varphi_s(i)$ is the given phase interval whereas synchronous particle pass modulation period.

The stability diagram are constructed for Mathie equation

$$\frac{d^2x}{d\tau^2} + \pi^2(2q\cdot\cos(2\pi\tau) + A)\cdot x = 0$$

in (q,A)-plane.

The pair of fundamental solutions are calculated under the diagram construction. They are used for transverse frequency calculation. The envelope modulation factor as well as matching beam radius are calculated.

The parameters q and A are connected with channel parameters by relations

$$q = \frac{eU}{2W_0}\cdot\left(\frac{\lambda}{\pi R_0}\right)^2, \qquad A = \frac{2}{\pi}\cdot\frac{eUT}{W_0\beta_s^2}|\sin(\varphi_s)|$$

Visual information: trajectory of operational point on stability diagram against a background of simple and parametric resonances lines, assumed beam envelop in the channel.

The transverse oscillations stability diagram is constructed taking into consideration space charge forces. The defocusing parameter A is absent in the considering equations.

$$\begin{cases} \dfrac{d^2x}{d\tau^2} + \left(2\pi^2 q \cdot \cos(2\pi\tau) - \dfrac{2\alpha}{x(x+y)}\right) \cdot x - \dfrac{1}{x^2} = 0 \\ \dfrac{d^2y}{d\tau^2} + \left(2\pi^2 q \cdot \cos(2\pi\tau) - \dfrac{2\alpha}{y(x+y)}\right) \cdot y - \dfrac{1}{y^2} = 0 \end{cases}$$

where $q = \dfrac{eU}{2W_0} \cdot \left(\dfrac{\lambda}{\pi R_0}\right)^2$, $\alpha = \dfrac{2IL}{I_0 \varepsilon (\beta\gamma)^2}$. Here I - ion current, I_0 - Alfven current for given ion type, L - modulation period length, ε - beam normalized emittance.

The diagram is constructed in the (q,α)-axes. The periodic solutions of the above equation system with period 1 have been find by iteration for diagram construction.

Visual information: operational point trajectory, assumed beam envelop in the channel.

The RFQ linac version evaluation has been made by beam simulation. During beam simulation the following effects must be calculated:
- longitudinal and transverse captures into accelerating and focusing regimes,
- effective emittance (an integration of phase portraits shared by cross-sections which have been injected in RFQ linac at different RF phase,
- a possible manifestation of simple and parametric resonances.

The mathematical model contains three particle types regarding the above tasks. The axis particles without velocity spread ($x_0 = y_0 = 0, x_0' = y_0' = 0, \beta = \beta_s$) have been used for the longitudinal capture calculation. For space charge field E_{zc} calculation: at the each integration steps the beam is assumed to be cylindrical with transverse size equals equilibrium crossection one.

The input cylindrical beam is subdivided into several crossections for calculation the transverse capture and the effective emittance. The model describes the each crossection motion and the current within this crossection is calculated with regard to beam bunching and particle losses

$$I_c = \dfrac{a(z)}{\theta^2(z)}$$

where $\theta(z)$ - the maximum excess of crossection transverse size above minimum bore radius at the linac part covered by the beam, $a(z)$ - the charge distribution having regard to beam bunching.

The same input $r_{xo}, r_{x0}', r_{yo}, r_{y0}'$ conditions are adopted for all crossections or the "ideal" conditions are adopted for each crossection.

The third particle type is putting into operation when the longitudinal and transverse oscillation frequencies are closely related and possible manifestation of simple or parametric resonances is expected. The third particle array includes non axis particles which have been injected into channel on the bound of equilibrium crossection parallel to the axis. In z-direction the third type particles experience the same space charge forces E_{zc} as the first type ones and in transverse directions the space charge forces are defined as in the

$$\mathbf{E}_{xc} \sim \frac{Ix}{r_x(r_x+r_y)}, \quad \mathbf{E}_{yc} \sim \frac{Iy}{r_y(r_x+r_y)}$$

where I, r_x, r_y are calculated by interpolation between two neighbouring crossections (of the second type) which are nearest to the particle under interest.

The information about matched beam parameters and about effective emittance expected growth is received long before the beam simulation is started. During process of beam simulation the phase portrait (W-Ws,Z-Zs) is observed on the background of current separatrix as well as fraction of particles outside of separatrix and envelopes of all crossections. The phase and energy spectrums for output beam are indicated after simulation is finished.

THE SECOND LEVEL CODES

The periods of vane modulation are calculated at the second stage taking into account the vane real shape. The functions U(n), m(n), Fs(n) and R(n) obtained at the first stage are used as input data. Electromagnetic fields distribution are calculated as well as period lengths in a case when synchronous phase is changes by value Fs(n+1)-Fs(n).. Electrostatic approximation is used in every cases where it is possible.

THE THIRD LEVEL CODES

For numerical experiments the 3-dimensional model based on Poisson's equation and macroparticles motion equations solving was developed (a variant based on full Maxwell's equations system solving is also available, but has no advantages for considered tasks). The code was written in C++ language. Its main characteristics are as follows:

a) The calculation area represents a parallelepiped $0 \leq x \leq x_{max}$, $0 \leq y \leq y_{max}$, $z_{min} \leq z \leq z_{max}$ covered with grid $i_m \times j_m \times k_m$ with steps $h_x = x_{max}/(i_m-1)$, $h_y = y_{max}/(j_m-1)$ and $h_z = z_{max}/(k_m-1)$.

b) Accelerated particles flow is simulated using macroparticles. Each macroparticle represents a parallelogram $h_x \times h_y \times h_z$ (PIC-model).

The macroparticle motion is described with the equation

$$dmv/dt = q(E+E_{ext})$$

where v - particle's velocity, m - mass, q - particle's charge, E - self electric field of the charged particles' bunch and E_{ext} - an external accelerating electric field with pre-defined spatial structure. The spatial structure of the external field may be defined as an analytical function or as a set of values on an additional 3-dimensional grid, calculated at the previous stage.

For the approximation of the forces acting on the particle the fields values in eight nearest grid cells are used.

The particles which touch a wall at any time moment are considered to leave the system.

c) For the electric field E definition Poisson's equation is solved in the cylindrical area using the iterations method with the Chebyshev's acceleration. For electrode modeling the values of potential in some grid cells may be forced to be constants. That is always true for boundary cells (except for z_{min} and z_{max} boundaries).

d) The values of z_{min} and z_{max} vary in time in such a way that the related phases $\phi_{min} = \int_0^{z_{min}} k dz - \omega t$ and $\phi_{max} = \int_0^{z_{max}} k dz - \omega t$ remain constant and therefore the accelerated particles' bunch is always placed inside the calculation area (which may be much less in longitudinal size than the accelerator length). The difference $f_{max} - f_{min}$ is always chosen to be an integer of 2π, this gives an opportunity to set periodical boundary conditions for z boundaries. Here w is the frequency of the external wave, k - the wave number.

e) To test the code the frequencies of plasma oscillations in a number of analytically studied situations (longitudinal waves in tube beams, transverse waves in plasma columns, etc.) were found in our computations and then compared with their theoretical values. All discrepancies were not greater than several percent.

f) The calculation results may be presented as plots of the configuration and phase spaces where the macroparticles positions are marked with dots and as plots of two-dimensional distributions of fields E_x, E_y, E_z, currents j_x, j_y, j_z, potential j and charge density r. All plots at different time moments may be saved on hard disk during the calculations and then may be watched consequently as a cartoon and drawn on printer.

A Matrix Representation of Lie Algebraic Methods for Design of Nonlinear Beam Lines

Serge N. Andrianov

St. Petersburg State University, St. Petersburg, 198904, Russia
E-mail:serge@asn.apmath.spb.su

Abstract. This report introduces the idea and concept of dynamic modeling paradigm for beamlines design. This approach is based on two aspects: matrix formalism for Lie algebraic methods and object-oriented programming approach. Using of these approaches permits to create a base of an expert system for studying of beamlines. For creation of corresponding databases and knowledge bases the computer algebra codes are used. Some aspects of employment of this approach are discussed. [1]

INTRODUCTION

This report presents key concepts of beamlines design. There are many works devoted to this problem (see, for example [1]). The authors of these works discuss approaches based on description pictures which are usual for present accelerator physics. These pictures deal with some set of different on their nature (mathematical one first of all) elements. This is one of the principal difficulty which interferes in building **knowledge bases** – the first part of any expert system. Indeed, these elements have different methodological description, operations and representations. In this case the knowledge engineering (i.e. the process of building expert systems) is hindered because we must use different kind of elements, define necessary connections between them (in the form of semantic nets, for instance), construct human interface for manipulation elements of the expert system. The analysis of the present status in the accelerator physics allow to make the following conclusions:

– the usage mathematical methods have to be adequate both to physical models and modern programming methods, to put it in another way the mathematical methods have to permit using of artificial intelligence methods and facilities;

– manipulated objects must maintain a natural extension of the set of processes and effects being investigated;

---

[1] This work is supported by the Russian Foundation for Fundamental Researches 96-02-17335-a

– manipulated objects must admit the most simple (from the a computational point of view) operations necessary for calculations.

These and other notes as well as author's experience demonstrated that one of the method satisfying these requirements is the matrix formalism of Lie algebraic methods [2]. In this paper we discuss how this approach can be used for beam physics and how the modelling process can be formulated. So, we formulate the principal properties of Lie algebraic tools [3] as a basic tools for our problems:

– the natural hierarchy of effects is investigated which can be included to the calculation schemes when necessary;

– the possibility of drawing of group-theoretical and algebraic methods which are highly developed and powerful mathematical ones [4];

– truncated Lie maps (Lie approximants) can be easily corrected for ensuring necessary properties (for example, symplectic properties for Hamiltonian systems);

– the simplicity of the mathematical nature of used objects – matrices;

– the possibility of unifying description of a subject of control (a beam line system) and an object of control (a particle beam) using the matrix tools including space-charge effects (see, for instance, [5]);

– the possibility of wide application of such modern computer approaches as object-oriented programming, computer algebra, intelligent interfaces, expert systems.

THE MATHEMATICAL CONCEPTS

Lie Transformations

It is known that time evolution in dynamical systems may be represented by one-parameter group of maps acting on the initial values of phase space variables $\mathcal{M} : X_0 \to X = \mathcal{M} \cdot X_0$. In the case of Hamiltonian systems it is a symplectic group of canonical maps – Lie maps. In the way we have to compute the action of this group for given dynamical systems. It can be done by solving the evolution equation for the operators that carry out the maps. One can write this equation in the form

$$\frac{d}{dt}\mathcal{M}(t|t_0) = \mathcal{M}(t|t_0) \cdot \mathcal{V}(t),$$

where $\mathcal{V}$ is a vector field of the dynamical system. The solution of this equation can be written in the form of chronological exponent operator (**T**-exponent):

$$\mathcal{M}(t|t_0) = \mathbf{T}\exp\left\{\int_{t_0}^{t} \mathcal{V}(\tau)d\tau\right\}.$$

For practical calculations, it is more convenient to move on to the Magnus representation for $\mathcal{M}$ with the help of a routine exponent operator

$$\mathcal{M}(t|t_0) = \exp\{\mathcal{W}(X;t|t_0)\},$$

where the vector field $\mathcal{W}$ can be calculated by using the well known Campbell-Baker-Haussdorf formula and its inverse variant – the Zassenhaus formula. We can write a vector field as a Lie operator associated with a function $G(X,t)$, for example, $\mathcal{W} = \mathcal{L}_G = G^*(X,t)\partial/\partial X$. In this case the latter equation can be rewritten in the form

$$\mathcal{M}(t|t_0) = \exp\{\mathcal{L}_{G(X;t|t_0)}\}$$

and the function $G(X;t|t_0)$ can be calculated from a function $F(X,t)$: $\mathcal{V} = \mathcal{L}_F$. This type of approach has found a wide range of applications to different topics. In particularly, it has been extensively used to solve beam dynamics in particle accelerators.

The Matrix Formalism

Let be $F(X,t) = \sum_{k=0}^{\infty} \mathbf{F}_k(t) X^{[k]}$, where $X^{[k]} = \underbrace{X \otimes \ldots \otimes X}_{k\, times}$ is the Kronecker power of a phase vector X of k-th order, $\mathbf{F}_k(t)$ is a $(n \times \binom{n+k-1}{k})$-dimensional matrix. This representation of the F generates an expansion of the function $G(X;t|t_0) = \sum_{k=0}^{\infty} \mathbf{G}_k(t|t_0) X^{[k]}$ and so we can write

$$\mathcal{M}(t|t_0) = \exp\left\{\sum_{k=0}^{\infty} \mathcal{L}_{G_k(X;t|t_0)}\right\}, \qquad G_k(X;t|t_0) = \mathbf{G}_k\, X^{[k]}.$$

This representation allows to write infinite products of exponentials of Lie operators

$$\mathcal{M} = \ldots \cdot \exp\{\mathcal{L}_{H_2}\} \cdot \exp\{\mathcal{L}_{H_1}\} = \exp\{\mathcal{L}_{V_1}\} \cdot \exp\{\mathcal{L}_{V_2}\} \cdot \ldots,$$

where $H_k = \mathbf{H}_k X^{[k]}$, $V_k = \mathbf{V}_k X^{[k]}$ are homogeneous polynomials of k-th order. The matrices $\mathbf{H}_k$ or $\mathbf{V}_k$ can be calculated with the help of the CBH- and Zassenhauss formulae and by using the Kronecker product and Kronecker sum for matrices. What is more using matrix representation for Lie operators one can write a matrix representation for a Lie map

$$\mathcal{M} \cdot X = \mathbf{M}\, X^{\infty} = (\mathbf{M}^{10}\mathbf{M}^{11}\mathbf{M}^{12}\ldots \mathbf{M}^{1k}\ldots) X^{\infty} = \sum_{k=0}^{\infty} \mathbf{M}^{1k}\, X^{[k]},$$

$$X^{\infty} = (1\ X\ X^{[2]} \ldots X^{[k]} \ldots)^*,$$

where the matrices $\mathbf{M}^{1k}$ (solution matrices) can be calculated according to the recurrent sequense of formulae of following type

$$\mathcal{M}_k \cdot X^{[l]} = \exp\{\mathcal{L}_{G_k}\} \cdot X^{[l]} = X^{[l]} + \sum_{m=1}^{\infty} \frac{1}{m!} \prod_{j=1}^{m} \mathbf{G}_m^{\oplus((j-1)(k-1)+l)} X^{[m(k-1)+l]},$$

where $\mathbf{G}^{\oplus l} = \mathbf{G}^{\oplus (l-1)} \otimes \mathbf{E} + \mathbf{E}^{[l-1]} \otimes \mathbf{G}$ –the Kronecker sum of l-th order. Using the generalized Gauss's algorithm we can write the matrix representation for the inverse map $\mathcal{M}^{-1}: X \to X_0 = \mathcal{M}^{-1} \cdot X$.

COMPUTER MODELLING

Dynamic Modeling Concept

The foregoing moments were put in the base of a new paradigm – **dynamical modeling** [6]. According to this paradigm we can go by the following way:
– to define an initial set of decisions (preliminary judgements of design problem);
– to applicate one or more mathematical models of various properties of the dynamical system (in our case of the beam line system). On this step in the frame of the matrix formalism we must select some of set of matrices from databases and include them to a calculation module. If there are no necessary matrices one can calculate them using computer algebra codes (in this work we usually use the *REDUCE* and *MAPLE* codes;
– to analyses of the solution and then accept or reject corresponding model, and in the later case, define a new set of decisions and return to the second step.

The running calculation experiments allows the filling our knowledgebases, for example in the form of recommendations and exclusions which can help for selections of models in future.

In the frame of this approach the mathematical model are regarded as objects, which have such very important characteristics as the inheritance and polymorphism.

From point of view of the matrix formalism the **bricks** of the corresponding databases are matrices of nonhigh dimensions calculated in the symbolic form with the help of computer algebra codes. These matrices are objects from which we build our models of a dynamical system - beam line system [7]. The extension of usual matrix algebra is realized on account of an introduction of the Kronecker sum and product. We have to note the possibility of using of parallel processing both in the symbolic and numeric modes of calculations.

Prototype of Expert Systems

First of all we must note that the definition "expert system" can be applied with some provisoes. Indeed, this definition can be used only after great number of calculations and information processing correspondingly. The calculation module is the main part intended for necessary calculations both in symbolic mode and in numeric mode. A multi-layer structure has been adopted. The interaction between calculation module and knowledge base (and databases correspondingly) can be realized with the help of oriented graphs methods. The calculation module invokes a procedure of multiple-key retrieval from the databases containing solution matrices for different models. The tutorial module supplies as a tutor and a system guide. In this module one can review of concepts, formulae, key parameters and other theoretical materials which are useful for solving a problem. The corresponding program library consists of programs related to both numerical and symbolic computations.

The special attention we can pay to the problem of synthesis of beam line systems with desired characteristics [8] and optimization of such systems. These problems are closely connected with the problem of calculation of invariants and symmetries (see [9]). The knowledge of invariants and symmetries can be used for solving the problem of synthesis of a control system (a beam line system) using an information about beam state. The usage of the matrix formalism allows to reduce the many problems of such kind to the problems of the linear algebra. The optimization module must maintain different methods of optimization which are included as required. In the frame of the matrix formalism in symbolic form such problems are the nonlinear programming ones. In particular, the known problems of a compensation of harmful nonlinear aberrations are reformulated as the problem of linear algebraic equations solving.

CONCLUSION

The above described approach was applied for ion-optical systems design including space charge problems, compensation of harmful nonlinear aberrations of different kinds [10]. For this purpose the package of programs for symbolic and numerical calculations which allow both to fill up the corresponding databases and to investigate beam evolution in nonlinear beam lines. As a demonstrating example of the package for modelling of chaotic behaviour of Hamiltonian systems is created (see [11]).

REFERENCES

1. Clearwater S.H., Lee M.J. *Prototype development of a beam line expert system.* IEEE PAC'87, **1**, p. 532, N.Y., 1987, Higo T., Shoaee H., Spancer J.E. *Some applications of AI to the problem of accelerator physics.* IEEE PAC'87, **1**, p. 701, N.Y., 1987. Osberg E.A., Ludgate G.A., Koscielniak S., Dohan D.A. *Dynamic object modelling as applied to the KAON control system.* Nuclear Instruments & Methods, **A293**, No 1-2, p.394, 1990. Schultz D.E., Brown P.A. The development of an expert system to tune a beam line. Nuclear Instruments & Methods, **A293**, No 1-2, p.486, 1990.

2. Andrianov S.N. *A Matrix Representation of the Lie Transformation.* Abstrs. of the Int. Congress on Computer Systems and Applied Mathematics, CSAM'93, St.Petersburg, July 19-23, 1993, St.Petersburg 1993, p.14.

3. Dragt A.J. *Lectures on nonlinear orbit dynamics.* Physics of High Energy Particle Accelerators, eds. by R.A.Carrigan, F.R.Huson, M.Month, AIP Conf. Proc., **87**, 1982, pp.147–311.

4. Andrianov S.N. *The Group-Theoretical and Algebraic Modelling of Particle Beam Control Systems .* Mechanics and Control Problems, issue 15, The Theory of Control Systems, St.Petersburg University, St.Petersburg, 1991, pp. 7–13 (in Russian).

5. Andrianov S.N. *Analytical simulation of space charge in ion-optical systems.* Proc. of First Int. Workshop Beam Dynamics & Optimization BDO'94, St.Petersburg (Russia), July 4–8 1994, St.Petersburg, SPSU, 1995, pp.19–29.

6. Hiroshi Nishimura *Dynamic accelerator modeling uses objects in Eiffel.* Comp. in Phys., **6**, No 5, 1992, pp.456–461.

7. Andrianov S.N. *Computer algebra and object-oriented programming*, Abstrs. of the Int. Conf. on Interval and Computer–Algebra Methods in Science and Engineering, INTERVAL'94, March 7–10, 1994, St.-Petersburg, Russia, pp.41–43.

8. Andrianov S.N. *Numerical-analytical calculations of symplectic maps for particles inside accelerators*, Proc. of IY Int. Conf. on Computer Algebra in Physical Research, Dubna, USSR,1990, World Sci., Singapore, 1991, pp.447–451.

9. Andrianov S.N., Yudin I.P. *Nuclear microprobe System with defined Characteristics* , Proc. of XII conf. on Charge Particle Accelerators, JINR, Dubna, Oct.13–15, 1992, Dubna (Russia), part 2, pp.305–310, 1993 (in Russian).

10. Andrianov S.N. *Construction of approximate symmetries and invariants for dynamical systems.//* Proc. of Second Int. Workshop Beam Dynamics & Optimization BDO'95, St.Petersburg (Russia), July 4-8 1995, St.Petersburg, SPSU, 1996, pp.16–24.

11. Andrianov S.N., Dvoeglazov A.I. *A Solving Module for Hamiltonian Dynamical Systems//* Abstrs. of Third Int. Workshop Beam Dynamics & Optimization BDO'95, St.Petersburg (Russia), July 1–5 1996, St.Petersburg, SPSU, 1996, p.3.

An Object Model For Beamline Descriptions

Barrey W. Hill, Hendy Martono and James S. Gillespie*

G. H. Gillespie Associates, Inc. P. O. Box 2961, Del Mar, CA 92014, USA

Abstract: Translation of beamline model descriptions between different accelerator codes presents a unique challenge due to the different representations used for various elements and subsystems. These differences range from simple units conversions to more complex translations involving multiple beamline components. A representation of basic accelerator components is being developed in order to define a meta-structure from which beamline models, in different codes, can be described and to facilitate the translation of models between these codes. Sublines of basic components will be used to represent more complex beamline descriptions and bridge the gap between codes which may represent a beamline element as a single entity, and those which use multiple elements to describe the same physical device. A C++ object model for supporting this beamline description and a grammar for describing beamlines in terms of these components is being developed. The object model will support a common graphic user interface and translation filters for representing native beamline descriptions for a variety of accelerator codes. An overview of our work on the object model for beamline descriptions is presented here.

INTRODUCTION

Our motivation is to develop an object model for beamline descriptions which supports a common graphic user interface incorporating a collection of accelerator simulation codes and facilitates translation between codes. This object model is purely descriptive. It is not used to simulate beam dynamics such as was described by L. Michelotti [6]. A hierarchical organization of basic beamline components, using sublines in combination with the ability to create "aliases", benefits translation and provides a rich design pattern for the development of a object oriented model for beamline descriptions. "Alias," as used here, refers to a component or subline which is represented by a persistent link to another component or subline and is also capable of storing deviations from the original data without duplication of redundant data. A beamline representation which also supports a flat model description, in which every element has an independent representation, greatly simplifies translation between different accelerator design and modeling codes, such as TRACE 3-D [1], TRANSPORT [2], ASM [3] and PARMILA [4]. A number of other benefits can be realized with this approach, including more efficient problem setup, compact views of very large models and elimination of redundant data storage within highly symmetrical beamlines [5]. The implementation of the beamline object model has been prototyped in the Particle Beam Optics Interactive Computer Laboratory, a cross-platform development project [7].

FODO CHANNEL EXAMPLE

The beamline object model will support a GUI in which sublines and aliases may be organized by the user in any representation that suits the problem. Automated tools will be provided which will organize the model into representations that optimize for particular needs, such as translation or size. Sublines, encapsulating a series of basic accelerator components, facilitate translation between accelerator modeling codes, which represent a series of components as one beamline element, and those codes which use a single component to describe the same beamline element. Figure 1 illustrates a trivial but concrete example of a FODO channel composed of four antisymmetrical Doublets separated by Drift elements.

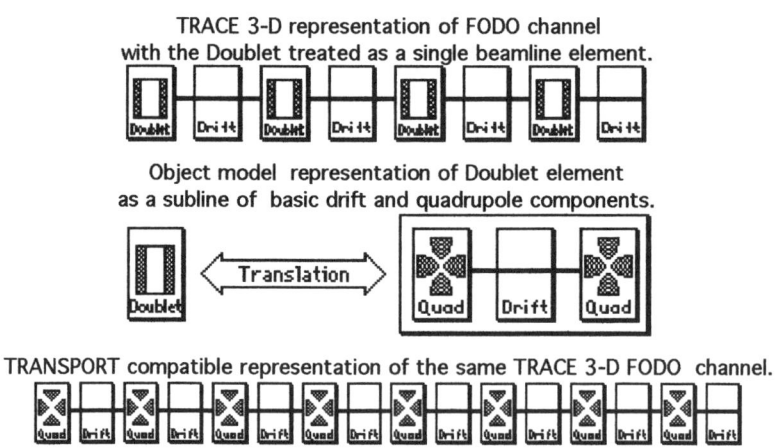

Figure 1. FODO channel translation between TRANSPORT and TRACE 3-D.

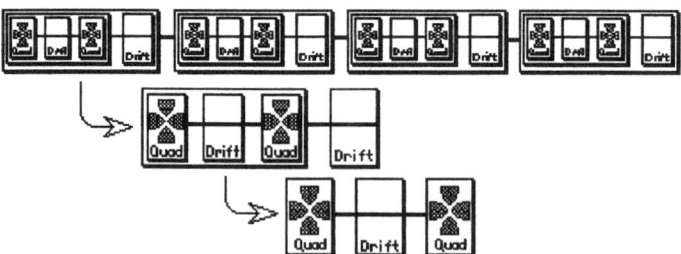

Figure 2. FODO channel model fragment composed with sublines and aliases.

Figure 2 illustrates the use of basic drift and quad components which are organized with nested sublines and aliases to exploit the inherent symmetry of the FODO channel example. The first subline contains a Doublet subline and a Drift component. The Doublet subline contains a Quadrupole component, and an alias to the Quadrupole, separated by a Drift component. This example will be referenced to describe the hierarchical beamline object model.

OBJECT MODEL IMPLEMENTATION

The implementation of the beamline object model can be described in terms of two different design patterns. The persistent hierarchical model representation is best described as a directed graph structure. Flat representations, which are created and disposed dynamically, are treated as tree structures. The actual manipulation of data is carried out on temporary tree structures generated from the directed graph. In the directed graph representation, nodes may have multiple parents and multiple children but the graph is only traversed from parent to child. A node may be both a parent and a child. There is one node with no parents which is the root node of the model. A node with multiple children is a subline and a node with exactly one child is an alias. A node with no children is a component. Figure 3A shows a directed graph representation for the FODO channel example.

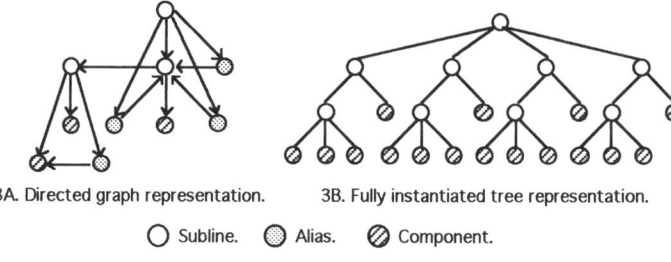

3A. Directed graph representation.   3B. Fully instantiated tree representation.

○ Subline.   ◉ Alias.   ⊘ Component.

Figure 3. Directed graph and tree structures for FODO channel.

When accessing specific parameter data, it is necessary to have a flat representation in which aliases to that data, and any deviations from the original data, are resolved. This flat representation is best viewed as a temporary tree structure with the nodes of the tree as sublines and the leaf nodes (nodes with no children) as components. Figure 3B illustrates a tree representation for a fully flat description of the FODO channel example. The entire tree in Figure 3B would never get instantiated at one time. However, operations on the model data would create temporary subtrees, following the directed graph and resolving aliases as required. Basically, subtrees are used as temporary flat structures which are created and disposed dynamically from any node in the directed graph as it is traversed recursively over sublines.

The hierarchical representation of the beamline model is persistent over time through a simple grammar and syntax. Bracketing and key words identify sublines, components, aliases and parameters. Any alias can specify a deviation from the original data. Returning to the FODO channel, an example of the file format for a hierarchical beamline description is presented in Figure 4.

```
{ 0 SUBLINE  "FODO"
  { 1 SUBLINE  ""
    { 5 SUBLINE  "Doublet 1"
      { 7 ELEMENT  "Doublet Quad 1"
        { (QUAD, L=0.077000000,
                 B=3.000000000,
                 A=0.010000000) }
      }
      { 8 ELEMENT  "Doublet Drift"
        { (DRIFT, L=0.010000000) }
      }
      { 9 ELEMENT ALIAS  "Doublet Quad 2"
        { 7 (B=-3.000000000) }
      }
    }
    { 6 ELEMENT  "Drift Space"
      { (DRIFT,  L=0.010000000) }
    }
  }
  { 2 SUBLINE ALIAS  "Doublet 2"
    { 1 (7:L=0.078500000)
        (6:L=0.096420000) }
  }
  { 3 SUBLINE ALIAS  "Doublet 3"
    { 1 () }
  }
  { 4 SUBLINE ALIAS  "Doublet 4"
    { 1 (6:"Final Drift") }
  }
}
```

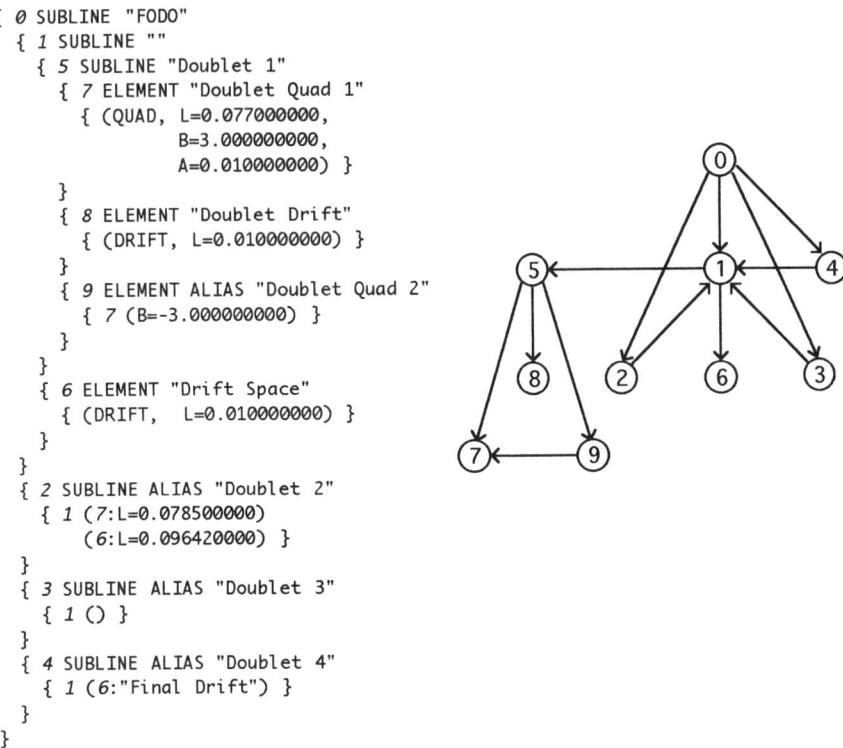

Figure 4. Persistent representation of the object model for the FODO example.

Each of the sixteen basic components represented in this example could be uniquely named, although they have not all been named here. Instead, the sublines have been named, as well as the three component pieces which contain the parameter data. The parameter data, for the remaining pieces, is generated dynamically from the alias links and the actual deviation data attached to specific component aliases. The Doublet subline (5) uses an alias to represent the antisymmetrical Quad components. The quad alias (9) uses the same data as the quad component (7) except for the magnetic field parameter, which has the opposite sign. The "Doublet 2" subline (2) is an alias to the subline (1) with parameter deviations for the lengths of the first Quad (7) and the Drift space (6) which precedes the next Doublet. The "Doublet 3" (3) subline is an identical alias to the subline (1). The "Doublet 4" (4) subline uses the same parameter data as subline (1), however the final Drift component has been renamed.

The node numbers on the directed graph and the corresponding numbers in the file example (shown in *Italics*) are only for the purpose of this discussion. The actual representation will use unique file copy identifiers. The file copy identifiers will be used to reference beamline descriptions which are defined in other files, so that a beamline model in one file can reference another file to describe particular sublines or components.

SUMMARY

The object model, for hierarchical beamline descriptions, supports a common user interface for a variety of accelerator modeling codes, and facilitates translation of the beamline data between the codes. Basic component definitions are being developed to support a variety of beamline elements. The object model implements aliases which can be extended to support remote linking over a local area network or the Internet. Filters can also be incorporated to support data translation from external sources such as a database or control system.

ACKNOWLEDGEMENTS

Portions of this work has been supported by the U. S. Department of Energy under SBIR grant number DE-FG03-94ER81767.

* Permanent address: Intuit, Inc., 6265 Greenwich Drive, San Diego, CA 92121.

REFERENCES

1. G. H. Gillespie and B. W. Hill, "Graphical User Interface for Trace 3-D Incorporating Some Expert System Type Features," in *Linear Accelerator Conference Proceedings* (Ottawa) AECL-10728, 1992, pp. 787-789.

2. G. H. Gillespie, B. W. Hill, N. A. Brown, and R. C. Babcock, "A Graphic User Interface for the Particle Optics Code TRANSPORT," to be published in the Proceedings of the 18th International Linac Conference, held August 26-30, 1996 in Geneva, Switzerland.

3. C. C. Paulson, A. M. Todd, M. A. Peacock, M. F. Reusch, D. Bruhwiler, S. L. Mendelsohn, D. Berwald, C. Piaszczyk, T. Meyers, G. H. Gillespie, B. W. Hill and R. A. Jameson, "Accelerator Systems Optimizing Code", in *Proceedings of the 1995 Particle Accelerator Conference*, 2, 1995, pp. 1164-1166.

4. G. H. Gillespie, B. W. Hill, J. S. Gillespie, "Making PARMILA Easy to Use - Really Easy to Use!," in *10th International Conference on High Power Particle Beams, BEAMS '94*, 1994, pp. 626-633.

5. N. Malitsky, R. Talman, F. Dell, S. Peggs, F. Pilat, T. Satogata, L. Schachinger, S. Tepikian, D. Trbojevic, C. G. Trahern and J. Wei, "A Proposed Flat Yet Hierarchical Accelerator Lattice Object Model," presented at LHC Workshop in CERN, Geneva, Switzerland, 1995.

6. L. Michelotti, "Towards C++ Object Libraries for Accelerator Physics," in *AIP Conference Proceedings*, 297, 1993, pp. 264-266.

7. G. H. Gillespie, B. W. Hill, N. A. Brown, R. C. Babcock, H. Martono and D. C. Carey, "The Particle Beam Optics Interactive Computer Laboratory," these proceedings.

Mapa - an Object Oriented Code with a Graphical User Interface for Accelerator Design and Analysis

Svetlana G. Shasharina and John R. Cary

Tech-X Corporation
4588 Pussy Willow Court, Boulder, CO 80301

Abstract. We developed a code for accelerator modeling which will allow users to create and analyze accelerators through a graphical user interface (GUI). The GUI can read an accelerator from files or create it by adding, removing and changing elements. It also creates 4D orbits and lifetime plots. The code includes a set of accelerator elements classes, C++ utility and GUI libraries. Due to the GUI, the code is easy to use and expand.

OBJECT ORIENTED APPROACH

C++ was chosen for the project (mapa) because it is the most commonly used object-oriented programming (OOP) language in scientific and programming communities. It is portable across many platforms and works elegantly with C procedures of X/Motif. This language provides data encapsulation, inheritance (including multiple inheritance) and polymorphism (thorough dynamic binding) [1]. It allows overriding and overloading of methods. The code, written in OOP language has a better chance to be clear, expandable, flexible and reusable.

Mapa uses many patterns (idioms) of OOP [2]. First of all, we used classical patterns of canonical classes, which emulate behavior of built-in types. Thus, we can treat vectors, Matrices, strings in a most convenient way (i.e. we can add them and perform other "natural" operations). We also used the Composite pattern for building primitive (single) and composite (beamline) accelerator elements, so that we can treat them equally. From behavioral patterns we used the Handle/Body pattern to build the garbage collection for strings, vectors and matrices. The Observer pattern was used for realization of Model/View-Controller structure, which allows to separate model from views and update the views upon changes in models. The Template method, which encapsulates logical parts of algorithms and leaves their definition to derived classes, saved us a lot of coding and made the code more transparent. We also used the Letter/Envelope pattern for providing polymorphic behavior for arithmetic classes (latter will be used for implementation of TPSA).

SYSTEM HIERARCHY AND MAPA'S CAPABILITIES

The code has two hierarchy trees describing the models of interest. One tree has to do with general systems, which have names (class System), parameters and options with unified I/O (class SimpleSystem). The map hierarchy describes systems with dynamic features (the Advance methid propagates dynamic variables though time). The two hierarchies meet to create (through multiple inheritance) the SimpleMap class (which is still an abstract class), from which most of mapa systems are derived. Thus, an abstract Element is derived from the SimpleMap, as well as Accelerator, set of classical nonlinear dynamics maps (Henon's, standard map etc.) and Torus (class for studying motion of particles on toroidal fusion devices). Concrete elements and composite element (Beamline) are derived from Element. Accelerator and Element are related through aggregation: Accelerator has a list of Element pointers.

GRAPHICAL USER INTERFACE

One of the main efforts in building mapa was to make the code user friendly. To reach the goal, we created a set of C++ classes for encapsulation and convenient use of X/Motif. These classes became the base of the GUI. The use the GUI of makes computing interactive. First of all, it allows the user to select the system of interest through a menu-like widget. Click of the mouse button brings up the controller of the system. The controller can read/write the system from/to files, list parameters and options, change them. It also brings up, start, stops and saves simulations relevant to the system (like now we have Monte Carlo simulations resulting in showing average behavior of dynamic variables of the system versus time). It also allows users to see the orbits in phase or real space (each orbit can be start by a click on the plot) and lifetime plots. The accelerator controller can change the beamline by visual adding and removing of elements from a table, whose parameters can be changed through the same interface. A special widget allows to find fixed points of different order by using various solvers, with the initial guess being being found and input graphically.

The set of controllers mirrors the abstract part of the system hierarchy, so that particular systems do not need a specialized controller. The GUI allows coexistence of many systems and simulations simultaneously, which makes the task of design and analysis more fast easy.

DISCUSSION

We are planning to (a) teach mapa to read from diffirent filc formats (like MAD (3) , SMF (4), CLASSIC); (b) implement TPSA for non-linear map analysis (5); (c) include possibilities of missalignments; (d) improve fitting; (e) make the set of accelerator elements richer; (f) create survival plots (g) try other (not mapa's) tracking engines (TEAPOT (6)).

REFERENCES

1. Stroustrup B., *The C++ progamming lanquage, second edition*, Massachusetts: Addison-Wesley, Reading, 1991.

2. Gamma E., Helm R. et. al, *Design Patterns*, Massachusetts: Addison-Wesley, Reading, 1995.

3. Grote H., Iselin F. C., *The MAD Program User's Reference Manual*, CERN report CERN/SL/9013, CERN, Geneva, 1991.

4. http://w4.lns.cornell.edu/~tpot.

5. Malitsky N., Reshetov A. and Y. Yan, ZLIB++: *Object-oriented numerical library for differential algebra*, Stanford Linear Accelerator Center: SSCL-659, 1994.

6. Talman R., TEAPOT: *A Thin-Element Accelerator Program for Optics and Tracking*, Particle Acceleretors, **22**, 35.

Recent Developments in the Accelerator System Model Code

S. Mendelsohn, D. H. Berwald, M. H. Hughes, T. J. Myers, C. C. Paulson,
M.A. Peacock, C. M. Piaszczyk, E. M. Piechowiak, J. W. Rathke
Northrop Grumman Corporation
Advanced Technology & Development Center, Bethpage, NY 11714, USA

George H. Gillespie and Barrey W. Hill
G. H. Gillespie Associates, Inc.
P.O. Box 2961, Del Mar, CA 92014, USA

Abstract. High-power linear accelerators have potential applications in fusion materials testing, intense spallation neutron production for neutron physics, production of nuclear materials and destruction of nuclear waste. Determining the optimal configuration and operating parameters for the accelerator early in the design process is an important step in minimizing development costs associated with these applications. A coordinated evaluation capability in the rf system and other major supporting subsystems would thus provide a fully integrated system design. Further, characterizing component redundancy where necessary to provide a specified facility/system on-line availability completes a highly representative engineering and costing configuration. The Accelerator System Model (ASM) code provides this unique capability for developing optimum designs. ASM has been under joint development of computer interface, physics and engineering activities for four years and now permits the detailed layout and evaluation of a wide variety of normal and superconducting accelerator and rf power configurations for the aforementioned devices and other accelerator applications. This paper describes the current capabilities of the latest version of ASM, details continuing concentration on an integral reliability/availability approach, new component development and support system models.

INTRODUCTION

The ASM code provides a new approach to computer tools for accelerator design in which engineering and beam dynamics are treated at comparable levels of fidelity. The initial goals of the ASM effort included the development of a modeling and analysis software package that was easy to use, automated many initial design calculations, supported trade studies used in assessing alternate accelerator designs and was flexible enough to readily incorporate new technology models [1,2]. These goals were achieved in a prototype version of ASM [3,4] and a collaboration was begun to significantly expand and improve upon the physics and engineering models, as well as to enhance the Graphical User Interface capabilities (developed by G. Gillespie and Associates[5]) in order to support new user tasks. The code uses a series of FORTRAN-based modules, linked to graphic objects, to provide for the "construction" of the accelerator directly on the computer screen. The effort has led to the development of significantly enhanced versions of ASM which have been applied to a number of accelerator system design problems [6,7,8]. This paper provides an overview of the current capabilities of the code.

THE ESTABLISHED DESIGN CODE

In the past three years Northrop Grumman has built on existing experience and technical capabilities to develop a simulation code which provides accelerator system configurations unique in detail and technical expanse. The approach develops detailed, self-consistent, accelerator component and system models addressing physics, engineering, subsystems, facilities, RAM and cost. Results have been benchmarked against existing designs stressing viability so that we can offer realistic trades. Although a systems code, it has evolved so far in engineering description that ASM can be used to establish detailed design configurations. It offers point-of-departure designs as a step-off to actual design, optimizes to specific parameters, and quickly evaluates configurations, performance, availability and cost for full-up systems within the context of a fully integrated accelerator system. The table below indicates the breadth of modeling capability available.

Beamline Physics and Engineering	System Engineering and Cost
(G) Ion Source (ECR, RF-driven)	(N) RF System Layout
(G) Low-energy Beam Transport (PMQ, EMS)	(N) Cryogenic System Layout
(G) Radio Frequency Quadrupole	(N) Dedicated Facilities
(G) Beam Funnel	(N) Accelerator Capital Cost
(G) Drift Tube LINAC	(N) System Life-cycle Cost
(G) Coupled Cavity LINAC	(N) RAM Models
(N) Drift Tube LINAC (Various Lattices, Magnet Types)	(N) System Integration
(N) Bridge Coupled DTL (APT-type)	(D) Target Systems
(N) CCL Cavity (APT-type SCL)	Miscellaneous Modules
(N) Coupled Cavity DTL (2 or 3 gaps)	(G) ASM Graphical Interface

(N) Superconducting LINAC
  - Low β Independent Cavities
  - High β Multi-elliptical Cell Cavities
	(N) System-independent Driver

(N) Beam Matching Logic (Interstructure, Funnel)	(D) Trade Study Optimizer
(N) RFQ (Const Aver Aperture, Tilted Field)	(D) Graphical Output
(D) High-energy Beam Transport	
(D) Cyclotron	

Source: (G) GHGA   (N) NGC   (D) NGC Development

The resulting analytic improvement is an ability to consider more configurations and technology trades, more quickly, using a more complete set of data and a more consistent and inclusive set of modeling algorithms. The applicability of trades currently supported by ASM relevant to current designs is indicated below.

	Linac Application		
Candidate Trade Study	IFMIF	ATW	NSNS
Beam Pulse Length	N/A	N/A	•
Alternative Accelerating Structures	√	•	•
Normal vs. Superconducting	√	•	•
Transition Energies & Matching	√	•	•
Beam Energy vs. Current	N/A	√	•
Accelerating Gradient	√	√	•
RF Frequency	√	•	•
Frequency Doubling	√	•	•
Current Funneling	N/A	•	N/A
Multiple vs. Single Beamlines	√	√	N/A
Multiple vs. Single Ion Injectors	√	•	N/A
Design Optimization vs. Plant Life	√	•	N/A
RF Amplifier Technology	√	√	•
RF Tanking	√	√	•
RF Pre-Amplifier Staging	√	•	•
RF Amplifier Redundancies	√	•	•
High Voltage Power Technology	N/A	•	•
RAM Trades	√	√	•

RECENT PHYSICS/ENGINEERING MODELS

In the past year significant effort has been expended to improve existing capabilities in representation and user interface. The following describes new modeling efforts to keep pace with applications, and new approaches to RAM and cost estimation.

Matching Section

From a physics point of view, the matching section is the most powerful element within the code. It provides the designer with the flexibility to determine approximate baseline designs and to explore operational and design variations around the base point. It presently is a generic module which uses the beam parameters from the immediately upstream element and matches the properties of the following element to those beam parameters. This allows the code to determine the optimal initial parameters for each element dependent upon the beam presented to that element. In this manner the design may proceed with minimal input. At present, off-nominal operation is obtained by allowing the beam to travel from element to element (the element parameters were determined previously using the matching section) without matching. This will soon be augmented with a second matching element that will modify the beam parameters to match the downstream accelerating element parameters (the inverse of the present matching section).

Couple Cavity Drift Tube LINAC

The Couple Cavity DTL is a relatively new accelerating structure developed at Los Alamos[9] that utilizes the best features of the DTL and the CCL to accelerate beams in the energy ranges normally reserved for DTLs. Its basic structure is that of a large CCL-like cavity with small (no enclosed magnet) drift tubes. The representation presently has the capability of modeling a one-gap (normal side coupled CCL), a two-gap and a three-gap structure. They are maintained as separate entities due to engineering and cost coding strictures. Device geometrical parameters and their derivatives (shunt impedance etc.) were obtained by scaling a series of SUPERFISH exercises and from data presented in the noted reference. Operational parameters may be input (constant or variable) or obtained from the input beam phase advances (matching section). The focusing lattice may be FoDo or FDo and the period may be constant or a multiple of the local beta.

Due to its design complexity the current APT design has been selected to baseline the CCDTL module. The model used a two-gap CCDTL between 7 and 8 MeV, a three-gap CCDTL to 20 MeV, and then back to a two-gap design between 20 and 100 MeV. Additionally, three different bore radii were used in the higher energy section. All sections used FoDo focusing. The code correctly reproduced the accelerator layout in terms of numbers of cells, section lengths, and power as well as accurately reproducing the beam parameters.

High Energy Quasi-Elliptical Superconducting LINAC Structure

These tank structures are normally designed to be identical over a rather large energy range. To provide the most flexible design tool possible a number of choices have been provided to the designer. He is able to choose the number of cells per cavity and cavities per cryostat (these can alternatively be chosen by the code to optimize the design); the length of the cryostat to be fixed or a multiple of the local beta; the phase advance to be fixed or vary with beta; and the design beta for the various designs (or as a specific value of the longitudinal acceptance).

The energy gain in a cavity is calculated in terms of the cell gaps, central and synchronous phases therefore also defining the longitudinal acceptance. The relationship between the central phase in two contiguous gaps is defined as a function of the cavity design velocity, the beam average velocity in the center of a designated cell, and a cell size modifier to account for field fringing. These relationships were iterated to obtain the phases for each cell predicated on the requirement that the cavity central phase is the synchronous phase. The cells have been parameterized using 238 separate SUPERFISH calculations. From these, a range of working values for the geometry - length, outer and cylinder radii, septum size - were chosen, providing closed-form parameterizations for surface resistance, peak electric and magnetic fields, peak power density and shunt impedance.

High Energy Beam Transport

A new physics section implements an HEBT model to describe beam transport from the end of the accelerator to a suitable target. This code has been constructed in a highly modular fashion and implemented such that geometrical configurations of interest can be accommodated. It is assumed that an arbitrary HEBT assembly can be constructed using only four distinct components. These are: periodic FoDo sections; achromatic bends; beam expander sections; and RF gaps. These four elements are modularly assembled into an HEBT configuration as required.

Typically, input includes the electromagnetic quadrupole (pole tip) field, the aperture and the betatron phase advance. Either the magnet geometry or the quadrupole distribution is subsequently calculated consistently. The HEBT physics is linear and includes first order transport of the beam along various components comprising the system. At present, space charge effects are omitted, though a linear model could readily be implemented if required. In addition, subroutines are included which perform detailed engineering calculations of both quadrupole and dipole magnets comprising the system. These calculate, for example, the volume and weight of the magnets, the volume of copper required and power requirements which are returned to ASM for subsequent costing and reliability analysis.

Diagnostics, Instrumentation and Control

The inclusion of a section devoted to diagnostics, instrumentation and control hardware represents further accountability of support systems. Based on the accelerator design build, a base set of diagnostics, sensor instrumentation, control hardware and computer substations are enumerated for each accelerating component: injector, LEBT, matching sections, RFQ, DTL, funnel, BCDTL, CCL and beamstop. Examples of this base set are Faraday cups and collimators, video profile and beam position monitors, flying-wire and harp profile systems predicated on component type and beam energies. This list is then expanded to include items the quantity of which can be estimated as relating to the number of individual parts in the component or the component geometry. Typical examples are: magnet power supplies, microstriplines or ionization chamber beamloss monitors, beam halo detectors based on the quantity of magnets by type; number of thermocouples based on the quantity of cavities or drift tubes in a DTL; number of beam transformers for beam current measurement or in-line vacuum valves based on physical tankage.

A computer control station is presently defined for the injector/LEBT, the LINAC (RFQ/DTL/other), and the HEBT. This includes computer and network hardware, data channel controlling hardware and attendant software. This composite list is cycled through the RAM module for evaluation and recommendation of spares and thence to the costing module for inclusion in overall facility totals.

Reliability, Availability, and Maintainability Module

The ASM RAM code was developed to follow the four major accelerator components: injector, LINAC, rf system, and HEBT. Support systems, such as electric power, cooling, heating, vacuum, air conditioning, computer control, and the target systems are usually modeled separately. However, the RAM models in ASM can be used to represent these systems as well. Each subsystem is modeled in multi-level detail for RAM purposes. For example, the accelerator system may be depicted as an overall RAM model of estimated reliability and availability dependent on the individual parameters for the injector, the LINAC, the RF system, the HEBT and support systems. Further, the injector performance may be derived from a lower level depiction including the Ion source and the LEBT. Yet a much more detailed depiction of the next, lower level of the ion source provides predictive performance characteristics as a function of the predicted reliability and availability of the RF antenna, RF generator, extractor, gas supply, HV power supply, turbomolecular vacuum pumps and the support structure. This detailed cumulative approach leads to greater predictive accuracy for the subsystem.

Just as detailed, the LINAC is represented by a system of major subsystems - the RFQ and DTL - and in work, the BCDTL and CCL. Each is further modeled at its own constituent level. The LINAC RF System is represented as three systems in series: the RFQ RF System, the DTL RF System, and the Global RF Instrumentation and Control which is primarily the phase reference system. Similarly, the High Voltage Power Supply, one per every RF Station in this usage, is represented for RAM purposes to a level of thyristors, transformers, capacitors, reactors, monitors, cooling and AC power distribution as an example. The High Energy Beam Transport can consist of a non-activated part, where the repairs are relatively easy, and the "hot" room, where repairs take longer because of the high radiation environment and the need to wait for the activated components to "cool" down.

Among the RAM module results produced are the average numbers of components that will require replacement over the period of a typical year and the average number of man-hours that will have to be expended for the repairs (corrective maintenance). These values are used in estimating the associated maintenance costs and to estimate the quantities of spares to be stocked as replacements parts. The results produced with the ASM RAM code (by executing it parametrically) can also be used to evaluate the sensitivities of the overall system availability to the variations in individual input failure and repair. Such information is useful in identifying the development items with the greatest payoff for system reliability.

Cost Module

Like the RAM calculation the costing is driven from the parts list. The capital cost of the accelerator and RF system hardware is comprised of both labor and material. A large database is used to store both types of costing information. The material database contains information on the unit cost of the item, and information concerning the quantity at which a lot savings occurs. Since the quantity of parts is determined by the engineering modules a material charge can be immediately estimated. The labor portion of the cost estimate is slightly different. The logic examines the number of tanks in the configuration and determines the total design labor charge. This is comprised of both non-recurring and recurring activities.

To develop fabrication costs, the number of major subassemblies for each type of accelerating structure is coded as well as expected savings from repetitive task "learning curve savings". Each designed accelerator component, as well as the RF system, are developed based on parts, fabrication and assembly, including material and labor. The capital costs also include the enclosures for the accelerator and RF systems. Since the size of the equipment going in the buildings is known, simple costs per unit meter of tunneling, or cubic meter of industrial building is used.

The module develops estimates of the annual expenditures required to operate the facility. The first of these is the AC power. This is coupled to the availability estimates and a yearly expenditure is developed. Thereafter estimates of staffing required to operate the machine, the facility, maintenance, management and indirect staffing are developed. Other aspects of the annual operating budget include allocations for consumables and a budget for the refurbishment and replace of components that wear out. A lifecycle cost is developed by calculating the net present value of each cost category in a consistent time-frame.

CONCLUSION

The ASM code is a highly representative modeling tool for the particle accelerator community enabling in-depth trade studies and a real basis for preliminary and final design efforts. Its strength lies in its modeling flexibility at different levels of detail, its comprehensive library of modeling capability, its consistent approach to an integrated evaluation of a full-up facility and most importantly its integrity in adhering to established physics design principles and requirements.

REFERENCES

1. G. H. Gillespie, B. W. Hill and R. A. Jameson, "A New Tool for Accelerator System Modeling and Analysis," Proc 1994 Inter'l Linac Conf 1, 110-112 (1994)
2. G. H. Gillespie, B. W. Hill and R. A. Jameson, "Progress in the Development of an Accelerator System Model," Proc of International Workshop on Beam Dynamics and Optimization (St. Petersburg), 93-101 (1994).
3. G. H. Gillespie, B. W. Hill and R. A. Jameson, "A New Approach to Modeling Linear Accelerator Systems," AIP Conf Proc 346, 597-603 (1995).
4. A. M. M. Todd, C. C. Paulson, M. A. Peacock and M. F. Reusch, "A Beamline Systems Model for Accelerator-Driven Transmutation Technology (ADTT) Facilities," AIP Conf Proc 346, 604-610 (1995).
5. G. H. Gillespie, "The Shell for Particle Accelerated Related Codes (SPARC) - A Unique Graphical User Interface," AIP Conf Proc 297, 576-583 (1993).
6. C. C. Paulson, G. H. Gillespie, R. A. Jameson, et al,"Accelerator Systems Optimizing Code," Proc of 1995 Particle Accelerator Conf 2, 1164-1166 (1995).
7. D. H. Berwald, et al, "Parametric Study of Emerging High Power Accelerator Applications Using Accelerator Systems Model (ASM)," to be pub in Proc of Eighteenth International Linac Conf 3 pp (1996).
8. G. H. Gillespie, G. E. McMichael, B. J. Micklich and G. R. Imel, "RFQ-Based, Transportable, High-Resolution Neutron Radiography System Concept," to be pub. in Proc of Fifth World Conf on Neutron Radiography, 8 pp (1996).
9. J.H. Billen, F.L. Krawczyk, R.L. Wood, L. Young, "A New RF Structure for Intermediate-Velocity Particles,"Proc 1994 Intern'l Linac Conf 1, 341-345 (1994)

HIGH-PERFORMANCE COMPUTING

PARALLEL BEAM DYNAMICS CALCULATIONS ON HIGH PERFORMANCE COMPUTERS

Robert Ryne* and Salman Habib[†]

*Accelerator Operations and Technology Division and [†]Theoretical Division
Los Alamos National Laboratory, Los Alamos, NM 87545

Abstract. Faced with a backlog of nuclear waste and weapons plutonium, as well as an ever-increasing public concern about safety and environmental issues associated with conventional nuclear reactors, many countries are studying new, accelerator-driven technologies that hold the promise of providing safe and effective solutions to these problems. Proposed projects include accelerator transmutation of waste (ATW), accelerator-based conversion of plutonium (ABC), accelerator-driven energy production (ADEP), and accelerator production of tritium (APT). Also, next-generation spallation neutron sources based on similar technology will play a major role in materials science and biological science research. The design of accelerators for these projects will require a major advance in numerical modeling capability. For example, beam dynamics simulations with approximately 100 million particles will be needed to ensure that extremely stringent beam loss requirements (less than a nanoampere per meter) can be met. Compared with typical present-day modeling using 10,000-100,000 particles, this represents an increase of 3-4 orders of magnitude. High performance computing (HPC) platforms make it possible to perform such large scale simulations, which require 10's of GBytes of memory. They also make it possible to perform smaller simulations in a matter of hours that would require months to run on a single processor workstation. This paper will describe how HPC platforms can be used to perform the numerically intensive beam dynamics simulations required for development of these new accelerator-driven technologies.

Introduction

Many countries are now involved in efforts aimed at developing high power linacs for transmutation of radioactive waste, disposal of plutonium, production of tritium, and as drivers for next-generation spallation neutron sources. For these projects, high-resolution modeling far beyond that which has ever been performed in the accelerator community will be required to reduce cost and technological risk, and to improve accelerator efficiency, performance, and reliability. Such accelerators will have to operate with extremely low beam loss (0.1-1 nA/m) in order to prevent unacceptably high levels of radioactivity. High resolution simulations using on the order of 100 million particles will be needed to help ensure that this requirement can be met. With the advent of HPC platforms such as massively parallel processors (MPPs) and clusters of shared memory processors (SMPs) such simulations are now possible. Near term HPC platforms will have memories of 100's of GBytes and performance of a few TFLOPs. Compared with high-end workstations (500 MFLOPs) and high-end PCs (100 MFLOPs), a 1 TFLOP HPC platform would outperform these by factors of 2000 and 10000, respectively.

Though HPC platforms have great potential, it is a challenge to use them effectively on a wide variety of problems. Some algorithms can only be parallelized with significant effort, while for some other problems it might be necessary to find a different method of solution, involving a different algorithm, that is more amenable to parallel computation. In general, care must be taken to ensure that all processors do roughly the same amount of work (the load balance issue), and that interprocessor communication does not seriously affect performance. Another challenge is to program parallel computers. Message passing libraries are quite mature and portable; also, languages such as High Performance Fortran (HPF) are maturing and becoming more widely available. But software challenges remain, particularly if platforms with memory hierarchies such as clusters of SMPs emerge as the most widespread HPC platforms, as many have predicted.

Approaches to Parallelization

There are two main approaches to using parallel computers. In the data-parallel approach, the style of programming is similar to that used in sequential computing, *i.e.* the programmer takes an essentially global view of the system and uses a single thread of control. This style of programming is well-suited to the Single-Instruction, Multiple-Data (SIMD) and Single-Program, Multiple-Data (SPMD) paradigms, in which all processors run the same program, synchronously or asynchronously, with their own data. Data-parallel programming is implemented using languages similar to traditional ones, but with extensions for explicitly parallel operations and directives. An example is High Performance Fortran (HPF). HPF includes Fortran 90 as a subset, but it also contains new commands such as the `FORALL` statement (which is a parallel `DO`) and compiler directives such as `distribute` (for distributing data across processors). For problems that are "embarrassingly parallel," such as modeling the dynamics of a number of noninteracting particles, the SIMD/SPMD approach is the natural choice. But, as will be seen later, purely data-parallel programming is extremely limiting, and to be useful for many problems it must be augmented with utility libraries and scientific software libraries.

A second approach to parallelization makes use of message passing libraries. In this approach the programmer takes a more local view of the problem, telling processors when to send and receive data, when to synchronize, *etc.* This style of programming accommodates the Multiple-Instruction, Multiple-Data (MIMD) paradigm. Though more difficult to use than the data-parallel approach, MIMD-style programming provides the flexibility needed to deal with problems that are not suited to the data-parallel model. Also, it often results in excellent performance.

It is possible to use a combination of these two approaches. For example, one can call non-HPF routines from inside an HPF program using a feature called HPF_EXTRINSIC. This style of parallel programming could become popular in the future since it allows people with traditional programming experience to work in a familiar environment that also allows them to call message-passing-based software, written by themselves or others, to deal with those parts of a calculation that are not amenable to data-parallel programming.

In the remainder of this paper we begin by presenting two simple examples. The first describes a way to run multiple copies of a single particle beam transport code on a parallel computer, and the second shows a data-parallel implementation of a drift subroutine. After these two examples, we discuss the need to perform internal data processing prior to writing the results of large scale parallel simulations. The remaining sections describe progress in developing Particle-In-Cell and Vlasov-Poisson codes for modeling intense beams on parallel computers.

Single Particle Tracking to Predict Dynamic Aperture

Dynamic aperture is an important consideration for large hadron colliders. One way to predict this is to separately track hundreds to thousands of particles for several million turns (assuming the particles do not interact with one another), and see whether they survive or strike the beam pipe during the simulation. This can be done by running a single-particle tracking code on several independent computers, or by using a parallel computer. In the latter case, an obvious approach would be to use message passing. A second, more tedious approach, applicable to codes that track many (non-interacting) particles simultaneously, would be to use data-parallel programming, replacing DO loops over particles in all the subroutines with FORALL statements. Within the confines of HPF a third approach is also possible that is extremely easy to implement and requires no knowledge of message passing. Consider a code called MYTRACK that tracks a single particle. One could write a main program using HPF that calls MYTRACK(V) as a subroutine, where V denotes the 6-vector of coordinates and momenta of the particle to be tracked. In addition to array declarations, compiler directives, etc., the main program would contain the following loop:

```
!HPF$ INDEPENDENT
      do N=1,NPARTICLES
      call MYTRACK(V(I))
      end do
```

The INDEPENDENT directive states that all iterations of the DO loop are independent of one another. The main program would also require an INTERFACE block specifying that MYTRACK is a "pure" subroutine, i.e. one having no side-effects.

Data-Parallel Implementation of a Drift Subroutine

Consider a subroutine to propagate the coordinates and momenta, of 100 million particles in a drift space. A crucial consideration, as with any parallel program, is the distribution of the data across processors. In this case the coordinates and momenta of a given particle should be on the same processor so that the above transformation requires no interprocessor communication. Suppose the data is stored in an array dimensioned a(6,100000000). The HPF compiler directive `distribute a(*,block)` would cause quantities specified by the first dimension to reside on the same processor, as desired, while the quantities specified by the second dimension would be stored in "blocks" across processors. For example, if 4 processors were in use, then the six coordinates of particles 1-25000000 would reside on processor 1; particles 25000001-50000000 would reside on processor 2; and so on. See Table 1.

Table 1: Layout of a two dimensional array distributed (*,block)

Processor 1	Processor 2	Processor 3	Processor 4
a(1-6,1)	a(1-6,25000001)	a(1-6,50000001)	a(1-6,75000001)
a(1-6,2)	a(1-6,25000002)	a(1-6,50000002)	a(1-6,75000002)
a(1-6,3)	a(1-6,25000003)	a(1-6,50000003)	a(1-6,75000003)
⋮	⋮	⋮	⋮
a(1-6,24999999)	a(1-6,49999999)	a(1-6,74999999)	a(1-6,99999999)
a(1-6,25000000)	a(1-6,50000000)	a(1-6,75000000)	a(1-6,100000000)

The following HPF subroutine implements the drift transformation:

```
      subroutine drift(a,b,t)
c a(1-6,n) = initial (x,px,y,py,z,pz) of nth particle
c b(1-6,n) =   final (x,px,y,py,z,pz) of nth particle
      real a,b,t
      dimension a(6,100000000),b(6,100000000)
!hpf$ distribute a(*,block)
!hpf$ distribute b(*,block)
      forall(i=1:100000000)b(1,i)=a(1,i)+a(2,i)*t
      forall(i=1:100000000)b(2,i)=a(2,i)
      forall(i=1:100000000)b(3,i)=a(3,i)+a(4,i)*t
      forall(i=1:100000000)b(4,i)=a(4,i)
      forall(i=1:100000000)b(5,i)=a(5,i)+a(6,i)*t
      forall(i=1:100000000)b(6,i)=a(6,i)
      return
      end
```

In the above subroutine, the `FORALL` statements could be replaced with array syntax. For example, the first exectutable statement could be written as

```
      b(1,:)=a(1,:)+a(2,:)*t
```

If the arrays had a seventh component to indicate whether or not a particle was lost, a masked `FORALL` statement could be used to transform only the desired particles:

```
      forall(i=1:100000000,a(7,i).eq.0.)b(1,i)=a(1,i)+a(2,i)*t
```

Internal Processing of Large Data Sets

Parallel calculations are of no value unless there is a way to make the resulting data available after the simulation is finished. But considering the size of arrays in large scale simulations, this is not a trivial issue. Consider, for example, a beam dynamics simulation with 100 million particles. A six-dimensional, double precision array would require 4.8 GB of memory, and writing out the data at a few hundred locations in the beamline would require TBytes of disk space. Furthermore, in many situations it is insufficient to simply write out a randomly chosen fraction of the data. Thus, it is often essential to do internal data processing in large scale parallel simulations. The following example describes a situation where the parallel data processing is more difficult to implement than the physics calculation itself.

Consider a simulation whose goal is to illustrate the dynamics of a chaotic system. The system is a one-dimensional double well potential with a sinusoidal drive. To illustrate the dynamics, 100 million particles are initially distributed around a ring of radius $r = 0.01$ in x-p_x space, and the system is allowed to evolve for roughly 12 drive periods. (For this simulation the use of 100 million particles is not excessive, since adjacent particles have trajectories that diverge exponentially due to the presence of chaos.) By the end of the simulation the ring has been repeatedly stretched and folded into the complicated shape shown in Figure 1.

This is another example of an "embarrassingly parallel" problem, since all 100 million particles evolve independently. But the coding to output the data for the figure is complicated (more complicated than the code to integrate the trajectories) and not easily parallelizable. It is obviously wasteful and unnecessary to print all 100 million points, which would require 1.6 GB. Instead, enough points are plotted so that, when connected by straight lines, the curve is smooth. To do this, the tangent vector of the curve is first computed at all 100 million points, and the angle between tangent vectors, θ, is computed. Next, a cumulative angle array, Θ, is computed, whose elements are given by $\Theta_n = \sum_{i=1}^{n} \theta_i$. This is done using a utility routine called SCAN_ADD, a type of parallel prefix operation generally included in parallel software libraries. Next, points are marked for printing based on the following criterion: Having chosen a point in the array to print, the next point to be printed is the next one in the array where Θ has changed by some threshold value (1 degree in this example). This is done using a SCAN_COPY routine. Finally, the points cannot simply be printed as follows,

```
      do 100 i=1,100000000
100   <if the ith point is marked for printing, print it>
```

since the DO loop is a scalar operation that would run far more slowly than anything else in the code. Instead, the points to be printed are moved to the start of a temporary array by ordering them using a mask of zeroes and ones as keys to indicate which points are marked or not marked for printing, respectively. The data is printed as shown,

```
      do 100 i=1,ntot
100   <print the ith point of the temporary array>
```

where ntot is the total number of points to be printed. Figure 1 contains approximately 62000 points, corresponding to a data reduction factor of 1600.

This example illustrates the fact that, in the data-parallel environment, it is crucial to have a library of utility routines (in this case SCANs and ORDER), to perform those operations that, by their nature, are not data-parallel. It also illustrates the fact that, though it will never be possible to take an arbitrary scalar code and have it converted automatically to an efficient parallel code, it would be extremely useful to have this capability for small segments of code.

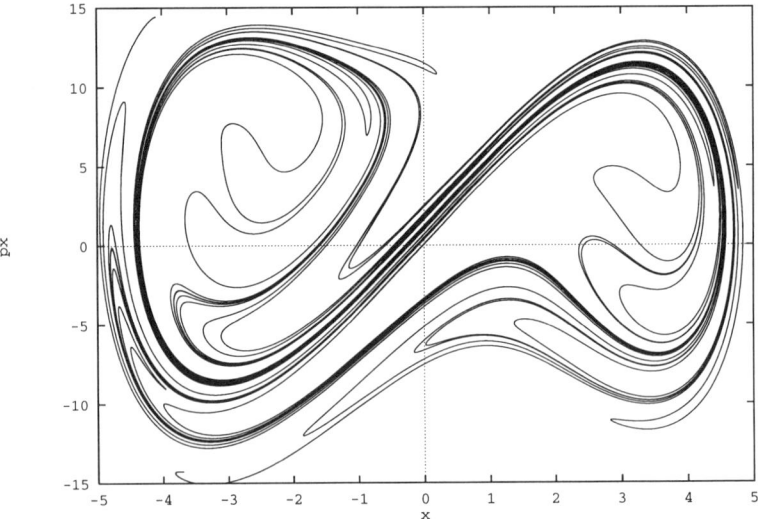

Figure 1: The final configuration of an initial circle of points of radius 0.01 after it has evolved in a driven, double well system. The plot contains approximately 62000 points out of 100 million that were initially distributed around the circle.

Particle-In-Cell Simulations

MPPs have had a significant impact in modeling beam halo in moderate-to-high average power linacs for the accelerator-driven technologies mentioned previously. Using resources provided by the Los Alamos Advanced Computing Laboratory, parallel beam dynamics codes have been written using CM Fortran (a precursor to HPF) that run on a 1024-node Thinking Machines CM5 computer. The codes use standard Particle-In-Cell (PIC) techniques, and the main difficulty in parallelization is related to the space charge routines, as described later.

Parallel Version of PARMILA

In addition to writing new codes, a parallel version of the beam dynamics code PARMILA (originally written in Fortran 77) has also been developed. To do this, a number of steps were required. First, all DO loops over large arrays, such as the particle array, were replaced with FORALL loops. This required rewriting large sections of the serial code between DO/ENDO statements, replacing scalar temporaries with parallel temporary arrays. Complications such as testing for lost particles inside of FORALL statements could be easily dealt with by using masked FORALLs. A more complicated situation arose when tests inside loops affected the program flow. Consider, for example, the serial code used to generate a 4D waterbag distribution:

```
      do 100 i=1,nptcls <loop over particles>
50    <generate 4 random numbers x1,x2,x3,x4>
      if(x1**2+x2**2+x3**2+x4**2.gt.1)goto 50
      <generate coordinates and momenta for this particle>
100   continue
```

This had to be replaced with code of the following form:

```
100   <generate four LARGE arrays x1,x2,x3,x4>
      <mask off if x1**2+x2**2+x3**2+x4**2.gt.1>
      <pack good data into final array>
      <if final array is not complete, goto 100>
      <generate coordinates and momenta for all particles>
```

This exemplifies another situation where a utility routine (namely PACK) is essential to perform an operation that cannot be performed within the confines of data-parallel programming.

Besides rewriting large sections of code associated with DO loops, some other simple tasks were required to port PARMILA. As mentioned above, it is necessary to insert compiler directives in subroutines to specify the layout of parallel arrays. Another task was related to subroutine calls and data reshaping. In data-parallel languages, parallel arrays cannot be reshaped through subroutine calls as they can in Fortran 90. Thus, a 2D array coord(6,ntot) cannot be used as in call mysub(coord(1)) and treated like a 1D array in subroutine mysub. Also, a 1D array x(ntot) cannot be used as in call mysub(x(ntot/2)) and treated as a 1D array of half the original length in the subroutine. These situations are straightforward to deal with, but it can be tedious to find all such occurrences and they can easily go unnoticed until the program fails to execute properly.

The major difficulty in developing a parallel version of PARMILA, as with any parallel PIC code, is the space charge calculation. This is discussed in the next.

Space Charge Calculation

In the PIC approach, the self-fields are computed in three steps: Charge deposition on a grid, field solution on the grid, and field interpolation at the particle positions.

These steps are not easily parallelizable. Consider, for example, charge deposition on a two-dimensional grid using area weighting. A serial routine would look like the following:

```
      do 100 n=1,np
      i=(x(n)-xmin)/hx
      j=(y(n)-ymin)/hy
      ab=xmin-x+i*hx
      cd=ymin-y+j*hy
      rho(i,j)=rho(i,j) + ab*cd
      rho(i+1,j)=rho(i+1,j)+cd*(hx-ab)
      rho(i,j+1)=rho(i,j+1)+ab*(hy-cd)
100   rho(i+1,j+1)=rho(i+1,j+1)+(hx-ab)*(hy-cd)
```

The equivalent parallel routine is shown below:

```
i=(x-xmin)/hx    ! i,j,x,y,ab,cd = arrays
j=(y-ymin)/hy    ! hx,hy,xmin,ymin = scalars
ab=xmin-x+i*hx
cd=ymin-y+j*hy
forall(n=1:np)rho(i(n),j(n))=rho(i(n),j(n))+ab(n)*cd(n)
forall(n=1:np)rho(i(n)+1,j(n))=rho(i(n)+1,j(n))+cd(n)*(hx-ab(n))
forall(n=1:np)rho(i(n),j(n)+1)=rho(i(n),j(n)+1)+ab(n)*(hy-cd(n))
forall(n=1:np)rho(i(n)+1,j(n)+1)=rho(i(n)+1,j(n)+1)+(hx-ab(n))*(hy-cd(n))
```

Unfortunately the above parallel routine has poor performance. First, the FORALL statements cause significant interprocessor communication. Second, if the density array rho is uniformly spread across processors, then the routine will not be load balanced. For example, if one deposited a Gaussian charge distribution on the grid, then processors associated with the tail of the distribution would finish accumulating charge sooner than processors associated with the core.

The performance of this algorithm can be improved using several approaches. One can use optimized utility routines that send data to processors based on index arrays to perform binary operations on the data (*e.g.* add, overwrite, min, max). This is easy to implement but the performance improvement is modest. Another approach is to first reorder the particles so that all those corresponding to the same grid point are contiguous, then use parallel prefix routines to efficiently perform the charge deposition. This approach is more difficult to implement than the one mentioned previously, but the performance improvement is significant. It has been used by Ferrell and Bertschinger in an N-body code for astrophysical simulations [1]. A third method for improving the performance uses routines written with message passing libraries. In this approach the programmer explicitly writes the code that includes logic to determine how to partition the data so that the load is balanced. Though this approach is the most difficult to implement, it yields the best performance improvement. This approach was developed within the plasma simulation community, and has been used as part of the Numerical Tokamak Project, a DOE-funded High Performance Computing and Communications project [2][3].

Besides charge deposition and field interpolation, the space charge calculation requires the solution of the field equations on the grid. The parallel beam dynamics codes developed at the Advanced Computing Laboratory use an FFT-based approach to convolve the charge density on the grid with a Green function defined on the grid. Using standard techniques it is possible to treat a bunch of charge in the presence of open boundary conditions [4]. We have also implemented a procedure that uses open boundary conditions transversely and periodic boundary conditions longitudinally.

Performance

We have used our parallel beam dynamics codes to perform linac simulations with 1-30 million particles. For example, the parallel version of PARMILA has been used to model a superconducting linac for the Accelerator Production of Tritium project. The code contains a new 3D space charge algorithm based on the method of Ferrell and Bertschinger as described previously. A 10 million particle simulation on a 128^3 grid, involving 900 space charge calculations, required 4.4 hours and 6.1 GBytes on the 256 node partition of the CM5. A sample output of the horizontal phase space for an initially mismatched beam shown in Figure 2. Symbols associated with the particles show the density in phase space, normalized to a maximum value of one.

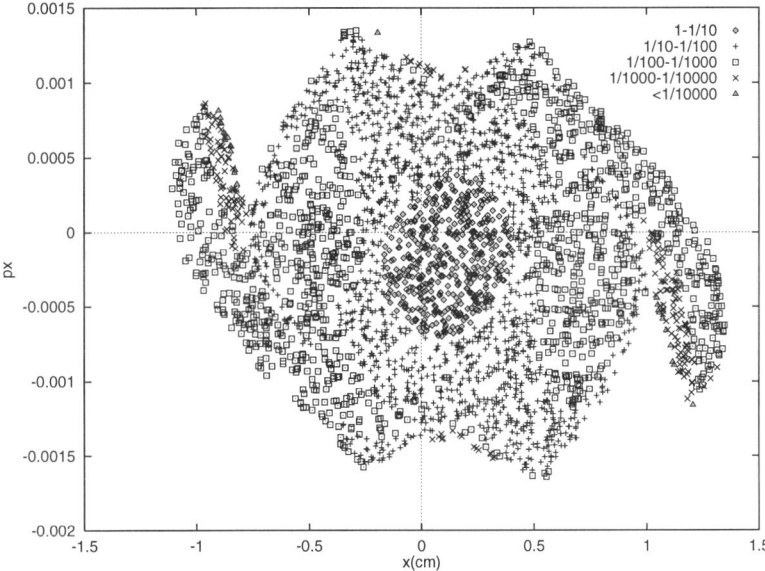

Figure 2: Results from a parallel version of PARMILA run on the CM5 at the Los Alamos Advanced Computing Laboratory. The simulation used 10 million particles.

The success of the parallel approach depends on scalability, i.e., the ability to run larger problems in the same amount of time using more processors, or the ability to run problems of a fixed size in less time using more processors. [1] The parallel version of PARMILA has excellent scalability as shown in Table 2:

Table 2: Scaling Results (3.75M particles, 64x64x64 grid)

Procs	CPU (min)	MEM (GB)
128	15.5	1.8
256	8.1	2.0
512	4.3	2.5

Direct Vlasov-Poisson Simulations

Direct Vlasov-Poisson solvers provide an alternative to PIC methods. In a direct solver, the beam distribution function is defined on a grid in phase space. Since no particles are needed there is no sampling noise in determining the distribution function at low densities. One difficulty with this approach is the enormous amount of memory required: a 3D simulation (with a 6D phase space) that uses 64 mesh points per dimension has a total of 68 *billion* mesh points.

In this approach one attempts a direct solution of the Vlasov-Poisson equations,

$$\frac{\partial f}{\partial t} + (\vec{p} \cdot \partial_{\vec{x}})f - (\nabla V \cdot \partial_{\vec{p}})f = 0, \qquad (1)$$

where $f(\zeta, t)$ is a distribution function on phase space ($\zeta = (\vec{x}, \vec{p})$). Here, the potential V is a sum of an externally applied potential and a space charge potential which is obtained self-consistently from Poisson's equation. A second-order accurate stepping algorithm for this system is given by

$$f(\zeta, t) = \mathcal{M}(t) f(\zeta, t = 0). \qquad (2)$$

where the mapping $\mathcal{M}$ is given by

$$\mathcal{M}(t) = e^{-\frac{t}{2}(\vec{p} \cdot \partial_{\vec{x}})} e^{t(\nabla V \cdot \partial_{\vec{p}})} e^{-\frac{t}{2}(\vec{p} \cdot \partial_{\vec{x}})}. \qquad (3)$$

This is analogous to the leap-frog algorithm used to advance particles in PIC codes, since the operators involving drifts and space charge "kicks" are done separately. In a split-operator spectral code, the mappings generated by these operators are evaluated using Fourier transforms. By transforming variables appropriately, the exponential operators turn into multiplications in Fourier space. It is worth noting

---

[1] Increasing the number of processors while keeping the problem size fixed cannot cause the execution time to decrease indefinitely since, if the problem size is too small, the processors will do too little calculation, and the execution time will be dominated by communication.

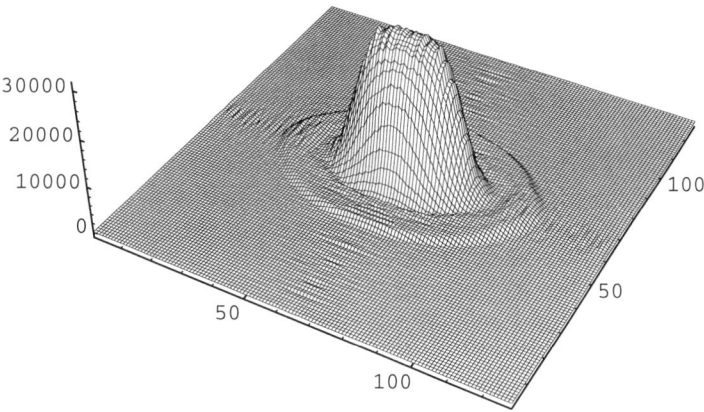

Figure 3: Charge density from a direct Vlasov-Poisson simulation with 268 million grid points in a 4-dimensional phase space.

that, using a method due to Yoshida, one can construct fourth order (and higher order) stepping algorithms [5].

Figure 3 shows the results of a 2D beam halo simulation using a direct Vlasov-Poisson solver. The figure shows the beam density, $\rho(x,y)$, of an initially mismatched Gaussian beam after 22 periods in a quadrupole channel. The simulation, which was performed on the CM5 at the Advanced Computing Laboratory, used a 128 mesh points in each phase space dimension, for a total of 268 million mesh points.

Conclusions

HPC platforms make it possible to perform the numerically intensive calculations needed to model future accelerators for accelerator-driven transmutation technologies and applications. As discussed in this paper, simulations of intense beams with up to 30 million particles are already being performed, and with the arrival of the next-generation of HPC platforms, simulations with 100 million particles are expected in the 1997-1998 time frame. Though the development of parallel beam dynamics codes would be a daunting task if done from scratch, the accelerator community is fortunate to be able to build on progress already made by researchers in the computational plasma physics and cosmology communities.

Though we have emphasized modeling intense charged particle beams, HPC platforms will have many other uses in the accelerator community. As mentioned in this paper, parallel tracking calculations to predict dynamic aperture in large circular machines can be done with ease, either through message passing or within the data-parallel framework of HPF. Also, HPC platforms will have a major impact in modeling complicated 3D electromagnetic structures with high accuracy. Parallel, 2D finite element codes have already been developed for this purpose, and development of 3D codes is underway [6].

In summary, HPC platforms will be needed to solve a variety of important simulation problems within the accelerator community. By the end of the decade we can expect to see many more accelerator physicists doing production computing on these platforms, thanks to maturing programming languages such as HPF, the availability of parallel scientific software libraries, and the availability of cycles at locations such as the Los Alamos Advanced Computing Laboratory and the National Energy Research Scientific Computing Center at Lawrence Berkeley National Laboratory.

Acknowledgements

The authors thank consultants of the National Energy Research Scientific Computing Center, including Youngbae Kim, for suggesting the use of the INDEPENDENT directive to run multiple copies of single particle tracking codes. This research was supported by the U.S. Department of Energy, Office of Energy Research, through the Division of Mathematical, Information, and Computational Sciences, the Division of High Energy and Nuclear Physics, and by the Office of Defense Programs, Accelerator Production of Tritium program. This research was performed in part using the resources located at the Advanced Computing Laboratory of Los Alamos National Laboratory, Los Alamos, NM 87545.

References

[1] R. Ferrell and E. Bertschinger, *Int. J. Mod. Phys. C* **5**, (1994) 933-956.

[2] V. K. Decyk, *Computer Physics Communications* **87** (1995), 87-94.

[3] J. Wang, P. Liewer, and V. Decyk, *Computer Physics Communications* **87** (1995), 35-53.

[4] R. W. Hockney and J. W. Eastwood, *Computer Simulation Using Particles*, (Adam Hilger, New York, 1988).

[5] H. Yoshida, *Phys. Lett. A* **150**, 262 (1990); Also, see E. Forest *et al*, *Phys. Lett. A* **158**, 99 (1991).

[6] X. Zhan and K. Ko, these proceedings.

Parallel Computation of Transverse Wakes in Linear Colliders [1]

Xiaowei Zhan and Kwok Ko[†]
SCCM Program, Stanford University, Stanford, CA 94305
[†] ATSP Department, Stanford Linear Accelerator Center, Stanford, CA 94309

Abstract

SLAC has proposed the detuned structure (DS) as one possible design to control the emittance growth of long bunch trains due to transverse wakefields in the Next Linear Collider (NLC). The DS consists of 206 cells with tapering from cell to cell of the order of few microns to provide Gaussian detuning of the dipole modes. The decoherence of these modes leads to two orders of magnitude reduction in wakefield experienced by the trailing bunch. To model such a large heterogeneous structure realistically is impractical with finite-difference codes using structured grids. We have calculated the wakefield in the DS on a parallel computer with a finite-element code using an unstructured grid. The parallel implementation issues are presented along with simulation results that include contributions from higher dipole bands and wall dissipation.

1. Introduction

In the Next Linear Collider (NLC) proposed by SLAC the detuned structure (DS) is being considered as a viable design to control the emittance growth of long bunch trains due to transverse wakefield effects in the linacs [1]. An off-axis beam transiting through the accelerator structure excites dipole modes which impart transverse kicks on subsequent bunches. This dipole wakefield can be greatly reduced if the modes are decohered by suitably tapering the cell dimensions to provide Gaussian detuning of their frequencies, while keeping the accelerating mode in synchronism with the beam.

At X Band, the NLC DS is a 1.8m long disk-loaded waveguide (DLWG) consisting of 206 cells each of which has the geometry shown in Fig. 1(a). The unit cell is formed between disks whose aperture a and thickness t vary along the structure according to the distributions shown in Fig. 1(b). The tapering profile for a leads to the Gaussian detuning of the 1st dipole band while that for t results in detuning of higher bands (3rd and 6th). The cell radius b allows the accelerating mode to be tuned to 11.424 GHz at 120 degree phase advance across the cell length L.

2. Transverse Wakefield Analysis

The wakefield at distances of the order of the bunch spacing is mainly determined by the resonant modes of the structure which are responsible for the long range effect. Hence,

---

[1] This work was supported by the U.S. Department of Energy, under contract No. DE-AC03-76SF00515.

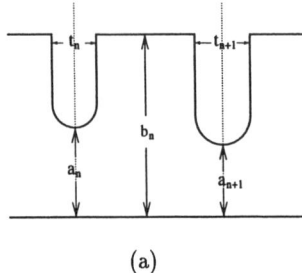
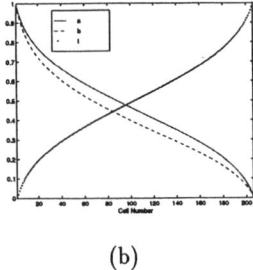

(a) (b)

Figure 1. (a) *The dimensions of a typical cell;* (b) *Variations of cell dimensions.*

the transverse wakefield at distance s behind the driving bunch is best calculated in the frequency domain using

$$W(s) = \frac{2}{206} \sum_{i=1}^{N} K_i \sin(\frac{\omega_i s}{c})$$

where ω_i, K_i are the frequency and the kick factor of mode i respectively. N is number of modes excited by the bunch and here N is large since there are 206 modes per dipole band. The kick factor is determined from the voltage

$$V_i = \int_{z_{min}}^{z_{max}} E_z(r_o, z) \exp\left(i\frac{\omega_i}{c} z\right) dz$$

and the stored energy

$$U_i = \frac{1}{2}\epsilon \int |E|^2 \, dV + \frac{1}{2\mu} \int |B|^2 \, dV$$

via the expression

$$K_i = \frac{|V_i|^2}{4(\omega_i/c) U_i \, r_o^2 \, L}$$

To lowest order, the DS can be considered as a closed axisymmetric cavity if the end cell effects due to the power input/output coupler and the entry/exit beampipes are ignored. Although the problem is now reduced to 2D, the challenge remains as to how one can solve for a large number of modes in a long heterogeneous structure efficiently and accurately.

Many theoretical efforts have been devoted to the assessment of the effectiveness of the DS in suppressing wakefields. The calculations to date have all been carried out with approximate methods primarily because of the size of the problem. They include equivalent circuits [2], mode matching [3], and the open-mode approach [4]. Also, a simplified geometry is assumed by treating the ends of the disks as flat than round. Detailed accounts of these analysis can be found in the literature, and the results are generally found to be in qualitative agreement with experimental data.

3. Grid-based Simulation

It is of interest to find the wakefield from a direct simulation of the entire DS on a numerical grid. Such an approach till now has not been possible for the following reasons. Most standard field solvers, based on finite-difference (FD) schemes, use structured grids which are not efficient for modeling gradual tapers [5]. To fit the tapering in a and b alone would require two mesh lines per cell for a total of 512. Given the cell to cell variation in a and b is few microns while the cell radius is about 1 cm, the number of mesh points in the radial direction can be in the thousands. It suggests that if even a moderate resolution is adopted in the axial direction, to model the whole length of the structure would require prohibitively large number of mesh points (estimated to be tens of millions). This is not to mention that structured grids are not suited to model the rounded ends of the disks. One concludes from these considerations that grid-based simulations with FD codes are not feasible.

The finite-element (FE) method has been shown to be highly accurate and efficient in modeling accelerator structures with the use of unstructured grids [6]. The advantage of the unstructured grid is the ability of the mesh to fit small differences in geometry, and to refine locally around curvatures without significant increase in the total number of elements or degrees of freedom (DOF). The number of elements can be further reduced without loss in accuracy through the use of quadratic elemnents. Even with such improvements, however, the memory and CPU time required for a reasonable FE mesh is still far being practical to compute on any high-end workstation. It becomes apparent that this limitation can only be overcome by way of parallel processing.

4. Domain Decomposition and Parallel Implementation

The DS geometry is highly amenable to domain decomposition for parallel processing because the partitioning is straighforward. Each processing node is assigned a cell bounded by half-disks (subdomain) as shown previously in Fig. 1. A total of 207 nodes is used for the DS (domain) since the two ends are terminated in half-cells. In the single program multiple data (SPMD) model, global data such as meshes and matrices are distributed to the nodes where global operations can be processed locally. The communication between nodes is performed through message passing interface (MPI) which, in the present case, applies solely to the data along the radial boundaries that separate a particular cell from its neighbours.

The finite element formulation of the 2D Maxwell Eqs. results in a generalized eigenvalue problem, $Kx = \lambda Mx$ where K is the stiffness matrix while M is the mass matrix. Both are large, sparse and symmetric. Eq. 5 can be solved using the Lanczos algorithm which has superior convergence properties, and is ideal for computing extremal eigenvalues of large problems [7]. In each Lanczos iteration, one solves the linear system $\tilde{K}x = b$ where the coefficient matrix $\tilde{K} = K - \sigma M$. σ is the shift introduced to accelerate convergence by better separating the eigenvalues.

To solve the linear system effectively on the parallel computer, one takes advantage of domain decomposition by separating the DOFs in the domain into "interior" DOFs that belong inside a subdomain, and those that reside on the boundaries bewteen subdomains, the "interface" DOFs. Since there is no coupling between the "interior" DOFs from different subdomains, and the "interface" DOFs are far less in number, $\tilde{K}$ can be reordered in the form

$$\tilde{K} = \begin{pmatrix} A & C^T \\ C & F \end{pmatrix},$$

in which the "interface" DOFs are ordered after the "interior" ones. Now A is a block diagonal matrix with A_i being the submatrix from the ith subdomain, F contains only the "interface" DOFs, and C supplies the coupling between the two groups.

The next step is to apply the direct method and factorize $\tilde{K}$ using the Crout decomposition

$$\tilde{K} = \begin{pmatrix} L_A & 0 \\ CL_A^{-T}D_A^{-1} & L_G \end{pmatrix} \begin{pmatrix} D_A & 0 \\ 0 & D_G \end{pmatrix} \begin{pmatrix} L_A^T & D_A^{-1}L_A^{-1}C^T \\ 0 & L_G^T \end{pmatrix}$$

where $A = L_A D_A L_A^T, G = L_G D_G L_G^T$ are Crout decompositions of A, G respectively, and $G = F - CA^{-1}C^T$. Remembering A is block diagonal the decomposition of its submatrices can be done in parallel at each individual node. Because it constitutes the major block of $\tilde{K}$, this parallelization results in substantial gain in computation efficiency.

5. Adaptive Mesh Refinement

In finite element analysis, the accuracy of the solution improves as the mesh is refined. The straightforward refinement is to subdivide each triangular element over the whole mesh so the increase in DOFs scales with the number of subdivisions. It is computationally more cost effective to refine adaptively to keep the number of extraneous DOFs to a minimum. The criteria being used to determine if refinement is necessary is based on the local gradient in the stored energy. The refinement stops when a specified tolerance value is reached. This error indicator turns out to be the most effective in generating the optimal mesh for high accuracy in a given solution. Figs. 2 show an initial mesh, a uniformly refined mesh, and an adaptively refined mesh. The two refined meshes yield the same accuracy but the adaptively refined mesh has two times less the number of DOFs, thereby producing a significant savings in memory requirement and run time.

6. Simulation Results for NLC Detuned Structure

A new FE field solver using quadratic, mixed elements, and incorporating both parallelization and adaptive refinement features, has been running on the Intel's Paragon XP/S 150

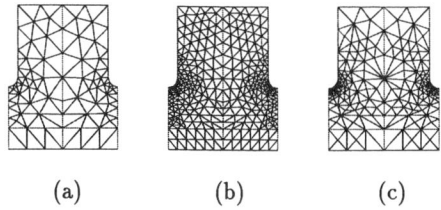

Figure 2. (a) *Initial mesh*; (b) *Uniform refinement*; (c) *Adaptive refinement.*

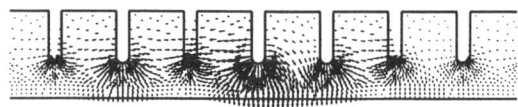

Figure 3. *Electric field partern (from cell 163 to 170) of mode 166 .*

at Oak Ridge National Laboratory. This massively parallel computer has 1,024 node with three 75-MFLOPS i860 XP processors and 64 MBytes memory per node. The code is written in C++ and is compatible with any mesh generator. For the NLC DS simulation, the mesh data is generated with MODULEF. Fig. 3 shows teh 166th mode in the first dipole band obtained from the simulation Fig. 4 compares the wakefield with the results from equivalent circuits and measurement including wall loss. The simulation data contains the contributions from 11 dipole bands for a total of 2266 modes versus 512 modes from the equivalent circuit analysis. The total CPU time consumed is about 15 hours which suggests that parallel computation can be practical for design and analysis.

7. Conclusion

A new 2D finite-element field solver is implemented on a parallel computer with adaptive mesh refinement. This enables the first-ever calculation of dipole modes in tapered accelerator structures using realistic geometries. The simulation results for the NLC DS include the contributions from higher bands and wall dissipation, and show good quantitative agreement with measured data for the wakefield. In terms of parallel computing efficiency the speedup with node increase is close to linear. The evaluation of the end-cell effects due to the input/output couplers and beampipes is in progress.

Acknowledgments

We are grateful to Drs. Cho Ng, Zenghai Li and Vinay Srivinas for their generous help and fruitful discussions. We also thank Professors Roger Miller and Gene Golub for their

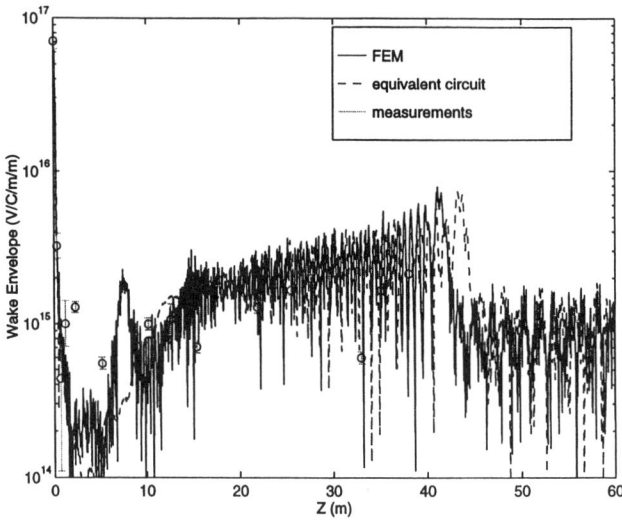

Figure 4. *Wake envelopes obtained from FEM and equivalent circuit calculations.*

valuable insights of the problem.

References

[1] The NLC Design Group. Zeroth-order design report for the next linear collider. *LBNL-PUB-5424, SLAC Report*, 474, UCRL-ID-124161, 1996.

[2] K. L. F. Bane and R. L. Gluckstern. The transverse wakefields of a detuned x-band accelerator structure. *SLAC-PUB-5783*, 1992.

[3] S. A. Heifets and S. A. Kheifets. Coupling impedance for modern accelerators. *AIP*, 249:151–235, 1992.

[4] M. Yamamoto. Study of long-range wake field in accelerating structure of linac. *KEK Report*, 94-9, 1995.

[5] The Mafia Collaboration. Mafia - the ecad system.

[6] Eric Nelson. *SLAC Report*, 92-16, 1992.

[7] Y. Saad. On the rates of convergence of the lanczos and the block lanczos methods. *SIAM Journal of Numerical Analysis*, 17:687–706, 1980.

Recent Fortran 90 developments in 3D electric fields calculations and applications related to the Spiral project at GANIL

Patrick Bertrand

GANIL, SPIRAL group,
PB 5027   14021 Caen Cedex   France

Abstract. In the frame of the new radioactive ion beam facility under construction at GANIL (the SPIRAL project), a new set of Fortran 90 codes has been developed. Among them, CHA3D is a Laplacian solver, which calculates 3D electric potentials using the 7 points finite difference scheme and the conjugate gradient algorithm in a vectorial-parallel version. It is interfaced with the EUCLID CAD software, so that the complex electrode geometries defined with it are automatically introduced, through Bezier surfaces. It is also interfaced with the AVS software allowing the visualization of the potential result together with the electrode shapes.

INTRODUCTION

The SPIRAL project, officially approved in October 1993, has been under construction since 1994 and will be achieved in 1998 [1]. This GANIL project is a new radioactive ion beam facility where the high energy primary beams will produce a wide variety of exotic elements in heated targets. These elements will be ionized using an ECR source [2], extracted, purified, injected in a new cyclotron called CIME, and finally conducted towards experimental areas.

In this paper, we focus our attention on the Laplacian solver CHA3D, which is now widely used in the SPIRAL project.

DESCRIPTION OF CHA3D

We want to calculate various 3D electric fields, solving the Laplace or Poisson equation :

$$-\Delta V = 0 \text{ (or } \rho/\varepsilon_0 \text{)} \quad \text{in } \Omega,$$
$$V = V_i \quad \text{on } \Gamma_i, \, i = 1, m_i,$$
$$\frac{\partial V}{\partial n} = 0 \quad \text{on } \Gamma_i, \, i = m_i + 1, m$$

where Ω is a bounded domain of R^3 and Γ_i surfaces corresponding to pieces of electrodes, or pieces of planes giving the symmetries of the problem.

To construct a parallel solver where the Dirichlet boundary conditions imposed by the electrodes are easy to introduce, we have made the following basic choices :

1. Use of Fortran 90 [3], which gives a natural way to "think parallel".
2. 7 points finite difference scheme in a very thin regular mesh.
3. Conjugate gradient algorithm allowing a good parallel implementation.
4. Description of each electrode with a set of parametrized surfaces which can be either user mathematical functions, or Bezier surfaces coming from the EUCLID CAD software.
5. Interface with the AVS software to visualize the electrodes and the results.

© 1997 American Institute of Physics

Finite Difference Scheme

To discretize the problem, we choose a rectangular 3D box Ω_h including Ω such that the symetry planes correspond to the 6 faces of the box, and a very thin mesh of size (mx+2, my+2, mz+2) with a constant step (dx, dy, dz) in each direction. The finite difference scheme on this regular mesh leads up to the linear system Hu = b with 7 non-zero diagonals in the sparse matrix.

In order to keep the natural parallelism of the algorithm, we declare these diagonals, the second member, the solution (and intermediate arrays) with 3D profiles as shown below, with initial values put as if the nodes were all free :

```
real,    dimension (0:mx+1, 0:my+1, 0:mz+1) :: hc, hx, hy, hz, u, b
logical, dimension (0:mx+1, 0:my+1, 0:mz+1) :: imposed
hc  = 2.*(1./dx/dx + 1./dy/dy + 1./dz/dz)
hx  = -1./dx/dx ;    hy = -1./dy/dy ;    hz = -1./dz/dz
imposed = .false.
```

Boundary Conditions Using User Functions

Concerning the Neumann symetry conditions, the array elements corresponding to the faces, edges and corners of the domain must be divided by 2, 4 and 8 in order to preserve the symetric definite positive property of H.

Each electrode is generally composed of several pieces of planes, cylinders, spheres or more sophisticated surfaces. Each of them can be defined in a common way with 3 specific formulae giving x, y and z depending upon two parameters s and t varying between 0 and 1. Once defined, these pieces of surfaces can be projected in the 3D mesh to block the nodes as Dirichlet conditions, by means of a very fine discretization of s and t. Here is an example with a simple half cylinder :

```
real,    dimension (n,n) :: s, t, x, y, z, indx, indy, indz
do i=1,n
  s(i,1:n) = (i-1.) / (n-1.)
  t(1:n,i) = (i-1.) / (n-1.)
enddo
x = r*cos(pi*t) ;    y = r*sin(pi*t) ;    z = h*s
indx = nint((x - xmin)/dx)                  ! same for indy indz
do i=1,n
do j=1,n
  u(indx(i,j), indy(i,j), indz(i,j))         = potential_value
  imposed(indx(i,j), indy(i,j), indz(i,j))   = .true.
enddo
enddo
```

The discretization of the parameters s and t, given by the dimension n, must be thinner than that of Ω_h to be sure that the projected surfaces generate connected points in the mesh, without hole. Then we update the matrices H and b to take into account the blocked nodes by :

```
where(imposed)
  hc = 0. ;    hx = 0. ;    hy = 0. ;    hz = 0. ;    b = 0.
endwhere
```

To avoid coding explicitly the definition of each piece of surface, we have written a routine containing the most frequently used shapes so that only the parameters are given in a file (axis, positions, angles, etc...). These electrode definitions by means of user functions is interesting when we want to obtain a rough result without using the CAD software. However this can be rather long if the electrode shapes are too complex.

Boundary Conditions Using Bezier Surfaces

EUCLID CAD software, from Matra Datavision, can produce automatically several files, in the normalized IGES representation, containing for each electrode a set of Bezier polynomials with their associated coefficents fitting the corresponding surfaces. Depending upon the complexity of the electrode, each polynomial can be of degree d_s and d_t up to 8 by 8, which means up to 81 coefficients. In practice, one complex electrode may produce a hundred of such polynomials.

We developed a software interface which transforms the IGES files into a unique global file where we add to the list of Bezier coefficients the particular potential value desired for each electrode. This file is then used exactly the same way than in the previous section. More precisely, this is made by using the Bernstein-Bezier representation parametrized in s and t to project the surfaces in the CHA3D mesh :

Let $P_x(d_s+1,d_t+1)$ be the matrix giving for one surface the x coordinates of its Bezier poles produced by EUCLID, and let B_t and B_s the Bernstein matrices of dimension (d_s+1,n) and (d_t+1,n) corresponding to the discretization of s and t. The matrix $x(n,n)$ is obtained by :

   x = matmul(transpose(B_s) , matmul(P_x,B_t))   ! idem for y and z,

where matmul and transpose are the Fortran 90 multiplication and transposition intrinsic procedures.

We notice also that depending upon the degree of each polynomial, the matrices P_x, P_y, P_z, B_s and B_t can be allocated in memory when necessary using the "allocate" Fortran 90 statement.

This operation is rendered automatic and takes a few minutes. Another advantage is that the calculated potential corresponds closely to the object which is to be later manufactered using EUCLID files as an input for machine tools.

Conjugate Gradient Algorithm

The conjugate gradient solver itself is very compact :

```
      subroutine gc3 (mx,my,mz,itemax,eps,hc,hx,hy,hz,r,v,y,b,u)
      real, dimension (0:mx+1,0:my+1,0:mz+1) ::  hc, hx, hy, hz, r, v, y, b, u
      criter   = eps*eps*sum(b*b)
         r     = b - hc*u - hx * (eoshift(u,dim=1,shift=1) + eoshift(u,dim=1,shift=-1)) &
     &                    - hy * (eoshift(u,dim=2,shift=1) + eoshift(u,dim=2,shift=-1)) &
     &                    - hz * (eoshift(u,dim=3,shift=1) + eoshift(u,dim=3,shift=-1))
         v     = r
         rn    = sum(r*r)  ;   ite   = 1
10       y     =     hc*v  - hx * (eoshift(v,dim=1,shift=1) + eoshift(v,dim=1,shift=-1)) &
     &                    - hy * (eoshift(v,dim=2,shift=1) + eoshift(v,dim=2,shift=-1)) &
     &                    - hz * (eoshift(v,dim=3,shift=1) + eoshift(v,dim=3,shift=-1))
         den   = sum(v*y)
         if(ite.ge.itemax .or. rn.lt.criter .or. den.eq.0)  return
         ro    = rn/den ;      rd   = rn       ;        ite  = ite + 1
         u     = u + ro*v
         r     = r - ro*y
         rn    = sum(r*r)      teta  = rn/rd
         v     = r + teta*v
      goto 10
      end
```

We observe that the 3-dimensional shapes of the H diagonals allow to use in a natural way the eoshift Fortran 90 function to shift arrays in the 3 directions, taking into account the 7 points finite difference scheme. Thanks to the update of H and b, the arrays r, v, y and u are not modified where the nodes are blocked.

All the statements are performed in parallel. Moreover, in a distributed memory computer like a CM5, the eoshift function generates parallel communications between processors through the "fat tree" network connecting them.

Performances

We consider the realistic case of a box containing 400*200*100 = 8 millions nodes, with millimetric steps. Such a problem needs roughly 10 arrays of that dimension, which represents 640 Megabytes of memory, using the 64 bit floating point Cray representaion. The convergence is reached with less than 2000 iterations. The main loop inside the conjugate gradient routine produces 22 floating point operations per iteration and per node, which gives a total number of :

$$N_{ops} = 22 * 2000 * 8 \cdot 10^6 = 352 \cdot 10^9$$

Using the 16 vectorial processors of our CRAY J9016 in parallel, the resolution takes about 7 minutes which corresponds to 840 megaflops per second. The performance is quite the same using a CM5 with 32 distributed processors.

Verifications Using AVS

It is crucial to check that the hundreds of surfaces coming from EUCLID (or from a list of user functions) are coherent and in particular that there is no one missing or in a wrong position. However, we must be aware that there is no possibility to check the validity of the boundary shapes with the mesh itself : of course, all the information is present in the logical array "imposed", but it is not easy to use it "visually" as a self-consistent verification.

The only way to operate is to take without any modification the Bezier input data and the projection routine used in CHA3D, and to connect them to a 3D viewer making the object appear on the screen.

Such a 3D viewer must give the possibility to put appropriate lights around the object, to rotate it easily in the space with the mouse, to zoom some parts and to click some pieces of surfaces making them transparent to look behind, etc...

AVS, from Advanced Visual Systems Inc., allows to construct a "network" composed of modules reading data, and exchanging them through data-flows towards the standard 3D viewer module called "geometry-viewer".

To simplify this process, we have constructed 2 special modules written in C so as to read automatically the hundreds of surfaces without clicking them one by one.

APPLICATIONS

Several electrostatic objects have been studied with CHA3D : in particular two electrostatic quadrupoles and an electrostatic deflector which are currently used in the SIRA experiments [4].

Concerning the SPIRAL project, three parts are being optimized : an Einzel lenses system for the future line following the ECR source, the central part of the RF electrodes of the Cyclotron CIME, and the spiral inflector for its axial injection.

Figure 1 represents a view of this rather complex object together with the 3D potential result shown in a particular horizontal plane.

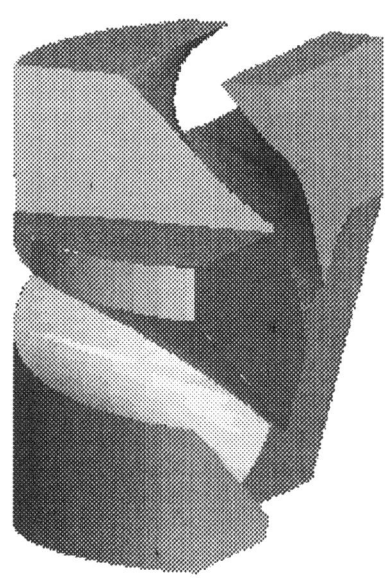

FIGURE 1. Spiral inflector

The two electrodes are constructed with 258 Bezier surfaces using the information given by EUCLID
The second figure shows the object from above with the potential values in a plane

CONCLUSION

We are convinced that this approach is interesting when we want to take into account the complete real geometry of electrostatic objects. Such a method needs memory and CPU performances, but these resources will be more and more available in multiprocessor computers.

We can also notice that the CHA3D solver may be used for heat transfert problems.

It could be also easily introduced in a space charge program using for example a PIC method applied to Vlasov equations. More precisely, this makes appear the following four Laplacian equations governing the collective effects :

$$-\Delta V = \frac{\rho}{\varepsilon_0}$$
$$-\Delta \vec{B} = \mu_0 \, \vec{rot}(\vec{j}) = \mu_0 \, \vec{rot}(\rho \vec{v})$$

In such a case, the presence of a regular finite difference mesh simplifies the particle-mesh and mesh-particle operations. Morover the use of an iterative conjugate gradient algorithm allows to decrease the number of necessary iterations by using the information coming from previous time steps.

It should be noticed however that although particle-mesh and mesh-paricle operations are not a problem with shared memory computers, they may be time consuming with distributed memory ones, depending upon the architecture of the network interconnecting the processors.

Finally we observe that CHA3D performances might be improved by using multigrid techniques.

ACKNOWLEDGMENTS

We would like to thank MM. Daudin, Dubois, Huguet and Ozille for their contribution in the use of EUCLID possibilities.

We would like to associate to this work the student Edorh Ananou, for the coding of the EUCLID-CHA3D interface.

We thank also Mrs. Bourgarel and MM. Chabert, Lieuvin and Ricaud for their constant encouragements.

REFERENCES

[1] Lieuvin M. and the SPIRAL group, The Status of SPIRAL, the RIB at GANIL, Cyclotron'95, Cape Town, South Africa (October 1995)

[2] Sortais P. et al., Developments of compac permanent magnet ECRIS, 12th International Workshop on ECR Ion Sources, Riken, Japan (April 1995)

[3] Brainerd W. S., Martin J. T., Smith B. T. and Wagener J. L., Fortran 90 Handbook, complete Ansi/Iso Reference , McGraw Hill (1992)

[4] Lecesne N. et al., Radioactive Ion Beam Production Tests for SPIRAL, 34th International Winter Meeting on Nuclear Physics, Bormio, Italy (January 1996)

AUTHOR INDEX

A

Abe, H., 343
Akers, W., 291
Andrianov, S. N., 355
Antonsen, T. M., 59

B

Babcock, R. C., 264
Balandin, V., 276, 282
Barts, T., 131
Bartsch, M., 65
Becker, R., 191
Becker, U., 65, 331
Bertrand, P., 395
Berwald, D. H., 369
Berz, M., 221, 253, 276
Bihn, M., 65
Blell, U., 65
Bondarev, B. I., 173, 349
Botton, M., 59
Bowling, B. A., 291
Braun, A., 297
Brown, N., 167
Brown, N. A., 264
Bruhwiler, D. L., 179

C

Cai, Y., 3
Callahan, D. A., 27, 125
Caporaso, G. J., 155
Carey, D. C., 264
Cary, J. R., 366
Catani, L., 309
Chan, K. C. D., 247
Chao, Y-C, 317
Chou, W., 131
Chowdhary, M., 317
Clemens, M., 65
Coats, R. S., 71
Colby, E. R., 45
Cooperberg, D., 83

D

Davies, W. G., 235
DeFord, J. F., 107
Dehler, M., 65
Dohlus, M., 65, 331
Douglas, S. R., 235
Drevlak, M., 65
Du, X., 65
Durkin, A. P., 173, 349

E

Ehmann, R., 65
Eufinger, A., 65

F

Freund, H. P., 143
Friedman, A., 27, 51, 125, 149

G

Gillespie, G. H., 247, 264, 369
Gillespie, J. S., 361
Golubeva, N., 276, 282
Grote, D. P., 27, 51, 125, 149
Gutschling, S., 65

H

Haber, I., 51, 125
Habib, S., 377
Hahne, P., 65
Hano, M., 197
Higo, T., 39
Hill, B. W., 247, 264, 361, 369
Holt, J. A., 270, 297
Hughes, M. H., 369

I

Irwin, J., 3
Iselin, F. C., 325
Iwashita, Y., 119

J

Jensen, K. L., 95
Johnson, W. A., 71

K

Kiefer, M. L., 71
Klatt, R., 65
Klein, W., 303
Ko, K., 9, 39, 389
Kodis, M. A., 95
Koiso, H., 215
Krietenstein, B., 65
Kroupa, M., 303

L

Langstrof, A., 65
Laskar, J., 15
Lee-Whiting, G. E., 235
Le Meur, G., 113
Lessner, E., 185
Levush, B., 59
Li, Z., 9, 39
Lin, X. E., 9
Liu, H., 203
Luger, G., 303
Lysenko, W. P., 247

M

Makino, K., 253
Malitsky, N., 337
Martens, M., 297
Martono, H., 264, 361
Mendelsohn, S., 369
Michelotti, L., 241, 270, 297
Min, Z., 65
Mix, L. P., 71

Myers, T. J., 369

N

Nelson, E. M., 77
Ng, C.-K., 9
Ng, W. C., 155
Nosochkov, Y., 3

O

Ohmi, K., 215
Oide, K., 215
Ostiguy, J.-F., 270

P

Pasik, M. F., 71
Paulson, C. C., 369
Peacock, M. A., 369
Piaszczyk, C. M., 369
Piechowiak, E. M., 369
Pinder, P., 65
Piquemal, A., 137
Podebrad, O., 65
Pointon, T. D., 71
Poole, B. R., 155
Pröpper, T., 65

Q

Quintenz, J. P., 71

R

Rathke, J. W., 369
Reusch, M. F., 179
Riley, C. P., 101
Riley, D. J., 71
Robin, D., 15
Rusthoi, D. P., 247
Ryne, R., 377

S

Schillinger, B., 161
Schmidt, D., 65
Schuett, P., 21
Schuhmann, R., 65
Schulz, A., 65
Schupp, S., 65
Schütt, P., 65
Seidel, D. B., 71
Shang, C. C., 155
Sharp, W. M., 27
Shasharina, S. G., 366
Shoaee, H., 291
Srinivas, V., 39
Steich, D., 155
Stern, C., 303
Symon, K., 185
Symon, K. R., 229

T

Talman, R., 337
Thoma, P., 65
Timm, M., 65
Touze, F., 113
Turner, D. J., 71

V

van Rienen, U., 65, 89
van Zeijts, J., 291
Verboncoeur, J. P., 83
Vinogradov, S. V., 349

W

Wagner, B., 65
Watson, W., 291
Weber, R., 65
Weiland, T., 65, 161, 331
Wipf, S., 65
Witherspoon, S., 291, 317
Wolter, H., 65

Y

Yan, Y., 3
Yan, Y. T., 259

Z

Zaidman, E. G., 95, 143
Zhan, X., 389
Zhang, M., 21

AIP Conference Proceedings

	Title	L.C. Number	ISBN
No. 295	The Physics of Electronic and Atomic Collisions: XVIII International Conference (Aarhus, Denmark, 1993)	93-74103	1-56396-290-X
No. 296	The Chaos Paradigm: Developments an Applications in Engineering and Science (Mystic, CT 1993)	93-74146	1-56396-254-3
No. 297	Computational Accelerator Physics (Los Alamos, NM 1993)	93-74205	1-56396-222-5
No. 298	Ultrafast Reaction Dynamics and Solvent Effects (Royaumont, France 1993)	93-074354	1-56396-280-2
No. 299	Dense Z-Pinches: Third International Conference (London, 1993)	93-074569	1-56396-297-7
No. 300	Discovery of Weak Neutral Currents: The Weak Interaction Before and After (Santa Monica, CA 1993)	94-70515	1-56396-306-X
No. 301	Eleventh Symposium Space Nuclear Power and Propulsion (3 Vols.) (Albuquerque, NM 1994)	92-75162	1-56396-305-1 (Set) 156396-301-9 (pbk. set)
No. 302	Lepton and Photon Interactions/ XVI International Symposium (Ithaca, NY 1993)	94-70079	1-56396-106-7
No. 303	Slow Positron Beam Techniques for Solids and Surfaces Fifth International Workshop (Jackson Hole, WY 1992)	94-71036	1-56396-267-5
No. 304	The Second Compton Symposium (College Park, MD 1993)	94-70742	1-56396-261-6
No. 305	Stress-Induced Phenomena in Metallization Second International Workshop (Austin, TX 1993)	94-70650	1-56396-251-9
No. 306	12th NREL Photovoltaic Program Review (Denver, CO 1993)	94-70748	1-56396-315-9
No. 307	Gamma-Ray Bursts Second Workshop (Huntsville, AL 1993)	94-71317	1-56396-336-1
No. 308	The Evolution of X-Ray Binaries (College Park, MD 1993)	94-76853	1-56396-329-9
No. 309	High-Pressure Science and Technology—1993 (Colorado Springs, CO 1993)	93-72821	1-56396-219-5 (Set)

	Title	L.C. Number	ISBN
No. 310	Analysis of Interplanetary Dust (Houston, TX 1993)	94-71292	1-56396-341-8
No. 311	Physics of High Energy Particles in Toroidal Systems (Irvine, CA 1993)	94-72098	1-56396-364-7
No. 312	Molecules and Grains in Space (Mont Sainte-Odile, France 1993)	94-72615	1-56396-355-8
No. 313	The Soft X-Ray Cosmos ROSAT Science Symposium (College Park, MD 1993)	94-72499	1-56396-327-2
No. 314	Advances in Plasma Physics Thomas H. Stix Symposium (Princeton, NJ 1992)	94-72721	1-56396-372-8
No. 315	Orbit Correction and Analysis in Circular Accelerators (Upton, NY 1993)	94-72257	1-56396-373-6
No. 316	Thirteenth International Conference on Thermoelectrics (Kansas City, Missouri 1994)	95-75634	1-56396-444-9
No. 317	Fifth Mexican School of Particles and Fields (Guanajuato, Mexico 1992)	94-72720	1-56396-378-7
No. 318	Laser Interaction and Related Plasma Phenomena 11th International Workshop (Monterey, CA 1993)	94-78097	1-56396-324-8
No. 319	Beam Instrumentation Workshop (Santa Fe, NM 1993)	94-78279	1-56396-389-2
No. 320	Basic Space Science (Lagos, Nigeria 1993)	94-79350	1-56396-328-0
No. 321	The First NREL Conference on Thermophotovoltaic Generation of Electricity (Copper Mountain, CO 1994)	94-72792	1-56396-353-1
No. 322	Atomic Processes in Plasmas Ninth APS Topical Conference (San Antonio, TX)	94-72923	1-56396-411-2
No. 323	Atomic Physics 14 Fourteenth International Conference on Atomic Physics (Boulder, CO 1994)	94-73219	1-56396-348-5
No. 324	Twelfth Symposium on Space Nuclear Power and Propulsion (Albuquerque, NM 1995)	94-73603	1-56396-427-9
No. 325	Conference on NASA Centers for Commercial Development of Space (Albuquerque, NM 1995)	94-73604	1-56396-431-7

	Title	L.C. Number	ISBN
No. 326	Accelerator Physics at the Superconducting Super Collider (Dallas, TX 1992-1993)	94-73609	1-56396-354-X
No. 327	Nuclei in the Cosmos III Third International Symposium on Nuclear Astrophysics (Assergi, Italy 1994)	95-75492	1-56396-436-8
No. 328	Spectral Line Shapes, Volume 8 12th ICSLS (Toronto, Canada 1994)	94-74309	1-56396-326-4
No. 329	Resonance Ionization Spectroscopy 1994 Seventh International Symposium (Bernkastel-Kues, Germany 1994)	95-75077	1-56396-437-6
No. 330	E.C.C.C. 1 Computational Chemistry F.E.C.S. Conference (Nancy, France 1994)	95-75843	1-56396-457-0
No. 331	Non-Neutral Plasma Physics II (Berkeley, CA 1994)	95-79630	1-56396-441-4
No. 332	X-Ray Lasers 1994 Fourth International Colloquium (Williamsburg, VA 1994)	95-76067	1-56396-375-2
No. 333	Beam Instrumentation Workshop (Vancouver, B. C., Canada 1994)	95-79635	1-56396-352-3
No. 334	Few-Body Problems in Physics (Williamsburg, VA 1994)	95-76481	1-56396-325-6
No. 335	Advanced Accelerator Concepts (Fontana, WI 1994)	95-78225	1-56396-476-7 (Set) 1-56396-474-0 (Book) 1-56396-475-9 (CD-Rom)
No. 336	Dark Matter (College Park, MD 1994)	95-76538	1-56396-438-4
No. 337	Pulsed RF Sources for Linear Colliders (Montauk, NY 1994)	95-76814	1-56396-408-2
No. 338	Intersections Between Particle and Nuclear Physics 5th Conference (St. Petersburg, FL 1994)	95-77076	1-56396-335-3
No. 339	Polarization Phenomena in Nuclear Physics Eighth International Symposium (Bloomington, IN 1994)	95-77216	1-56396-482-1
No. 340	Strangeness in Hadronic Matter (Tucson, AZ 1995)	95-77477	1-56396-489-9

Title	L.C. Number	ISBN
No. 341 Volatiles in the Earth and Solar System (Pasadena, CA 1994)	95-77911	1-56396-409-0
No. 342 CAM -94 Physics Meeting (Cacun, Mexico 1994)	95-77851	1-56396-491-0
No. 343 High Energy Spin Physics Eleventh International Symposium (Bloomington, IN 1994)	95-78431	1-56396-374-4
No. 344 Nonlinear Dynamics in Particle Accelerators: Theory and Experiments (Arcidosso, Italy 1994)	95-78135	1-56396-446-5
No. 345 International Conference on Plasma Physics ICPP 1994 (Foz do Iguaçu, Brazil 1994)	95-78438	1-56396-496-1
No. 346 International Conference on Accelerator-Driven Transmutation Technologies and Applications (Las Vegas, NV 1994)	95-78691	1-56396-505-4
No. 347 Atomic Collisions: A Symposium in Honor of Christopher Bottcher (1945-1993) (Oak Ridge, TN 1994)	95-78689	1-56396-322-1
No. 348 Unveiling the Cosmic Infrared Background (College Park, MD, 1995)	95-83477	1-56396-508-9
No. 349 Workshop on the Tau/Charm Factory (Argonne, IL, 1995)	95-81467	1-56396-523-2
No. 350 International Symposium on Vector Boson Self-Interactions (Los Angeles, CA 1995)	95-79865	1-56396-520-8
No. 351 The Physics of Beams Andrew Sessler Symposium (Los Angeles, CA 1993)	95-80479	1-56396-376-0
No. 352 Physics Potential and Development of $\mu^+ \mu^-$ Colliders: Second Workshop (Sausalito, CA 1994)	95-81413	1-56396-506-2
No. 353 13th NREL Photovoltaic Program Review (Lakewood, CO 1995)	95-80662	1-56396-510-0
No. 354 Organic Coatings (Paris, France, 1995)	96-83019	1-56396-535-6
No. 355 Eleventh Topical Conference on Radio Frequency Power in Plasmas (Palm Springs, CA 1995)	95-80867	1-56396-536-4
No. 356 The Future of Accelerator Physics (Austin, TX 1994)	96-83292	1-56396-541-0
No. 357 10th Topical Workshop on Proton-Antiproton Collider Physics (Batavia, IL 1995)	95-83078	1-56396-543-7

	Title	L.C. Number	ISBN
No. 358	The Second NREL Conference on Thermophotovoltaic Generation of Electricity	95-83335	1-56396-509-7
No. 359	Workshops and Particles and Fields and Phenomenology of Fundamental Interactions (Puebla, Mexico 1995)	96-85996	1-56396-548-8
No. 360	The Physics of Electronic and Atomic Collisions XIX International Conference (Whistler, Canada, 1995)	95-83671	1-56396-440-6
No. 361	Space Technology and Applications International Forum (Albuquerque, NM 1996)	95-83440	1-56396-568-2
No. 362	Two-Center Effects in Ion-Atom Collisions (Lincoln, NE 1994)	96-83379	1-56396-342-6
No. 363	Phenomena in Ionized Gases XXII ICPIG (Hoboken, NJ, 1995)	96-83294	1-56396-550-X
No. 364	Fast Elementary Processes in Chemical and Biological Systems (Villeneuve d'Ascq, France, 1995)	96-83624	1-56396-564-X
No. 365	Latin-American School of Physics XXX ELAF Group Theory and Its Applications (México City, México, 1995)	96-83489	1-56396-567-4
No. 366	High Velocity Neutron Stars and Gamma-Ray Bursts (La Jolla, CA 1995)	96-84067	1-56396-593-3
No. 367	Micro Bunches Workshop (Upton, NY, 1995)	96-83482	1-56396-555-0
No. 368	Acoustic Particle Velocity Sensors: Design, Performance and Applications (Mystic, CT, 1995)	96-83548	1-56396-549-6
No. 369	Laser Interaction and Related Plasma Phenomena (Osaka, Japan 1995)	96-85009	1-56396-445-7
No. 370	Shock Compression of Condensed Matter-1995 (Seattle, WA 1995)	96-84595	1-56396-566-6
No. 371	Sixth Quantum 1/f Noise and Other Low Frequency Fluctuations in Electronic Devices Symposium (St. Louis, MO, 1994)	96-84200	1-56396-410-4
No. 372	Beam Dynamics and Technology Issues for + - Colliders 9th Advanced ICFA Beam Dynamics Workshop (Montauk, NY, 1995)	96-84189	1-56396-554-2
No. 373	Stress-Induced Phenomena in Metallization (Palo Alto, CA 1995)	96-84949	1-56396-439-2

Title	L.C. Number	ISBN
No. 374 High Energy Solar Physics (Greenbelt, MD 1995)	96-84513	1-56396-542-9
No. 375 Chaotic, Fractal, and Nonlinear Signal Processing (Mystic, CT 1995)	96-85356	1-56396-443-0
No. 376 Chaos and the Changing Nature of Science and Medicine: An Introduction (Mobile, AL 1995)	96-85220	1-56396-442-2
No. 377 Space Charge Dominated Beams and Applications of High Brightness Beams (Bloomington, IN 1995)	96-85165	1-56396-625-7
No. 378 Surfaces, Vacuum, and Their Applications (Cancun, Mexico 1994)	96-85594	1-56396-418-X
No. 379 Physical Origin of Homochirality in Life (Santa Monica, CA 1995)	96-86631	1-56396-507-0
No. 380 Production and Neutralization of Negative Ions and Beams / Production and Application of Light Negative Ions (Upton, NY 1995)	96-86435	1-56396-565-8
No. 381 Atomic Processes in Plasmas (San Francisco, CA 1996)	96-86304	1-56396-552-6
No. 382 Solar Wind Eight (Dana Point, CA 1995)	96-86447	1-56396-551-8
No. 383 Workshop on the Earth's Trapped Particle Environment (Taos, NM 1994)	96-86619	1-56396-540-2
No. 384 Gamma-Ray Bursts (Huntsville, AL 1995)	96-79458	1-56396-685-9
No. 385 Robotic Exploration Close to the Sun: Scientific Basis (Marlboro, MA 1996)	96-79560	1-56396-618-2
No. 386 Spectral Line Shapes, Volume 9 13th ICSLS (Firenze, Italy 1996)		1-56396-656-5
No. 387 Space Technology and Applications International Forum (Albuquerque, NM 1997)	96-80254	1-56396-679-4 (Case set) 1-56396-691-3 (Paper set)
No. 388 Resonance Ionization Spectroscopy 1996 Eighth International Symposium (State College, PA 1996)	96-80324	1-56396-611-5
No. 389 X-Ray and Inner-Shell Processes 17th International Conference (Hamburg, Germany 1996)	96-80388	1-56396-563-1
No. 391 Computational Accelerator Physics (Williamsburg, VA 1996)	97-70181	1-56396-671-9